Note to readers

Checkbooks are designed for students seeking technician or equivalent qualification through the courses of the Business and Technician Education Council (BTEC), the Scottish Technical Education Council, Australian Technical and Further Education Departments, East and West African Examinations Council and other comparable examining authorities in technical subjects.

Checkbooks use problems and worked examples to establish and exemplify the theory contained in technical syllabuses. *Checkbook* readers gain real understanding through seeing problems solved and through solving problems themselves. *Checkbooks* do not supplant fuller textbooks, but rather supplement them with an alternative emphasis and an ample provision of worked and unworked problems, essential data, short answer and multi-choice questions (with answers where possible).

Preface

This Checkbook now covers the main topics of the new BTEC guide syllabus for Electronics NIII (unit U86/332). This has meant a radical revision: only the chapters on feedback, noise and oscillators remain from the first edition, other topics having been transferred to *Electronics 2 Checkbook* since they are now studied at level NII.

The new chapters deal with decibel measurement, operational amplifiers, DA and AD converters, controlled rectifiers, triggering devices, optoelectronic devices, fibre optics and power amplifiers.

As before, each chapter contains the relevant theory and definitions in summarised form, together with any essential formulae. Additional information is provided by way of worked examples which, it is hoped, will also demonstrate the best way to answer examination questions. Conventional and multi-choice problems for self-study are also included.

S. A. Knight

1 Using decibels

A MAIN POINTS CONCERNED WITH DECIBELS

1 There are a number of advantages to be gained from expressing power, voltage and current ratios in logarithmic units. One of the most important of these is that the enormous ranges of voltage and power amplification encountered in electronics and communication work can be expressed in figures conveniently small instead of astronomically large. For example, an amplifier may show large gain variations against frequency input. The gain ratios encountered may well lie between 10 and 1,000,000. The problem of plotting a graph showing these variations will be evident if this range of values is attempted on a piece of ordinary linearly scaled paper. The use of logarithmic units enables the overall gain of a system to be calculated by addition or subtraction rather than laborious multiplication or division.

2 The **decibel** (dB) is one-tenth of the bel, a unit in which we may express power ratios and the gain or loss ratios of the related quantities of voltage and current. The decibel is based on common logarithms to base 10.

Two power levels, P_1 and P_2 are said to differ by N decibels when

$$N = 10 \log \frac{P_1}{P_2}$$

With these units, if the signal power delivered to a load at the receiving end of a cable, say, is one-tenth at the sending end, the *loss* is 10 dB. With two identical cables in series, the received power is one-hundredth the sending power and the loss is 20 dB. Power loss is usually expressed with a negative sign or stated to be 'so many dBs down'.

If in any amplifier the output power is 100 times the input power, the *gain* is $10 \log 100 = 20$ dB. With two such amplifiers in cascade, the gain in power ratio will be $100 \times 100 = 10^4$ which is equivalent to $10 \log 10^4 = 40$ dB. Notice that the total power ratio of two (or more) systems in cascade involves the product of the individual power ratios but only the sum of their decibel equivalents.

3 In most cases, two powers P_1 and P_2 are compared by observing either the voltage developed across a known impedance, or the current flowing through the impedance. If then the input and output impedances of, say, an amplifier are *equal*, the power ratio will be proportional to the square of the voltage (or current) ratio, since $P = I^2 Z$ or V^2/Z. Hence

$$\frac{P_1}{P_2} = \left[\frac{V_1}{V_2}\right]^2 = \left[\frac{I_1}{I_2}\right]^2$$

$$\therefore \quad N = 10\log\left[\frac{V_1}{V_2}\right]^2 = 20\log\frac{V_1}{V_2}$$

$$\text{or} \quad N = 20\log\frac{I_1}{I_2}$$

Because the input and output impedances of any system are rarely equal, the voltage or current ratio is frequently 'misused' to describe a power ratio without regard to the necessity of having equality between the input and output impedances. It is not necessarily correct, for example, to say that if an amplifier has a voltage gain of 100, it has a power gain of 40 dB. This would only be true if the power gain was $(100)^2$ or 10^4. However, it is quite common practice to express voltage (and current) gain in this fundamentally incorrect manner, that is, if I know that a system has a voltage gain of 1000, I say its amplification is 60 dB, irrespective of the input and output conditions.

4 The adoption of logarithmic units to describe power ratios was not simply because of the convenience of substituting small numbers for large ones; it is found that the human ear responds logarithmically to sounds of different intensities, i.e. powers. So that if an increase of sound intensity by 10 times produces a given sensation at the ear, a further increase of 10 times, that is to say, a total increase of 100 times, will produce *twice* the sensation at the ear. The science of acoustics, which in many ways is related to communications signalling, therefore, shares a common unit relating to power ratios.

B WORKED PROBLEMS ON DECIBELS

Problem 1 Express as decibel gains or losses, power ratios of (a) 200, (b) 40, (c) 0.05.

For power gains or losses the expression we need is

$$\text{Gain } N = 10\log(\text{power ratio})$$

We notice that if the power ratio is less than 1, that is, we have a power loss, then the logarithm is negative.

(a) Gain $= 10\log 200$
$= 10 \times 2.301 = \mathbf{23.01\ dB}$

(b) Gain $= 10\log 40$
$= 10 \times 1.602 = \mathbf{16.02\ dB}$

(c) Gain $= 10\log 0.05$
$= 10 \times -1.301 = \mathbf{-13.01\ dB}$

It would be usual to round off these solutions to 23 dB, 16 dB and −13 dB respectively. The power loss in Example (c) is automatically indicated by the negative sign.

Problem 2 An amplifier has a power gain of 18 dB. What is the ratio of output to input power?

Here we have $10 \log (\text{ratio}) = 18$

$\therefore$ $\log (\text{ratio}) = 1.8$

So $\text{ratio} = 10^{1.8}$ (or antilog 1.8)

$= \mathbf{63.1}$

Problem 3 Explain how two power ratios expressed in dB can be expressed in terms of *absolute* power.

It is clear from the previous example that the decibel is a unit of power *ratio* only. When we say that the power gain of the amplifier is 18 dB we know that the output power is 63.1 times the input power, but what these powers actually are is not known. If we choose some definite reference for our measurements, however, a power of so many dBs above or below that level will acquire for us a definite meaning in quantitative terms.

In general communications practice the reference power level is taken to be 1 mW (0.001 W); then any power gain or loss P can be expressed as

$$\mathbf{10 \log \frac{\mathit{P}}{1\ mW}\ dB}$$

referred to 1 mW. Thus a power gain of 23 dB (see *Problem 1*) which represents a power ratio of 200 times will, referred to a reference 1 mW as input, represent an output of 200 mW. In the same way, a power loss of −23 dB referred to 1 mW as input will represent an output power of 1/200th of 1 mW, or 0.005 mW. The expression 'dB referred to 1 mW' is abbreviated to **dBm**. The reference level 0 dBm is then equal to 1 mW.

Problem 4 What power, in watts, is represented by 15 dBm?

If the required power is P watts, then

$$\frac{P}{1\ \text{mW}} = 10^{1.5} \text{ (or antilog 1.5)}$$

$$= 31.63$$

$$\therefore \quad P = 31.63\ \text{mW or } \mathbf{0.0316\ W}$$

Problem 5 Six networks are wired in cascade as shown in *Figure 1.1*, the decibel gains and losses being as indicated for each network. If 1 mW is applied at the input, what will the output power level be?

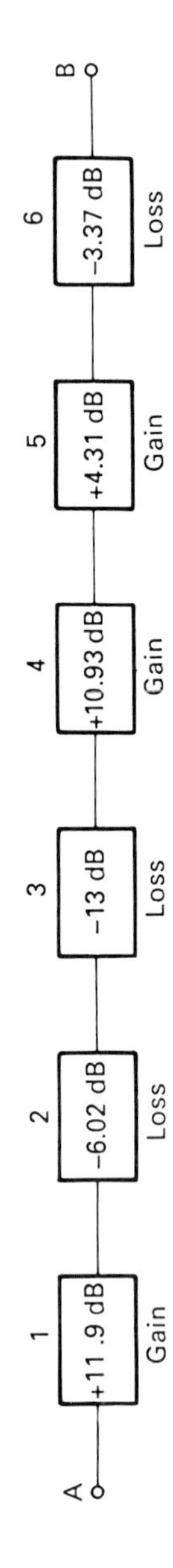
A
1
+11 .9 dB
Gain
2
−6.02 dB
Loss
3
−13 dB
Loss
4
+10.93 dB
Gain
5
+4.31 dB
Gain
6
−3.37 dB
Loss
B

Figure 1.1

The input in this problem is 1 mW or 0 dBm. Since $\log(P_1 \times P_2) = \log P_1 + \log P_2$, the total gain is the algebraic sum of the individual network gains which is

$$+11.9 - 6.02 - 13 + 10.93 + 4.31 - 3.37 = +4.75 \text{ dB}$$

The output level is therefore

$$0 + 4.75 \text{ dBm} = 4.75 \text{ dBm}$$

an overall gain. For an input power of 1 mW, therefore, the output power is

$$\frac{P}{1 \text{ mW}} = 10^{0.475} \text{ (or antilog } 0.475)$$

$$P = \mathbf{2.98\ mW}$$

Problem 6 An amplifier has a voltage gain of 20 at low frequencies and 55 at high frequencies. What is the relative response at low frequencies to that at high frequencies, in dB?

$$\text{Relative response} = 20 \log \left[\frac{\text{output at low frequency}}{\text{output at high frequency}}\right]$$

$$= 20 \log \frac{20}{55}$$

$$= 20 \log 0.364$$

$$= 20 \times -0.44 = \mathbf{-8.8\ dB}$$

Problem 7 In a transistor amplifier, an input of 10 mV produces an output current of 2.0 mA in a 600 Ω output load resistor. If the input resistance of the amplifier is 300 Ω, find the voltage, current and power gains in dB.

$$\text{Output voltage } V_o = 2.0 \times 10^{-3} \times 600 = 1.2 \text{ V}$$

$$\text{Voltage gain } A_v = \frac{V_o}{V_i} = \frac{1.2}{10 \times 10^{-3}} = 1200$$

$$\therefore A_v = 20 \log 1200 = \mathbf{61.6\ dB}$$

$$\text{Input current } I_i = \frac{V_i}{R_i} = \frac{10 \times 10^{-3}}{300}$$

$$= 33.4 \times 10^{-6} \text{ A} = 33.4\ \mu\text{A}$$

$$\text{Current gain } A_i = \frac{I_o}{I_i} = \frac{2 \times 10^{-3}}{33.4 \times 10^{-6}} = 60$$

$$\therefore \quad A_i = 20 \log 60 = \mathbf{35.6\ dB}$$

$$\text{Power gain} = A_v A_i = 1200 \times 60 = 72000$$

$$= 10 \log 72000 \text{ dB}$$

$$= \mathbf{46.6\ dB}$$

Problem 8 Show that a 3 dB voltage reduction is equivalent to a fall in the voltage to 0.707 of its original value. What is important about this result?

$$\text{Relative gain} = 20 \log 0.707 = -3 \text{ dB}$$

If we interpret this result in terms of power, we find that the power ratio corresponding to -3 dB is $10^{-0.3} = 0.5$. Hence the response is 3 dB down at the **half-power point**. This is a measure of the selectivity of a tuned circuit where the 3 dB points (one on each side of the resonant frequency) give a measure of the bandwidth. It is also the point where circuit reactance is equal to the resistance. The same reference points are taken for the response levels of amplifier systems and transmission circuits.

C FURTHER PROBLEMS ON DECIBELS

(a) SHORT ANSWER PROBLEMS (answers on page 152)

1 Define the decibel. Explain why the decibel is used in communications engineering.
2 Why is voltage or current gain measured in dB, double the figure obtained for a power gain?
3 Why is a reference level necessary when measurements are being made in dB? Why is the phrase 'relative to 1 mW' used after some statements of dB?
4 Why is the expression for voltage and current gain ratios of a system in dB only strictly correct if the input and output impedances of the system are equal?
5 Why does a negative sign appear whenever a power or voltage ratio less than unity is expressed in dB units?
6 What is unity gain expressed in dB?
7 Two amplifiers connected in cascade have individual power gains of 15 dB and 25 dB respectively. What is the overall power gain?
8 If in the previous question the input to the amplifier system is 0 dBm, what will be the output power?
9 An attenuator introducing a voltage loss of 6 dB is connected between a 1 V source and a load resistor. What is the voltage across the load?
10 An operational amplifier has an open-loop gain of 10^6 times. What is this in dB?

(b) MULTI-CHOICE PROBLEMS (answers on page 152)

1 A power ratio of 100 times is equivalent to (a) 10 dB, (b) 20 dB, (c) 40 dB, (d) none of these.

2 A 3 dB power loss is the same as a power ratio of (a) $1/\sqrt{2}$, (b) $\sqrt{2}$, (c) 1/2.

3 A 3 dB voltage drop is equivalent to a ratio of (a) $1/\sqrt{2}$, (b) $\sqrt{2}$, (c) 1/2.

4 Two systems are connected in cascade. One has a voltage gain of 15 dB, the other a voltage gain of −9 dB. The overall voltage ratio is (a) 6 dB, (b) 10 dB, (c) 2, (d) 4, (e) none of these.

5 Four cascaded networks each have a voltage gain of 1.25. The overall gain is (a) 10 dB, (b) 14 dB, (c) 18 dB, (d) none of these.

6 A certain attenuator introduces a voltage loss of −5 dB. Three of these attenuators are wired up in series and 1 V is applied at the input. The output voltage will be (a) 0.178 V, (b) 0.067 V, (c) 0.707 V, (d) none of these.

(c) CONVENTIONAL PROBLEMS

1 Express as decibel gains or losses, voltage ratios of 100, 40, 5 and 0.5 times. What power gains correspond to 23 dB, 6 dB, −30 dB?

[40 dB, 32 dB, 14 dB, −6 dB; 200, 4, 0.001]

2 What do you understand by the term dBm? What special advantage does it have in the calculation of power ratios? What powers are represented by (a) 17 dBm, (b) −24 dBm?

[(a) 50 mW, (b) 4 μW]

3 Calculate the overall gain or loss, in dB, of the networks shown in *Figure 1.2*. If the power input is 30 mW, determine the output power in dB relative to 1 mW.

[26.5 dBm]

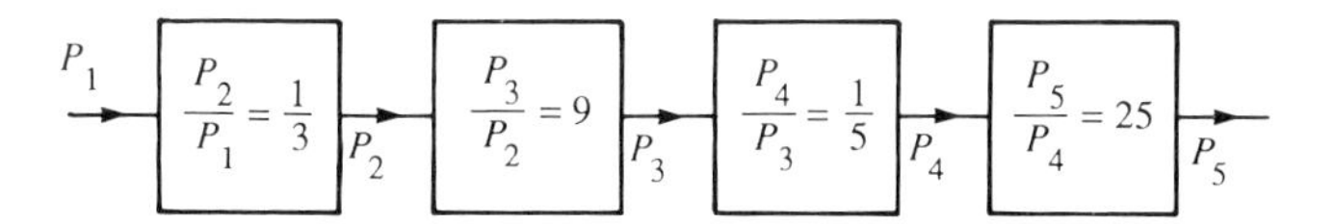

Figure 1.2

4 Calculate the current in, and the voltage drop across, a 600 Ω resistance in which the power dissipated is (a) 5 dB above, (b) 10 dB below, a level of 1 mW.

[4.1 mA, 2.45 V; 0.13 mA, 0.077 V]

5 In *Figure 1.3*, an a.c. source maintains a constant 500 mV r.m.s. across the terminals AB. Calculate the r.m.s. voltage across the load terminals CD.

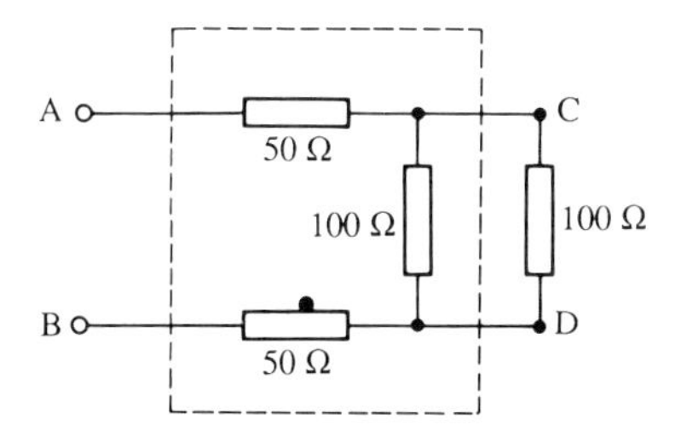

Figure 1.3

Calculate, in dB, the reduction in power output attributable to the network within the dotted frame.

[0.167 V; 9.54 dB]

6 A cable 30 km in length has an attenuation of 1.2 dB per km. What output power is received if the sending end input power is 100 mW? If an amplifier is placed midway along the line, what should be its power gain if the output required is 1 mW?

[0.025 mW; 16 dB]

7 Voltmeters used by communications engineers have scales calibrated in volts and also in dB referred to 1 mW dissipated in a 600 Ω resistance load. If the power in a line is measured by one of these meters and found to be 1 mW, what is the corresponding line voltage?

[0.775 V]

8 A low-frequency amplifier has a power gain of 56 dB. The input resistance is 600 Ω and the output load is a resistance of 10 Ω. What will be the load current if a 1 V r.m.s. signal is applied at the input?

[8.15 A]

9 An amplifier has a response at 100 Hz which is 8.5 dB down on its response at 1 kHz. If the voltage gain at 100 Hz is 15, what is its voltage gain at 1 kHz?

[40]

10 An amplifier has its input and output resistances equal at 75 Ω each. An input signal of 500 mV produces an output of 10 V. Calculate the voltage and power gain ratios and the overall gain in dB. If the output resistance is doubled, the input resistance and voltage gain ratio being unaltered, find the power gain ratio and the gain in dB.

[20, 400, 26 dB; 200, 23 dB]

11 The input signal to a transmission line varies between 23.5 mW and 1.25 W. Express each of these powers in dB relative to 1 mW and calculate the fluctuation in the level of the signal in dB.

[13.7 dB, 31 dB; 17.3 dB]

12 A common-emitter amplifier has an input resistance of 900 Ω. When the input signal level is 10 mV, a current of 2 mA flows in a 500 Ω collector load resistor. Find the voltage, current and power gains of this amplifier in dB.

[40 dB, 45 dB, 42.5 dB]

13 An electronic system has an automatic gain control circuit made up of three amplifying stages in each of which the gain changes by 3 dB for each volt of control signal applied. A change in the system input from 0.1 mV to 100 mV causes a 4 dB change in the output voltage. Find the corresponding change in the control signal voltage.

[6.22 V]

14 Equal powers are being dissipated in two resistances of value 600 Ω and 1200 Ω. What are the ratios of the voltages and the currents acting on each of the resistances? What would these ratios be expressed in dB?

[$\sqrt{2}$:1, 1:$\sqrt{2}$; −3 dB, +3 dB]

15 A neper is a communications unit which basically expresses a current ratio on a transmission line. If 1 neper = 8.686 dB, express the dB in nepers. Express a cable loss of 1 neper per kilometre as dB per mile.

[0.115 dB; 13.9 dB/mile]

2 Feedback

A MAIN POINTS CONCERNED WITH NEGATIVE FEEDBACK

1 A feedback amplifier is an amplifier where the input is dependent to a greater or lesser extent on the signal at the output. When the input signal is effectively reduced in magnitude by the addition of feedback, the method is known as **degenerative** or **negative feedback** (n.f.b.). If there is an increase in the total input due to feedback, the method is known as **regenerative** or **positive feedback**. Negative feedback significantly modifies the characteristics of an amplifier and enables a predictable performance to be achieved free from the effects of variations in circuit parameters and the environmental conditions in which the amplifier may be used.

2 *Figure 2.1* represents any amplifier having an input voltage v_i and an output voltage v_o. The voltage gain of the amplifier is

$$A_v = \frac{v_o}{v_i} \quad \text{and} \quad v_o = A_v \cdot v_i$$

Figure 2.2 shows the same amplifier but now a voltage proportional to the output voltage is fed back to the input terminals. The basic amplifier gain is still A_v but the input voltage available at the **terminals** of the amplifier is no longer v_i but v, where

$$v = v_i + v_f$$

Here, v_f is the signal voltage fed back from the output and, being proportional to v_o, can be expressed as βv_o, where β is the feedback fraction. Hence

$$v = v_i + \beta v_o$$

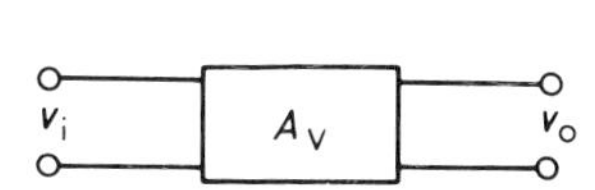

Figure 2.1

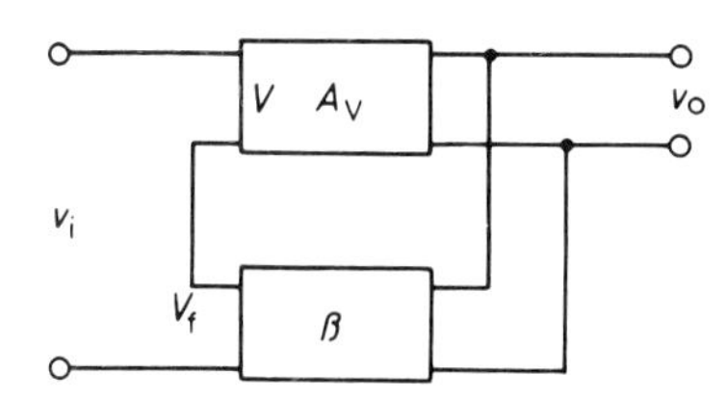

Figure 2.2

But now $$A_v = \frac{V_o}{v} \text{ or } v = \frac{v_o}{A_v}$$

$$\therefore \quad \frac{v_o}{A_v} = v_i + \beta v_o \text{ or } v_o = a_v v_i + A_v \beta v_o$$

$$\therefore \quad v_o - a_v \beta v_o = A_v v_i \text{ or } v_o(1 - \beta A_v) = A_v v_i$$

$$\therefore \quad \frac{v_o}{v_i} = \frac{A_v}{1 - \beta A_v}$$

This ratio v_o/v_i is now the **overall** gain of the complete circuit with feedback applied, and is symbolised A_v'.

$$\therefore \quad A_v' = \frac{A_v}{1 - \beta A_v}$$

This is the general feedback equation.

3 The quantity βA_v in the feedback equation is known as the **loop gain** because it represents the total gain measured round the feedback loop from v to v_f, that is, it is the product of the individual gains of the amplifier, A_v, and the feedback network, β. In the absence of feedback, $\beta = 0$ and the amplifier exhibits its basic gain of A_v as in *Figure 2.1*. This is the **open-loop** gain. The greatest possible value that β can have is 1, as it is not possible to feed back a signal that is greater than v_o

4 Three possible ranges of value that the loop gain βA_v may have are of interest.
   (a) If βA_v lies between 0 and 1 (but not equal to 1), that is, if βA_v is positive and less than unity, the denominator of the feedback equation will be less than unity and the gain with feedback A_v' will be **greater** than A_v. This is a case of positive feedback because the gain of the system has been increased by the feedback.
   (b) If βA_v is negative, the denominator will be greater than unity, since $1 - (-\beta A_v) = 1 + \beta A_v$, and consequently A_v' will be less than A_v. This is a case of negative feedback because the gain of the system has been reduced.
   (c) If $A_v = 1$, the denominator becomes zero and A_v' is theoretically infinite. This implies that the circuit will provide a finite output with no signal input. When the condition occurs the amplifier is unstable. This situation, when **deliberately** introduced into an amplifier system, provides an **oscillator**.

5 For the case of negative feedback, βA_v is negative, hence either A_v or β must be negative. Either the amplifier or the feedback network, or both, may introduce a phase shift between their input and output terminals. These phase shifts have been indicated by the angle θ and ϕ respectively in *Figure 2.3*. If the feedback network consists simply of a resistance divider (as it often does in practice) there will be no phase shift and ϕ will be zero. For βA_v to be negative in this case, A_v must be negative and so the amplifier phase shift θ must be 180°. The total phase shift round the loop is then 180° and the voltage fed back to the input, v_f, will be antiphase to the existing input. If the amplifier phase shift is zero, the phase shift

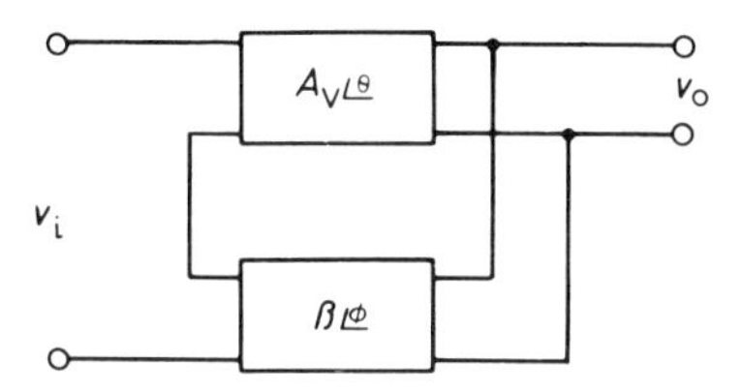

Figure 2.3

in the feedback network will have to be 180° in order for the feedback to be negative.

6 The general feedback equation for negative βA_v becomes

$$A_v = \frac{A_v}{1-(-\beta A_v)} = \frac{v_o}{1+\beta A_v}$$

It is often wisest, however, to work always to the equation having the negative sign in the denominator and let the overall resultant sign be determined by the signs of β and A_v. In more advanced work it is not always possible to tell whether the feedback is positive or negative.

Suppose the basic gain of an amplifier, A_v, is very large or $\beta A_v \gg 1$. Then $(1+\beta A_v) \simeq \beta A_v$ and

$$A_v' = \frac{A_v}{\beta A_v} = \frac{1}{\beta}$$

But β is a constant for a particular feedback system, hence the gain with feedback becomes independent of the basic gain A_v. Undesirable variations in A_v, therefore, leave A_v' unaffected. The gain in other words has been stabilised.

B WORKED PROBLEMS ON NEGATIVE FEEDBACK SYSTEMS

Problem 1 In a certain voltage amplifier the open-loop gain $A_v = 200$. Find the overall gain when feedback is applied if (a) $\beta = 0.004$, (b) $\beta = -0.02$. Comment on the results.

(a) Overall gain $$A_v' = \frac{A_v}{1-\beta A_v} = \frac{200}{1-(0.004\times 200)}$$

$$= \frac{200}{0.2} = \mathbf{1000}$$

(b) $$A_v' = \frac{200}{1-(-0.02\times 200)} = \frac{200}{5}$$

$$= \mathbf{40}$$

Case (a) represents positive feedback because the gain has been increased from 200 to 1000. The amplifier may not remain stable with this value of overall gain. Case (b) represents negative feedback because the gain has been reduced from 200 to 40. This may seem to be a considerable disadvantage of n.f.b. Indeed it is, when looked at simply in terms of voltage gain, but other advantages of n.f.b. are so numerous that it pays to design the original amplifier with sufficient gain so that after feedback has been introduced, the overall gain will not be less than the desired figure.

Problem 2 An amplifier with a voltage gain of −20 000 is used in a feedback circuit where $\beta = 0.02$. Calculate the overall gain with feedback. If the gain of the amplifier dropped to one-half of its inherent value because of a reduction in the supply voltage, what would the overall circuit gain become?

With the full gain available,

$$A_v' = \frac{-20\,000}{1-(0.02 \times -20\,000)} = -\frac{20\,000}{401} = \mathbf{-49.9}$$

When the basic gain falls to −10 000,

$$A_v' = \frac{-10\,000}{1-(0.02 \times -10\,000)} = -\frac{10\,000}{201} = \mathbf{-49.75}$$

This problem illustrates that important property of negative feedback mentioned earlier – gain stability. Although the inherent gain of the amplifier without f.b. fell by 50%, the overall gain with f.b. fell by less than 0.5%.

Problem 3 Describe, using circuit diagrams, how a feedback voltage may be derived from the output of an amplifier. Show how the way in which such feedback is derived affects the output resistance of the amplifier.

Figure 2.4 (a) and *(b)* shows possible arrangements for obtaining a fraction of the output voltage of an amplifier for feedback purposes. *Figure 2.4(a)* shows two resistors connected in series across the output terminals of the amplifier. This is direct voltage feedback because v_f is directly proportional to v_o and the feedback fraction β is clearly given by $R_2/(R_1+R_2)$. The total resistance of R_1 and R_2 in series is very much **greater** than the normal load R_L so that the introduction of feedback does not modify the loading.

Figure 2.4(b) illustrates an alternative method, current-derived feedback. Here the load current is passed through a resistor R_1 in series with R_L. The voltage developed across R_1 is now proportional to the output current i_o and the feedback fraction β is now $R_1/(R_1+R_L)$. The value of R_1 is very **small** compared with R_L so that the introduction of the feedback does not modify the output loading. Under this condition β approximates to the ratio R_1/R_L.

The output resistance R_o of an amplifier is the resistance measured at the output terminals when the input is zero. The introduction of feedback modifies the output resistance. Consider *Figure 2.4(a)* again. Suppose v_o increases for some reason but v_i remains constant. The feedback voltage $v_f = \beta v_o$ will increase and the effective input voltage will be reduced (negative feedback). This will tend to reduce the output voltage in turn. The net effect is to keep v_o constant, hence the output circuit behaves as a device with a **small** internal resistance, i.e. a constant-voltage generator. The diagram illustrates this form of output equivalent generator.

Conversely, looking at *Figure 2.4(b)*, if i_o increases for any reason, v_f increases and the effective input quantity (current or voltage) will **decrease**. The net effect is to keep i_o constant. The output circuit behaves this time as a constant-current generator, i.e. a device with a **large** output resistance R_o. The equivalent circuit is shown in the diagram.

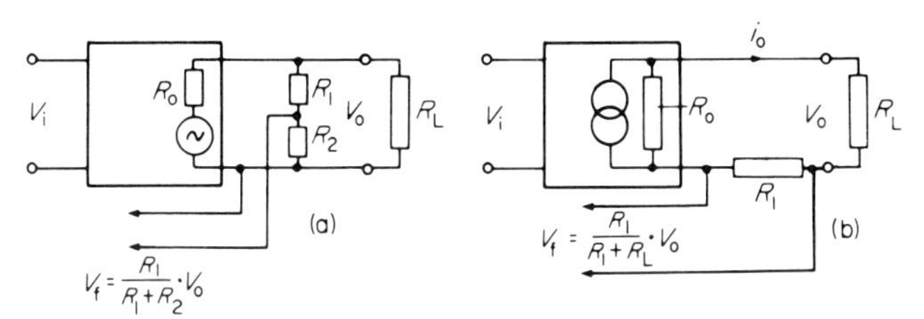

Figure 2.4

We deduce that voltage feedback (parallel connection) **reduces** the output resistance and current feedback (series connection) **increases** the output impedance. If R_o is the output resistance of an amplifier without feedback and R_o' is the output resistance with feedback added, it can be proved that for parallel connection as in *Figure 2.4(a)*,

$$R_o' = \frac{R_o}{1 + \beta A_v}$$

and for the series connection as in *Figure 2.4(b)*,

$$R_o' = R_o(1 + \beta A_v)$$

Problem 4 Using suitable diagrams, show how negative feedback may be applied to the input terminals of an amplifier. Show the effect each of the methods you describe has on the input resistance of the amplifier.

Feedback can be applied either in series or in parallel with the input terminals of an amplifier. In *Figure 2.5(a)* feedback is applied in series with the input and in *Figure 2.5(b)* feedback is applied in parallel with the input.

For the case of series applied feedback, v_f is opposing v_i and the effective input magnitude at the amplifier terminals will be $v_i - v_f$. This will be less than v_i and hence, as seen from the input terminals, the input current will be **reduced** from i_i to i_i'. This means that the input resistance as seen from the **source** terminals has apparently **increased**.

Now consider the situation in *Figure 2.5(b)* where the feedback is applied in parallel with the input. Here we are concerned with current addition. The effective input current i_i' will be $i_i - i_f$ or rearranging, $i_i = i_i' + i_f$. Hence i_i has increased and so, as seen from the **source** terminals, the input resistance has apparently fallen.

We deduce that with series applied feedback the input resistance is **increased** and with parallel applied feedback the input resistance is **decreased**. If R_i is the input resistance without feedback and R_i' is the input resistance with feedback added, it can be proved that for the series connection of *Figure 2.5(a)*.

$$R_i' = R_i(1 + \beta A_v)$$

and for the parallel connection of *Figure 2.5(b)*,

$$R_i' = \frac{R_i}{1 + \beta A_v}$$

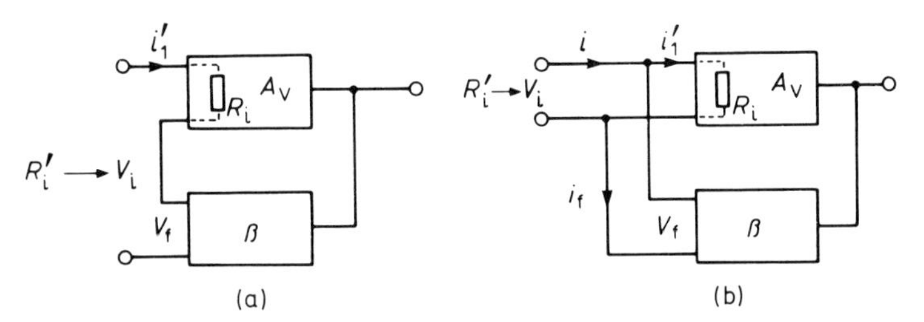

Figure 2.5

Problem 5 An amplifier has an open-loop gain of 2×10^5 and an input resistance of 50 kΩ. When n.f.b. is applied in series with the input, the gain is reduced to 5×10^2. Find (a) the feedback fraction β, (b) the input resistance with feedback.

$$A_v' = \frac{A_v}{1 + \beta A_v}$$

Rearranging gives $\quad A_v = A_v' + \beta A_v A_v'$

$$\therefore \quad \beta = \frac{A_v - A_v'}{A_v A_v'} = \frac{(2 \times 10^5) - (5 \times 10^2)}{10^8}$$

$$= \frac{1.99 \times 10^5}{10^8} = \mathbf{1.995 \times 10^{-3}}$$

Strictly, this value of β is negative of course.

(b) For series applied feedback, input resistance with feedback

$$R_i' = R_i(1 + \beta A_v)$$

$$R_i' = 50 \times 10^3[1 + (1.995 \times 10^{-3} \times 2 \times 10^5)]$$
$$50 \times 10^3[1 + (3.99 \times 10^2)]$$
$$20 \times 10^6\,\Omega = \mathbf{20\,M\Omega}$$

Problem 6 What effect does n.f.b. have on the bandwidth of an amplifier?

For large amounts of negative feedback, $A_v' \simeq 1/\beta$, hence if β can be made independent of frequency, the overall gain will be independent of frequency. Bandwidth is defined as that range of frequencies over which the gain does not fall below $1/\sqrt{2}$ of its maximum or mid-band value. *Figure 2.6* shows the open-loop response of an amplifier having a bandwidth of $f_2 - f_1$ Hz and average mid-frequency gain A_o. When n.f.b. is applied, the gain is reduced to

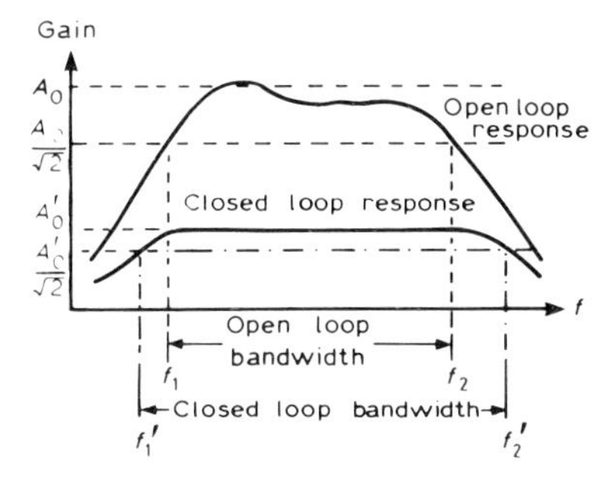

Figure 2.6

$A_o' = A_o(1 + \beta A_o)$ and is stable at this value over the new mid-frequency range. A study of the diagram makes it clear that the bandwidth has now increased to $f_2' - f_1'$. It can be proved that

$$f_2' = f_2(1 + \beta A_o) \text{ and } f_1' = \frac{f_1}{f_1(1 + \beta A_o)}$$

As f_1' is often a very low frequency, to a good approximation the overall bandwidth becomes f_2'. Although n.f.b. reduces the gain of an amplifier, the bandwidth is increased, so that the product gain × bandwidth is usually unaffected.

Problem 7 Describe an amplifier in which the whole of the output signal is fed back in opposition to the input signal. Deduce the gain of this amplifier with feedback and suggest an application for the amplifier.

The common-collector amplifier or **emitter-follower** circuit shown in *Figure 2.7* is an example of 100% feedback. The load resistor R_L is in the emitter circuit, resistors R_1 and R_2 are the usual bias components and the collector is returned directly to the V_{cc} line where, from the signal point of view, it is effectively at earth potential. Suppose a sinusoidal signal is applied at the input terminals.

As v_i rises towards its positive peak value, emitter current increases and the

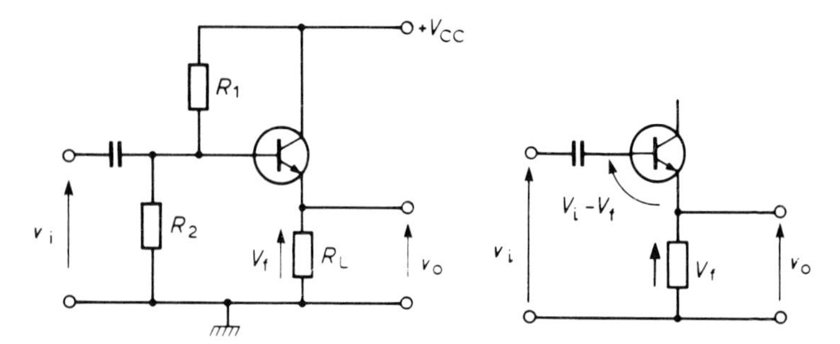

Figure 2.7

Figure 2.8

emitter voltage rises relative to earth. The converse happens when the input falls towards its negative peak value. Hence the emitter voltage **follows** the input signal variations and v_i and v_o are in phase **relative to the earth line**. The effective base-emitter input signal to the transistor, however, is the difference between v_i and v_f as *Figure 2.8* shows, and around the input loop these voltages are in phase opposition. Notice that v_o is in phase with v_i and hence A_v $(=v_o/v_i)$ is positive; but v_f is phase opposed to v_i, **relative to the emitter**, and hence to v_o. So **all** of the output voltage developed across R_L is fed back in series with v_i. Substituting $\beta = -1$ into the feedback equation gives

$$A_v' = \frac{A_v}{1 + A_v}$$

This result must be less than unity; hence the voltage gain of the emitter-follower is always slightly less than 1 and there is no voltage phase reversal.

The usefulness of the emitter-follower stems from the effect that 100% n.f.b. has on the input and output resistances. As the feedback voltage is derived in **parallel** with the output load and fed back in **series** with the input, the input resistance will be very high and the output resistance very low. Hence this amplifier is ideal as an electronic 'transformer' for matching a stage of high output resistance into one of low input resistance without voltage reduction or phase reversal. Feeding pulse signals into a capacitive load is one obvious application.

C FURTHER PROBLEMS ON NEGATIVE FEEDBACK

(a) SHORT ANSWER PROBLEMS

1 Define the terms (a) open-loop gain, (b) feedback fraction, (c) gain stability, (d) series-derived feedback, (e) parallel-derived feedback.

2 Say whether each of the following statements is true or false: (a) Series-connected feedback increases the input resistance. (b) Series-connected feedback reduces the voltage gain. (c) Parallel-connected feedback increases the output resistance.

(d) Parallel-connected feedback reduces the voltage gain. (Take all these feedbacks to be negative.) (e) The emitter-follower is an example of 100% n.f.b. (f) If β is positive, the circuit is certain to oscillate. (g) Negative feedback has no effect on bandwidth.

3. An amplifier has a voltage gain of -1000 and is used with a feedback network in which $\beta = 0.01$. Calculate the gain with feedback applied. [−90.9]
4. An amplifier has an open-loop gain of 10^4. Calculate the gain when 0.9% of the output voltage is fed back to the input. [109.9]
5. An amplifier is required to have a final gain of 80. What should be the open-loop gain of the basic amplifier if β is to be fixed at -0.01? [400]
6. An amplifier having an open-loop gain of 2×10^4 has n.f.b. applied which reduces the gain to 50. What is the value of β? [−0.02]
7. Complete the spaces in this table:

A_v	β	β	A_v'
50	−0.004		
180			18
	−0.5		160
25		−6.25	

8. In the block diagram of *Figure 2.9* feedback is shown being taken from the output terminals of the amplifier by way of resistors R_1 and R_2. If β is to be 0.1 and $R_1 = 10\,000\ \Omega$ what should be the value of R_2? [1111 Ω]
9. In the block diagram of *Figure 2.10* calculate the value of the feedback fraction β, (a) approximately, (b) exactly. [0.01, 0.0099]
10. In the block diagram of *Figure 2.11*, what is the overall voltage gain of the system v_o/v_i? [8.69]

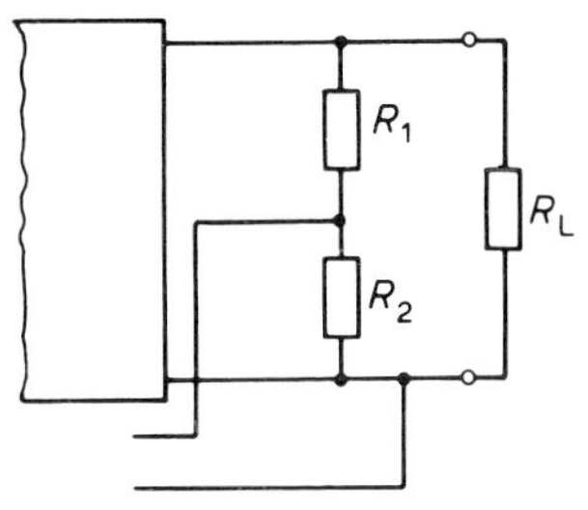

Figure 2.9

Figure 2.10

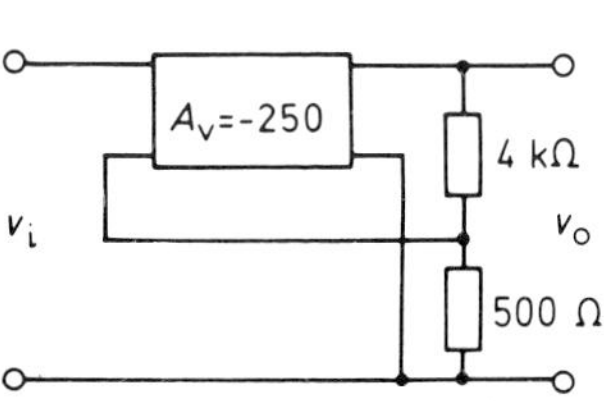

Figure 2.11

(b) MULTI-CHOICE PROBLEMS (answers on page 152)

1 If the loop gain βA_v is positive and less than unity, the gain of the system is (a) reduced, (b) increased, (c) unaffected.
2 If the loop gain βA_v is equal to unity, the amplifier (a) will be stable, (b) will be unstable, (c) may or may not be unstable.
3 The effect of negative feedback being derived from the circuit in *Figure 2.9* is to reduce the output resistance because (a) the feedback network loads the output by way of R_2, (b) R_1 and R_2 are in parallel with the load, (c) the amplifier output acts as a constant-voltage generator, (d) the output acts as a constant-current generator.
4 In the block diagram of *Figure 2.12*, the resistors R_1 and R_2 are $\pm 10\%$ tolerance. The extreme values of β will be (a) 0.091 and 0.109, (b) 0.076 and 0.109, (c) 0.090 and 0.091, (d) 0.100 and 0.091.
5 *Figure 2.13* shows negative feedback being applied in parallel with the amplifier input terminals. The **voltage** gain of this amplifier is (a) reduced by the feedback, (b) increased by the feedback, (c) unaffected by the feedback.
6 The voltage gain of the circuit of *Figure 2.14* is (a) 30, (b) 28.6, (c) 27.7, (d) 600.
7 A feedback amplifier with $A_v = 10^4$ and $\beta - 0.3$ will have an overall gain of about (a) 3.3, (b) 33, (c) 3300, (d) 10^4.
8 The amplifier of *Problem 6* has an input resistance of 2000 ohms before feedback is applied. After feedback the input resistance is (a) 95 kΩ, (b) 105 kΩ, (c) 38 kΩ, (d) 42 kΩ.

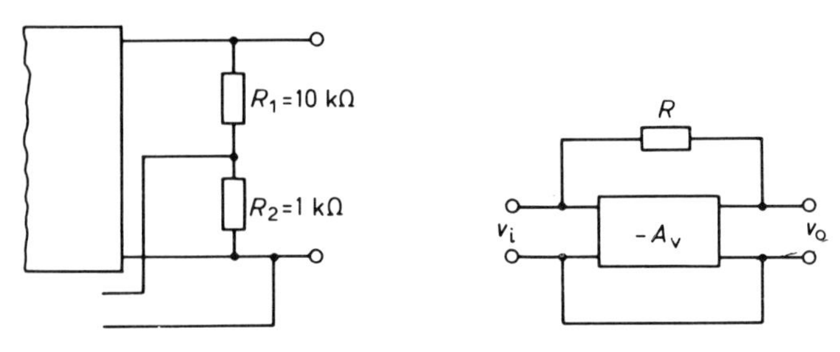

Figure 2.12

Figure 2.13

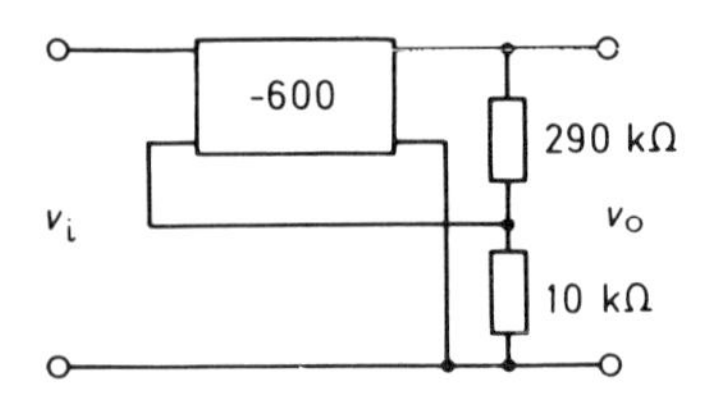

Figure 2.14

9 Given that current i_2 in the block diagram of *Figure 2.15* is 2 mA, current i_1 is (a) 18 mA, (b) 22 mA, (c) 10 mA, (d) 40 mA.

10 The rato i_o/i_1 is the overall current gain of the amplifier of *Problem 9*. This is approximatey (a) 11.1, (b) 9.1, (c) 4.5, (d) 102.

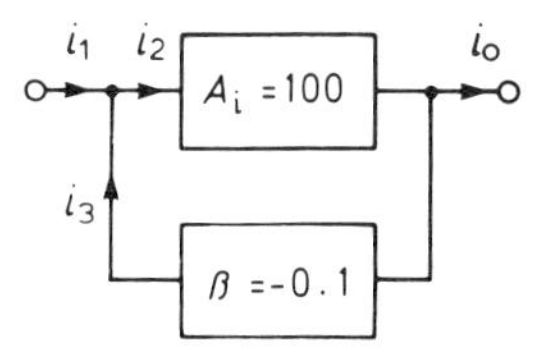

Figure 2.15

(c) CONVENTIONAL PROBLEMS

1 An amplifier has a gain of −500 and is used in a feedback circuit where $\beta = 0.05$. Calculate the overall voltage gain. If the gain of the amplifier drops to −400 because of a reduction in supply voltage, what will the overall gain become?

2 In the feedback circuit shown in *Figure 2.16*, the amplifier gain is −1000 when the 330 Ω resistor is short-circuited. What is the gain when the short-circuit is removed? If the resistors used have a tolerance of ±5%, calculate the maximum and minimum values that the gain might have with feedback.

[30.3] [27.6, 33.34]

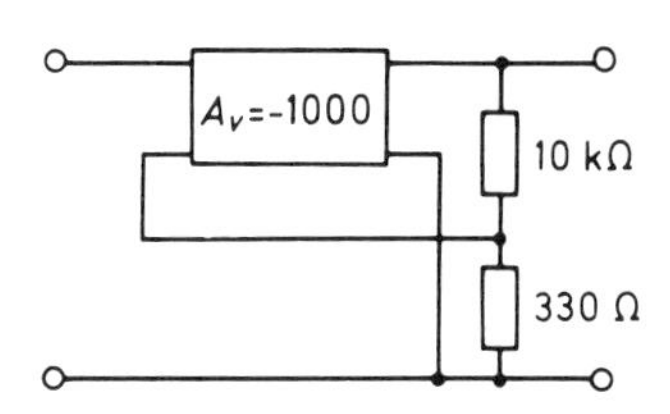

Figure 2.16

3 Negative feedback is taken from a potential divider connected across the output terminals of an amplifier and applied in series with the input signal. If the feedback reduces the open-loop gain of the amplifier by one-half, by what factor will the input and output resistances change?

[Input resistance doubled, output resistance halved]

4 An amplifier is to be designed to have an overall gain of 20 ± 0.1%. The basic amplifier from which the final circuit is to be evolved has a gain A_v ± 10%. Calculate the required value of β and the open-loop gain A_v.

[0.049, 2220]

5 An amplifier has an open-loop gain of 300 which is found to fall by 20% due to supply variations. If the gain is to be stabilised so that it falls by only 1%, calculate the amount of n.f.b. required. If the original 3 dB bandwidth of the amplifier was

500 Hz to 50 kHz, calculate the 3 dB frequencies when the feedback is applied.

[0.08, 20 Hz, 1.24 MHz]

6 An amplifier A in the absence of feedback has a gain that is liable to fall by 40% of its rated value as a result of external factors. By the application of n.f.b. an amplifier is to be provided that has a rated gain of 100 and with the requirement that this gain should not fall below 99 using the amplifier A. Find the necessary gain of A and the feedback fraction β. [6600, 0.0098]

7 An amplifier has an open-loop gain of 500, a bandwidth of 100 kHz and an input resistance of 1000 ohms. By the application of n.f.b. wired in series with the input, the bandwidth is increased to 5 MHz. What will be the new values of gain and input resistance? [10, 50 kΩ]

8 An amplifier with n.f.b. operative is found to have a voltage gain of 25. When the feedback leads are interchanged at one end, the gain rises to 85. Calculate (a) the feedback fraction, (b) the amplifier open-loop gain.

[(a) 0.041, (b) 38.5]

9 An amplifier without feedback has an input impedance equal to a shunt capacitance of 30 pF. If the open-loop gain is 20 and 10% n.f.b. is added in series with the input, what will be the apparent input capacitance? (Hint: capacitive reactance is inversely proportional to frequency.) [10 pF]

10 An amplifier having an output resistance of 1 kΩ has an overall voltage gain of −10 000. The gain is to be reduced to 200 by simultaneously applying current and voltage feedback in series with the input so that the output resistance is unaffected. Calculate the required current feedback resistor to be connected in series with the 1 kΩ load resistor (refer to *Figure 2.11*) and the percentage voltage feedback. [2.55 Ω, 0.0051]

11 Sketch a circuit diagram of an emitter-follower amplifier. Explain how negative feedback operates in this circuit. Is it voltage or current feedback?

3 Unwanted signals: noise

A MAIN POINTS CONCERNED WITH THE NATURE AND EFFECTS OF NOISE

1 Noise is defined as any spurious or unwanted electrical signal set up in or introduced into an electronic system. Noise can be divided into two general classes:

(a) Internal noise generated within the electronic system itself as the result of the random movement of charge carriers in resistors and connecting wires and in active devices such as transistors or valves.

(b) External noise caused by atmospherics, aircraft reflections, X-ray and diathermy apparatus and any spark-producing systems such as electric motor commutators and car ignition circuits.

For the signal output of any system to be of use, the signal power must be greater than the noise power. The ratio of the wanted signal power to the unwanted noise power is known as the **signal-to-noise ratio** (S/N ratio). If the signal level falls below the value giving the minimum signal-to-noise ratio, further amplification can only increase the level of both signal and noise equally, and the signal-to-noise ratio is unaffected.

2 The random movement of charge carriers in any electrical circuit conductor generates noise voltages. This noise is known as **thermal noise** or Johnson noise. Free electrons wander about within the atomic structure of the conductor, first in one direction and then another, but at any particular instant there will be more electrons moving in some directions than others. Since the movement of electrons constitutes an electric current, an e.m.f. of a certain polarity will be induced instantaneously between the ends of the conductor. The time duration of each of these states is extremely short and so the polarity and magnitude of the generated e.m.f. fluctuates rapidly. Over a period that is long compared with the mean time an electron spends moving in a particular direction, the total e.m.f. between the ends of the conductor is zero.

Thermal noise voltage is measured in terms of r.m.s. and can be calculated from the formula

$$V_n = \sqrt{(4kTRB)}$$

where k is a constant $= 1.38 \times 10^{-23}$ J/K

T = temperature of the conductor in K (= °C + 273°)

R = resistance of the conductor in ohms

B = bandwidth in Hz of the circuit at whose output the noise appears or the bandwidth over which the noise is measured, whichever is the smaller.

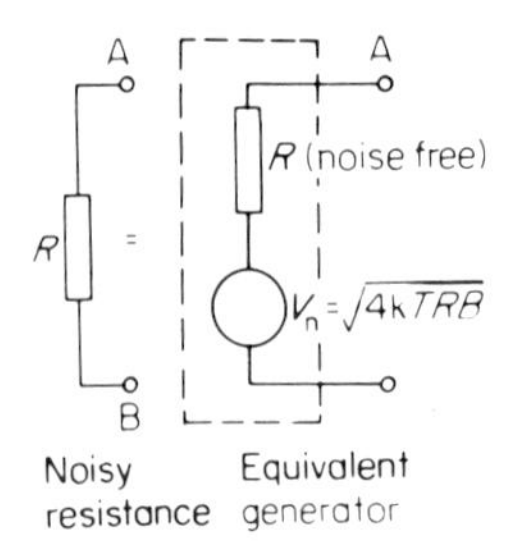

Figure 3.1

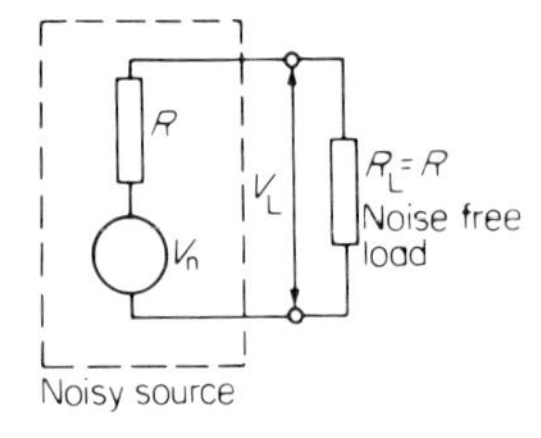

Figure 3.2

Strictly the **total** bandwidth of a system should be considered but the usual 3 dB bandwidth gives sufficiently accurate results in most cases.

Noise which contains equal magnitude of all frequencies is known as **white noise**. Thermally generated noise is white noise.

3 Thermal noise can be represented as a Thévenin equivalent generator having an e.m.f. equal to the noise e.m.f. connected in series with a noiseless resistance R having a value equal to that of the noisy resistor, see *Figure 3.1*. The open-circuit voltage across the termals A–B is then V_n. No polarity is associated with this hypothetical generator.

Let a noiseless load resistor R_L be connected to the generator (see *Figure 3.2*) such that $R_L = R$. This condition ensures the maximum transfer of power from the generator to the load. The load voltage V_L will be $\frac{1}{2}V_n$ and so

$$V_L = \tfrac{1}{2}\sqrt{(4kTRB)} = \sqrt{(kTRB)}\text{ volts}$$

Then the load power

$$P_L = \frac{V_L^2}{R_L} = \frac{kTRB}{R} = kTB\text{ watts}$$

Thus the maximum or **available** noise power that a thermal noise source can deliver to a load is **kTB** watts. This power is independent of R but is directly proportional to both the temperature and the bandwidth. For an unmatched load, the noise power is reduced.

4 Noise is a function of temperature. The temperature of a resistor is not simply that of the ambient but that resulting from its own generated heat. The noise equation applies strictly to metallic conductors and hence to wire-wound resistors. Carbon resistors introduce additional noise because of the random variation in the contact resistance between the granules making up the body of the resistor. This **contact** (or current) noise increases as the value of the resistance and the current flowing in the resistor increases. The use of low-noise resistors such as metal-oxide types reduces this kind of noise. The total noise voltage developed in a carbon resistor is given by

$$V_n = \sqrt{[(\text{thermal noise})^2 + (\text{contact noise})^2]}$$

The same form of equation related the total noise generated by two or more noise sources in series:

$$V_n = \sqrt{[(V_{n1})^2 + (V_{n2})^2 + \ldots]}$$

5 Noise generated in active devices like transistors results from several different effects and is superimposed on the ordinary direct and alternating signal currents that the device is normally carrying.

Shot noise is the result of fluctuations when charge carriers cross potential barriers. In a transistor, shot noise is caused by the random arrival and departure of carriers by diffusion across a p–n junction. A transistor has two such junctions and hence two such sources of noise. This noise is usually more significant than the thermal noise of associated components and can mask very low level signals.

Partition noise comes about because of the random way in which the emitter current divides between base and collector paths. Electron-hole recombinations in the base give rise to this noise. Partition noise is generally negligible in modern transistors.

Flicker noise is believed to be due to emitter surface leakage and conductivity variations in the semiconductor material. It is predominantly low-frequency in nature and becomes insignificant above a few kilohertz. Unlike white noise, flicker (or 'pink') noise is consequently frequency dependent. For this reason it is often known as $1/f$ noise.

Like the bipolar transistor, the field-effect transistor exhibits shot noise, thermal noise and $1/f$ noise. However, the FET is inherently a lower noise device than the bipolar because (a) there is only one p–n junction involved and so shot noise is reduced, (b) the channel resistance is small, hence thermal noise is small, and (c) carriers move directly along the channel from source to drain and so partition noise is absent.

6 External noise relates to noise originating outside any electronic system. Such noise can be either man-made or occur naturally. Man-made noise is, to a certain extent, under his control. Natural noise is rarely controllable.

7 The noise generated in an amplifier or introduced along with the input signal determines the smallest wanted signal level that can be amplified usefully. An important measure of the performance of an amplifier or receiver is the **signal-to-noise ratio**, which is defined as the ratio of the wanted signal power to the unwanted noise power, so

$$\text{S/N ratio} = \frac{\text{signal power}}{\text{noise power}}$$

As it is usual to express this ratio in decibels, we have

$$\text{S/N ratio} = 10 \log \left[\frac{\text{signal power}}{\text{noise power}} \right] \text{dB}$$

Since $P = V^2/R$, we can substitute for the case where the signal voltage is developed in series with a total resistance R. Then

$$\frac{\text{signal power}}{\text{noise power}} = \frac{(\text{signal voltage})^2}{R} \times \frac{R}{(\text{noise voltage})^2}$$

$$= \left[\frac{\text{signal voltage}}{\text{noise voltage}} \right]^2$$

$$\text{Hence S/N ratio} = 10 \log \left[\frac{\text{signal voltage}}{\text{noise voltage}}\right]^2$$

$$= 20 \log \left[\frac{\text{signal voltage}}{\text{noise voltage}}\right] \text{dB}$$

B WORKED PROBLEMS ON NOISE

Problem 1 Discuss briefly the main contributors to natural noise.

The bulk of natural noise is contributed by atmospheric or static noise resulting from lightning discharges. The effect of lightning on radio reception is well known, resulting in violent crackling sounds or, in the presence of 'sheet' lightning, in an almost continual 'frying' background to the wanted programme. Atmospheric noises of this sort can be picked up over very great distances and static interference is often detectable even when no thundery weather is apparent.

At high frequencies, above some 20 MHz or so, the effect diminishes but other sources of noise appear. Solar disturbances lead to background noise and fading at frequencies above some 50 MHz, and thermal radiation from the surface of the earth, particularly during the summer months, generates noise which becomes prominent at frequencies over some 200 MHz. Variations in the ionised layers in the upper atmosphere lead to what is known as quantum noise, though only very high frequency bands are affected by this. Galactic or cosmic noise, originating in outer space, can also be troublesome above some 1000 MHz.

Problem 2 Mention two examples of man-made noise and briefly discuss how they may be eliminated or reduced in their effects on electrical apparatus such as amplifiers or receivers.

Two common examples of man-made noise interference are mains hum and spark-induced radiation.

Any electronic system that receives its power supply from rectified a.c. mains is liable to have hum voltages introduced into its circuitry. Alternating magnetic flux from wires carrying alternating current or from the iron cores of transformers and chokes may link with the input circuits of the amplifying devices and induce in them an unwanted e.m.f. Similarly, stray capacitance between active terminals and some other conductor around which there is an alternating electric field will lead to the flow of unwanted current in the active leads. Electric and magnetic screening, together with a sensible layout and orientation of such components as transformers, will eliminate most of the problems associated with hum.

Spark-induced noise arises whenever an electric circuit is interrupted, whether by switch, motor commutator, car ignition or welding equipment, to name a few possibilities. The resulting sparks generate high-frequency currents in the circuit wires and these in turn act as transmitting aerials, radiating radio signals which cover a very wide band of frequency. Voltage pulses are also set up which make their way into the mains wiring system, from which they enter other mains operated systems. Screening the equipment will prevent a lot of direct radiation from the source of the sparking. In the same way, screening any susceptible equipment will prevent any direct pick-up of the offending radiation. Mains-borne pulses can be reduced or eliminated by the use of choke-capacitance suppressors wired into the mains supply leads close to the apparatus. Aerials cannot be screened, of course, but they can be orientated away from any source of interference or repositioned so that some sort of localised screening is achieved, such as the wall of a house or a line of trees.

Problem 3 What is meant by the noise factor of an electronic system?

Noise factor (or noise figure as it is sometimes called) is a measure of the noise quality of an electronic system. It is expressed as the ratio of the **total** noise at the output of the system to **that part** of the output noise which is due to the input noise alone. Hence

$$\text{Noise factor } F = \frac{\text{total noise power at the output}}{\text{output noise power due to input noise power}}$$

It is therefore a measure of the amount of internal noise generated by the system per unit gain. If the system introduced no noise, F would be unity. In a practical system, F should be kept as small as possible. F may, of course, be expressed in decibels.

Problem 4 Show that the noise factor can be expressed as the ratio of the input and output signal-to-noise ratios of a system.

Let an amplifier have a power gain A_p and an internally-generated noise power P_{na}, see *Figure 3.3*. Let the amplifier be fed from a matched input source which provides an input noise power P_{ni}. The total noise power at the output is P_{no}.

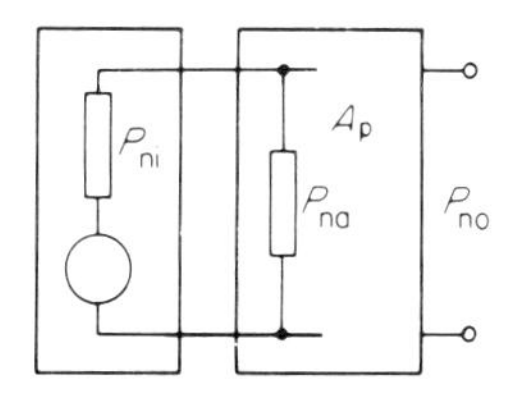

Figure 3.3

Now from the definition of noise factor

$$F = \frac{P_{no}}{P_{ni} \times A_p}$$

$$\text{But} \quad A_p = \frac{\text{signal power at the output}}{\text{signal power at the input}} = \frac{P_{so}}{P_{si}}$$

$$\therefore \quad F = \frac{P_{no}}{P_{ni} \times \dfrac{P_{so}}{P_{si}}} = \frac{P_{si}/P_{ni}}{P_{so}/P_{no}}$$

$$= \frac{\text{input S/N ratio}}{\text{output S/N ratio}}$$

Problem 5 What is the noise voltage generated by a 100 kΩ resistor at 27°C if the bandwidth is 1 MHz?

Using the formula

$$V_n = \sqrt{(4kTRB)}\, V$$

$$V_n = \sqrt{(4 \times 1.38 \times 10^{-23} \times 10^5 \times 300 \times 10^6)}\ \text{V}$$

$$= \sqrt{(1656 \times 10^{-12})}$$

$$= 40.7 \times 10^{-6}\ \text{V} = \mathbf{40.7\ \mu V}$$

Make a note that 27°C = 273 + 27 = 300 K, 100 kΩ = 10^5 Ω and 1 MHz = 10^6 Hz. Care must always be taken when substituting values into this formula.

Problem 6 If the signal level at the input of an amplifier is 105 μV and the noise level is 15 μV, what is the input signal-to-noise ratio in dB?

$$\text{S/N ratio} = 20 \log \left[\frac{\text{signal voltage}}{\text{noise voltage}} \right] \text{dB}$$

$$= 20 \log \frac{105}{15} = 20 \log 7$$

$$= 20 \times 0.85 = \mathbf{17\ dB}$$

Problem 7 Sketch frequency spectra diagrams showing the distributions of white noise and pink noise and their overall resultant distribution.

The distribution of white noise, made up of thermal, shot and partition noise is shown in *Figure 3.4(a)*. This noise has a constant energy per unit bandwidth. Pink or $1/f$ noise is frequency dependent and its distribution is shown in *Figure 3.4(b)*. Low-frequency amplifiers are particularly prone to this noise which is most evident below a few kilohertz. In this respect, an a.c. amplifier is preferable to a d.c. amplifier as the falling gain of the former type at low frequencies minimises the amount of $1/f$ noise passing through the system.

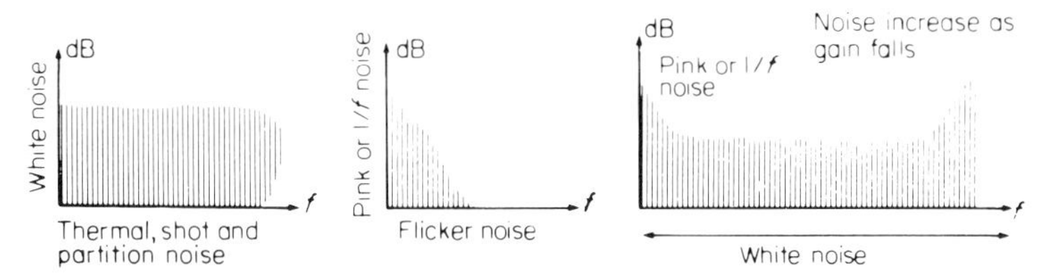

Figure 3.4

Figure 3.4(c) shows the overall resultant noise spectrum. At high frequencies the noise increases above the white noise level because of the fall in the gain of the system. This applies to both a.c.- and d.c.-coupled amplifiers.

C FURTHER PROBLEMS ON NOISE

(a) SHORT ANSWER PROBLEMS (answers on page 152)

1 Give an example of (a) a low frequency noise, (b) a high frequency noise, (c) a man-made noise, (d) a natural noise.
2 Explain what is meant by (a) white noise, (b) pink noise.
3 Why is a FET less inherently noisy than a bipolar transistor?
4 Define the term signal-to-noise ratio.
5 What is meant by the noise factor of a system?
6 What is the noise voltage generated by a 470 kΩ resistor at a temperature of 17°C if the bandwidth involved is 100 kHz? [22.4 μV]
7 The signal level at the input of a receiver is 25 μV and the noise level is 2 μV. What is the S/N ratio in dB? [22 dB]
8 Are these statements true or false? (a) A wideband amplifier is noisier than a narrow band amplifier, whatever the operating frequencies may be. (b) A piece of wire resting on a bench is not generating any thermal noise voltage. (c) The transfer of noise power from a source to a load is reduced when the source and load are mismatched. (d) The noise generated by a carbon resistor is wholly thermal in nature. (e) The maximum noise power that can be delivered by a resistance depends upon the value of the resistance. (f) A FET has lower shot noise than a bipolar transistor. (g) Partition noise cannot exist in a field-effect transistor. (h) An amplifier with a poor low-frequency response is less likely to suffer from $1/f$ noise than an amplifier with a good low-frequency response. (j) If the temperature doubles, the noise voltage generated by a wirewound resistor will double.

(b) MULTI-CHOICE PROBLEMS (answers on page 152)

1 If temperature and bandwidth are both doubled, the noise voltage generated by a wirewound resistor will be (a) doubled, (b) halved, (c) quadrupled, (d) unaffected.
2 Two resistors each having identical noise voltages are joined in series. The total noise voltage will be (a) twice the individual voltages, (b) half the individual voltages, (c) $\sqrt{2}$ times the individual voltages, (d) the same as either individual voltage.
3 Contact noise in resistors is affected by (a) the amount of thermal noise being generated, (b) the strength of the current flowing in the resistor, (c) the circuit bandwidth.
4 Shot noise in a transistor is caused by (a) fluctuations in the emitter current, (b) variations in the base-emitter potential, (c) the random injection of carriers across the p–n junctions, (d) variations in the leakage current.
5 Flicker noise in transistors is (a) significant at all frequencies, (b) significant only at very high frequencies, (c) significant only at very low frequencies, (d) non-existent.
6 If the collector current of a bipolar transistor is reduced, the shot noise (a) increases, (b) decreases, (c) is unaffected.
7 There is no partition noise in a FET because (a) there are no minority carriers, (b) there is only one p–n junction, (c) the channel has a low resistance, (d) the source current does not divide before reaching the drain.
8 Interference pick-up on an aerial system can often be reduced by (a) screening the aerial, (b) re-positioning the aerial, (c) using an indoor aerial, (d) fitting a suppressor in the aerial down lead.
9 An amplifier has a noise factor of 2, so it is introducing (a) negligible noise, (b) a lot of noise.
10 At frequencies above about 30 MHz the only significant source of noise is (a) atmospheric noise due to lightning, (b) thermal noise in resistors, (c) current noise in transistors, (d) car ignition systems.
11 White noise is noise for which the available power in a state bandwidth is (a) greatest above 30 MHz, (b) greatest below a few kilohertz, (c) the same at all frequencies.
12 The maximum available noise power obtainable from a resistor R is (a) proportional to R, (b) proportional to R^2, (c) proportional to $\sqrt{R}$, (d) independent of R.
13 A voltage source is matched to a load and both are at a temperature of 290 K. The signal-to-noise ratio in the load is 50. If the temperature doubles, the signal-to-noise ratio will be (a) unaffected, (b) 25, (c) 100, (d) 200.
14 The maximum noise power which can be delivered to a load of resistance R is (a) kTB, (b) $kTBR$, (c) $4kTBR$, (d) $\sqrt{(4kTBR)}$.
15 An amplifier has a power gain of 100. The noise power measured at its output is 100 times the noise power at its input. Its noise factor is (a) 0 dB, (b) 20 dB, (c) 40 dB, (d) 100 dB.

(c) CONVENTIONAL PROBLEMS

1 If the signal level at a tape recorder head output is 1 mW and the signal-to-noise ratio is 30 dB, what is the output noise power? [1 μW]

2 What is the available noise power from a resistance at a temperature of 27°C over a bandwidth of 2 MHz? [8.28×10^{-15} W]

3 Two resistors generating noise voltages of 100 μV and 64 μV respectively are joined in series. What is the total noise voltage? [118 μW]

4 The thermal noise generated in a 1 MΩ resistor at 20°C is 254 μV. What is the circuit bandwidth? [4 MHz]

5 An amplifier has a power gain of 20 dB. The signal power at the input is 0.2 mW and the noise power at the output is 0.2 μW. What is the signal-to-noise ratio of the system in decibels? [59 dB]

6 An amplifier has a flat passband characteristic extending from 100 kHz to 500 kHz and a voltage gain of 40 dB. What will be the noise voltage at its output resulting from a 500 kΩ resistor at 20°C connected across its input terminals? [5.68 mV]

7 An aerial has a noise power equal to 0.1 μW and the noise factor of the receiver matched to this aerial is 10. Calculate (a) the total noise power at the output of the receiver if it has a power gain of 19 dB, (b) the proportion of the output noise power contributed by the receiver. [(a) 80 μW, (b) 90%]

8 Prove that the power delivered to a matched load by a noise source of internal resistance $R\Omega$ is kTB watts.

9 Explain briefly the following types of noise occurring in low-level amplifiers: (a) shot noise, (b) partition noise, (c) flicker noise, (d) thermal noise.

10 A receiver is receiving a signal and measurements show that the S/N ratio at the receiver output is 30 dB. Assuming that all the noise occurs in the first stage of the receiver, calculate the output S/N ratio when (a) the output stage gain of the receiver is increased by 6 dB, (b) the incoming signal fades so that the received power falls to a quarter of its previous value. [(a) 30 dB, (b) 24 dB]

11 A tuned amplifier has a 3 dB bandwidth B Hz, a noise factor F, and is fed by a voltage source V_i having an internal resistance R_g. Prove that the output signal-to-noise ratio can be expressed as

$$\frac{V_i}{4kTR_g BF}$$

12 Deduce that in a superheterodyne receiver using a v.h.f. carrier, the signal-to-noise ratio at the receiver output will be practically unaffected by increasing the sensitivity of the intermediate-frequency amplifier.

4 Oscillators

A MAIN POINTS CONCERNED WITH SINUSOIDAL AND RELAXATION OSCILLATORS

1 An oscillator is fundamentally a power converter in the sense that its only input is the d.c. from a power supply and its output is a periodic time varying signal which may or may not be sinusoidal in form. An oscillator produces this output without requiring an externally applied input signal as an amplifier does. The input signal is, in other words, self supplied. In the circuit diagram of an oscillator, therefore, there is no evidence of a terminal which is specifically an input signal terminal. But there is an output terminal from which the generated waveform can be obtained.

2 Oscillator circuits can be divided into **negative-resistance** or **feedback** types. In turn, the output obtained may be
   (i) sinusoidal in form, obtained from **sinusoidal** or **harmonic** oscillators;
   (ii) non-sinusoidal in form, generally rectangular or triangular, obtained from **relaxation** oscillators.

   These outputs, whatever their precise form, always have some definite repetition rate or frequency. This repetition rate is determined by an *LC* or *RC* circuit arrangement.

3 The block diagram of a feedback oscillator is given in *Figure 4.1* and shows an amplifier with a feedback network connecting output to input. The amplifier

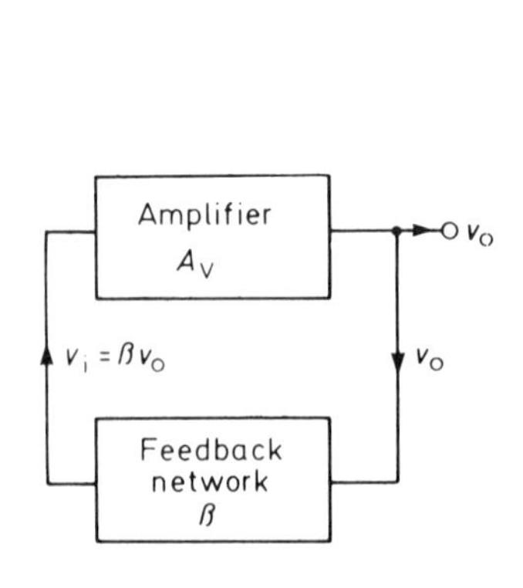

Figure 4.1

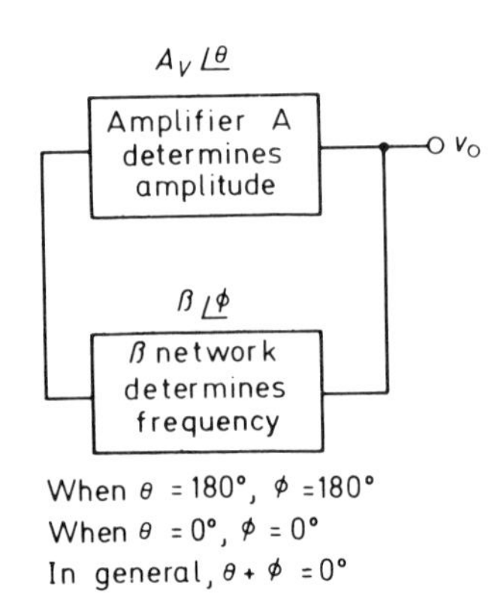

Figure 4.2

forms the active part of the circuit providing signal energy and the feedback network forms the passive part of the circuit whose function is to determine the frequency of operation. The feedback must be positive and the loop gain must be equal to unity. This means that there must be an overall zero phase shift round the loop and the amplifier gain must be such that it is capable of supplying sufficient energy to compensate for that which is dissipated in the resistance of the loop plus that required at its own input terminals to maintain oscillations.

In any feedback circuit the overall gain A'_v is given by the general feedback equation

$$A'_v = \frac{A_v}{1 - \beta A_v}$$

The condition for instability is $\beta A_v = 1$, making A'_v theoretically infinite. For oscillations to be possible therefore, $A_v = 1/\beta$. In making $\beta A_v = 1$, there are two implied conditions for the circuit to be an oscillator:

(i) The magnitude of the gain round the loop shall be unity to maintain oscillations; and

(ii) the phase shift round the loop shall be zero, or 180° different from the condition normally required for negative feedback.

The strict criterion for oscillation is therefore that $\beta A_v = 1 \angle 0°$. In practice, the magnitude of the gain needs to be greater than unity. Variations in supply voltage, temperature and component value changes, all vary the values of A_v and β slightly. It could be possible for βA_v to fall below unity with a particular combination of circumstances. The oscillation could then either not begin at switch on or, having begun, could collapse. When βA_v is adjusted to be slightly greater than unity, more signal is fed back than is needed for oscillations to occur and a definite build-up of signal level around the loop is ensured. This increase is limited in practice by non-linearities within the amplifier section and by the finite value of the supply voltage, but the additional gain is immediately available to counter any tendency for the oscillations to collapse.

4 Two performance features of oscillators need to be stable in most cases: (i) the generated frequency, (ii) the generated amplitude. At the desired frequency only, the phase shift is zero. This condition may be achieved by any arrangement where the phase shift in the amplifier is equal in magnitude but opposite in sign to the phase shift in the feedback network. Hence, for 180° phase shift in A (see *Figure 4.2*), the β network must provide a further shift of 180°; and for 0° shift in A, the β network must provide 0° (or 360°) shift. This condition will be met at some frequency ω_1 in the β-network phase characteristic. Thus it is the β network which determines the **frequency** of the generated oscillation.

At this frequency, the magnitude of the loop gain must be rather greater than unity for small signal levels, and should become exactly equal to unity for the desired output voltage amplitude. Thus it is a feature of the amplifier circuit and not the β network which maintains the oscillation **amplitude** substantially constant. These two requirements can be designed for separately. All oscillators, except those using the negative-resistance characteristics of a circuit device, are **phase-shift oscillators**, although they are often known under a variety of other names.

B WORKED PROBLEMS ON OSCILLATOR SYSTEMS

Problem 1 Describe an oscillator using a single transistor stage as amplifier, with feedback applied by way of an LC circuit. How could the feedback fraction be adjusted in an oscillator of this sort?

Figure 4.3 shows the circuit of what is known as a **tuned-collector** oscillator. In this oscillator there is a 180° phase shift through the common-emitter transistor amplifier and a further 180° is introduced by the appropriate direction of connection of the coupling coil L_1. The total phase shift around the loop (from base back to base) is 360° and so oscillation is possible. The magnitude of the feedback fraction must be sufficient to make the product $\beta A_v > 1$. This can be ensured by having sufficient gain in the amplifier, and then by adjusting the degree of mutual coupling between L and L_1 (by alteration of the physical spacing of the coils) to set β equal to $1/A_v$. This circuit can be tuned readily over a range of frequencies by making C variable. The frequency of the generated oscillation is that of the tuned circuit resonance, given closely by $1/2\pi\sqrt{(LC)}$.

There are a great number of different arrangements involving a single transistor amplifier and a resonant LC circuit, but if very low frequency operation is required, the values of both L and C become large and resistive losses increase. There is also the problem of physical bulk. Feedback networks which eliminate the inductor are then preferred.

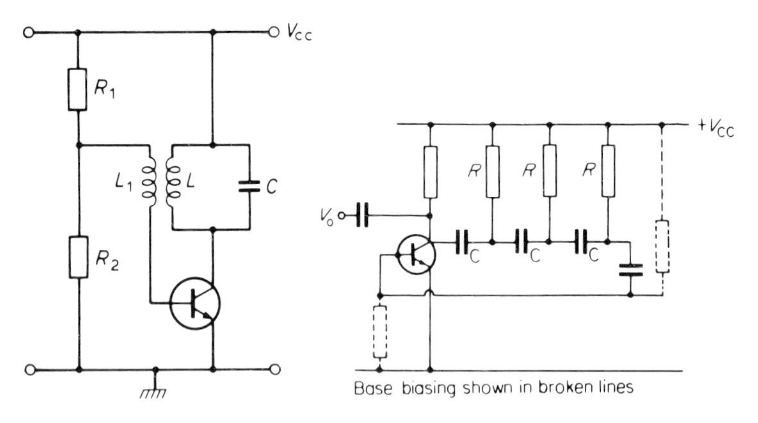

Figure 4.3 **Figure 4.4**

Problem 2 Describe an oscillator using C and R elements in its feedback network and mention any advantages or disadvantages this circuit might have.

Figure 4.4 shows an oscillator using three resistors and three capacitors in the feedback network. The transistor amplifier is connected in common-emitter mode and the base biasing arrangement is drawn in broken lines for clarity. In this circuit, 180° phase shift occurs through the transistor, hence a further 180° phase shift must occur in the feedback network if oscillations are to be established. Three *RC* phase shifting stages are necessary if the attenuation is not to be excessive. A single stage is illustrated in *Figure 4.5*. This is a phase advance network because the voltage across R, V_R, leads the input voltage, V_i, by some angle ϕ which can lie between 0° and 90°. As the angle approaches 90°, the output voltage V_o falls rapidly and becomes zero at 90°. Two such stages will produce a 180° phase shift but again with infinite attenuation. Three stages, each contributing a 60° phase shift, will give the desired total of 180° without excessive attenuation.

As for all oscillators, the gain of the amplifier multiplied by the attenuation (the feedback fraction) of the *RC* network must be at least equal to unity. This type of oscillator is known usually as a phase-shift oscillator in its own right. It is not an easy matter to vary the frequency of this oscillator, as either all three resistors **or** all three capacitors have to be adjusted simultaneously. It is, however, capable of generating reasonably low frequency sinusoidal outputs. It can be proved that the frequency at which the circuit oscillates is

$$f = \frac{1}{2\pi\sqrt{(6)}CR}$$

and that at this frequency

$$A_v = -29 = \frac{1}{\beta}$$

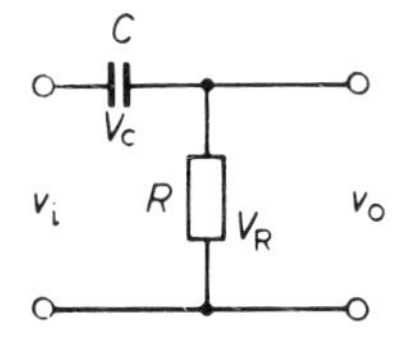

Figure 4.5

Problem 3 When is zero phase shift in the amplifier section of an oscillator necessary? How can a zero phase shift amplifier be achieved?

Zero phase shift in the amplifier is necessary when the phase shift through the feedback network is also zero. The amplifiers discussed in the previous two problems were single-stage amplifiers and introduced a 180° phase shift. The feedback systems were then designed to provide a further 180°. If the feedback system has zero phase shift, then the amplifier must also have zero phase shift. In other words, if βA_v is to be positive, β and A_v must both be positive or both negative.

A two-stage common-emitter amplifier has zero phase shift and so could be used for this purpose. An emitter-follower also has zero phase shift but its gain is less than unity, hence the oscillation criterion that $\beta A_v = 1$ could not be realised. However, a single-stage common-base amplifier could be used in conjunction with a transformer coupling feedback system. The common-base amplifier has zero phase shift but a high voltage gain; if therefore the transformer is connected to introduce zero phase shift also, the circuit will oscillate. The transformer will also provide a suitable match from the high impedance output to the low impedance input that the common-base amplifier exhibits.

Problem 4 Describe a form of oscillator employing a zero phase shift amplifier. Enumerate the advantages that this circuit has over those described previously.

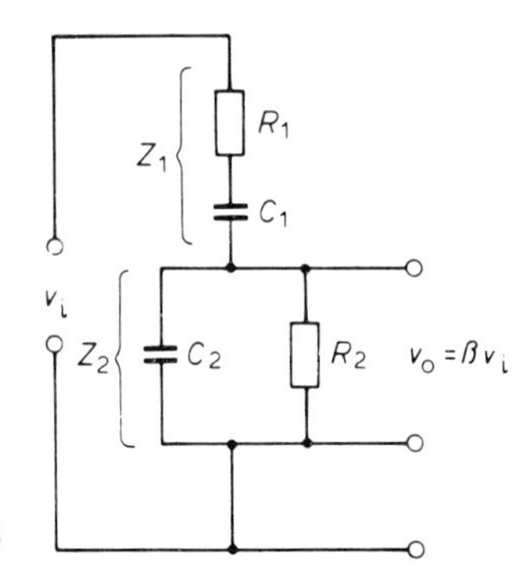

Figure 4.6

Clearly we require a zero phase shift feedback network for this oscillator. A very well-known feedback system is the **Wien network** shown in *Figure 4.6*. This is a frequency selective potential divider and the attenuation of the network is

$$\frac{V_i}{V_o} = \frac{Z_1 + Z_2}{Z_2}$$

It can be proved that V_o is in phase with V_i at only **one** frequency determined by the component values used in the circuit and that at this frequency the attenuation ratio is 3. Hence, if this network is used to connect the output to the input of a zero phase shift amplifier, a total phase shift of zero will exist around the circuit at the specified frequency. Since the ratio $V_o/V_i = 1/3$, the amplifier must have a gain >3 in order to maintain oscillations.

Figure 4.7(a) shows an elementary transistor oscillator making use of the Wien network. To adjust the frequency, either the two Wien capacitors or the two resistors may be ganged together as variables. As the gain required from the amplifier is only 3 when oscillations are established, a large amount of negative feedback may be applied within the amplifier. This will stabilise the gain and lead to a constant output amplitude. It will also lead to the high input resistance and the low output resistance that the Wien network ideally requires if it is not to load the amplifier output at one end or be shunted by the amplifier input at the other. A circuit using an integrated amplifier is shown in *Figure 4.7(b)*. Here the negative feedback is applied by way of a thermistor Th. The thermistor decreases its resistance when the temperature rises and conversely. Any tendency for the output amplitude to change, therefore, leads to a corresponding change in the value of the thermistor resistance (by internal heating). This in turn affects the amplifier gain, so that ideally

$$A_v = \frac{Th + R_3}{R_3} = 3,$$

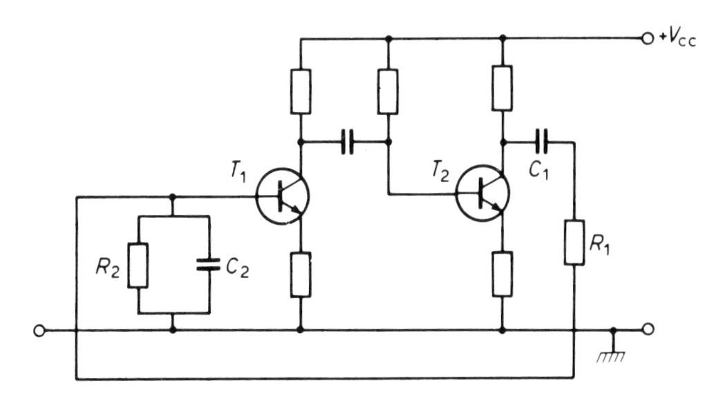

Figure 4.7(a)

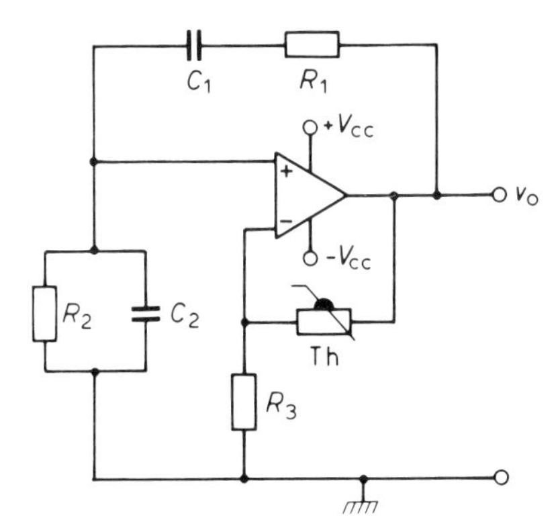

Figure 4.7(b)

and the stable condition is reached. It can be proved that the frequency of oscillation is given by

$$f = \frac{1}{2\pi\sqrt{R_1 R_2 C_1 C_2}} \text{ which reduces to}$$

$$= \frac{1}{2\pi CR}$$

when $R = R_1 = R_2$ and $C = C_1 = C_2$.

The Wien oscillator has several outstanding advantages when compared with the phase-shift oscillator and the tuned-collector (type) oscillators:

(i) It produces very low frequencies with negligible distortion down to a fraction of the Hertz if required.
(ii) It is easily tuned with twin capacitors or twin resistors.
(iii) It has a wide frequency range for a given variation in the capacities since the frequency is proportional to $1/C$ and not $1/\sqrt{C}$ as in an oscillator using tuned LC systems.
(iv) There is no inductance involved.

Problem 5 A Wien phase shift network is designed for 20 kHz zero phase shift. If $R_1 = R_2 = 1\ \text{k}\Omega$, find C_1 and C_2 taking $C_1 = C_2$.

If this network is connected to an amplifier that has an output resistance of 500 Ω and an input capacitance of 0.002 μF, find the frequency at which oscillations will actually occur.

When $R_1 = R_2 = R$, and $C_1 = C_2 = C$, the frequency at which the phase shift is zero is

$$f = \frac{1}{2\pi CR}$$

$$\therefore \quad 20 \times 10^3 = \frac{1}{2\pi \times 10^3 \times C}$$

$$\therefore \quad C = \frac{10^6}{2\pi \times 20 \times 10^6}\ \mu\text{F}$$

$$\therefore \quad C_1 = C_2 = 0.0079\ \mu\text{F}$$

(say **0.008 μF**)

When connected to the amplifier, the **effective** form of the Wien network becomes as shown in *Figure 4.8*. R_1 now equals 1.5 kΩ and C_2 equals 0.01 μF. The frequency for zero phase shift is now

$$f = \frac{1}{2\pi\sqrt{(R_1 R_2 C_1 C_2)}}$$

$$f = \frac{1}{2\pi\sqrt{(1.5 \times 10^3 \times 10^3 \times 0.008 \times 0.01 \times 10^{-12})}}$$

$$f = \frac{10^3}{2\pi\sqrt{(1.5 \times 0.008 \times 0.01)}}\ \text{Hz}$$

$$f = \mathbf{14.53\ kHz}$$

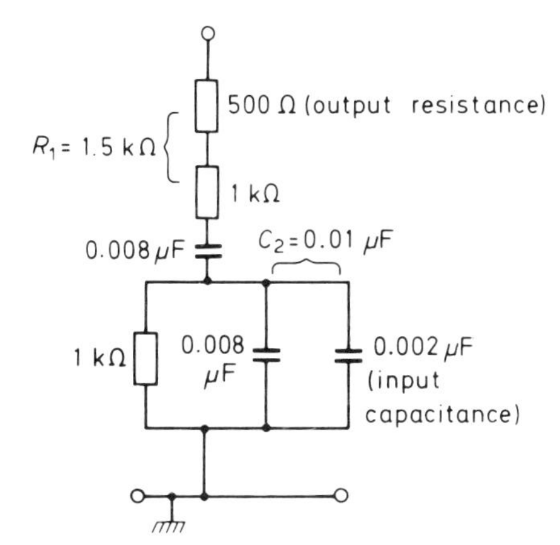

Figure 4.8

Problem 6 Describe one form of non-sinusoidal oscillator. Derive the frequency of operation and sketch the waveforms appearing in various parts of the circuit.

Non-sinusoidal or relaxation oscillators are fundamentally those types of oscillator in which the transistor or transistors switch periodically on and off. This contrasts with sinusoidal oscillators where the transistor or transistors operate without interruption and the output waveform is continuous.

The circuit of *Figure 4.9* shows a two-stage common-emitter amplifier in which each transistor output is coupled through capacitors to the input of the other. If $R_1 = R_2$ and $C_1 = C_2$, the circuit is said to be symmetrical. Asymmetrical arrangements do not affect the basic operation. The circuit generates a rectangular waveform and is known as an **astable multivibrator**.

Suppose that at some instant the transistors have equal collector currents. This condition will not persist for any slight disturbance in the current conditions will

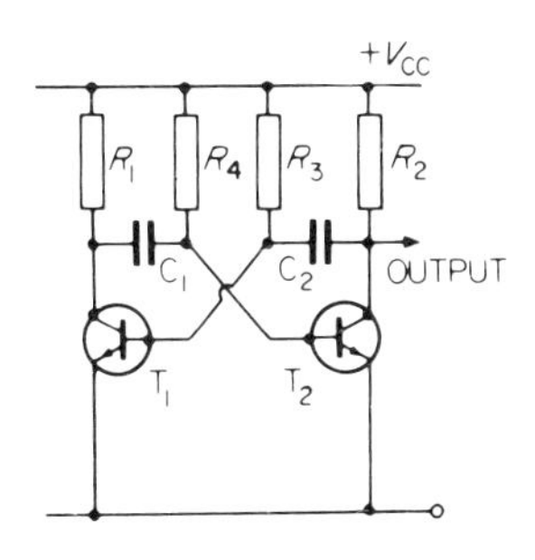

Figure 4.9

upset the balance. A sequence of events is now set in motion which may be tabulated as follows:

(a) The collector current of T_1 (say) rises very slightly.
(b) The collector voltage of T_1 will fall because of the increased voltage drop across R_1.
(c) This voltage drop will be transmitted to the base of T_2 by way of capacitor C_1.
(d) T_2 will be biased back and its collector current will fall.
(e) The collector voltage of T_2 will rise because of the reduced voltage drop across R_2.
(f) This voltage rise will be transmitted to the base of T_1 by way of capacitor C_2.
(g) T_1 will be forward biased and its collector current will rise.

This condition now augments the increase in the collector current of T_1 which initiated the sequence. The overall effect is cumulative and in an extremely short time the base voltage of T_2 has gone low enough to cut off this transistor while T_1 is conducting sufficiently to saturate it. This brings the collector voltage of T_1 to almost earth potential, with the collector voltage of T_2 at V_{cc}. This condition lasts only as long as the charge built up on C_1 allows the base of T_2 to remain low enough to hold T_2 off. When C_1 has discharged through R_4 and T_1 sufficiently to allow the base voltage of T_2 to return to its value fixed by R_4 alone, T_2 will start to conduct and the sequence of events described above will repeat but with T_2 taking the place of T_1. The circuit therefore switches successively between two **quasi-stable** states.

(a) T_2 cut off and T_1 saturated.
(b) T_1 cut off and T_2 saturated.

The frequency of operation can be deduced by considering the discharge path of C_1 by way of R_4 and the saturated T_1, see *Figure 4.10(a)*. For simplicity, the circuit is redrawn in *Figure 4.10(b)* where C_1 is carrying the charge it will have at the instant the sequence is completed, i.e. V_{cc} volts across it. This voltage will decay exponentially according to the equation

$V_{R4} = K \cdot \exp(-t/C_1R_4)$ where K is a constant.

When $t = 0$, $V_{R4} = 2V_{cc}$ (be sure you see how this happens), hence $K = 2V_{cc}$ and

$$V_{R4} = 2V_{cc} \cdot \exp(-t/C_1R_4)$$

T_2 will remain cut off until $V_{R4} = V_{cc}$; let this occur after a time t_1 seconds. Then

$$V_{cc} = 2V_{cc} \cdot \exp(-t_1/C_1R_4)$$

$$\therefore \exp(-t_1/C_1R_4) = 2$$

$$\therefore \quad t_1 = 0.693C_1R_4 \text{ sec}$$

In a similar manner, assuming symmetrical component values, the time T_1 in turn is cut off is given by

$$t_2 = 0.693C_2R_3$$

and since $t_1 = t_2$ under this condition, the frequency of the waveform will be $1/2t_1$ or

$$f = \frac{1}{1.39C_1R_4} = \frac{1}{1.39C_2R_3}$$

The waveforms are illustrated in *Figure 4.11*.

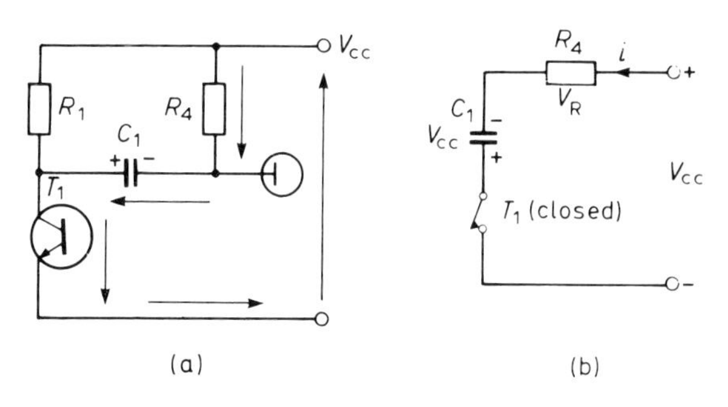

Figure 4.10

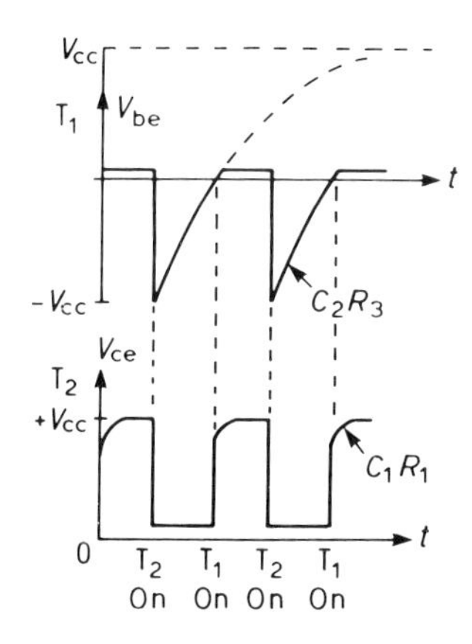

Figure 4.11

Problem 7 List the most likely causes which are liable to change the output frequency of an oscillator, and explain briefly how they may be eliminated.

The output frequency of an oscillator is likely to change if
(i) there is a change in the supply voltage,
(ii) there are changes in the tuned circuit reactances due to thermal or other variations,
(iii) there is a variation in the loading on the oscillator.

Variations in the supply voltage are best dealt with by using a stabilised supply. In most instances a simple series stabiliser is sufficient, or a shunt zener diode may be used.

Thermal changes and mechanical variations (small movements of circuit wires, for example) are allied in many cases. Tuning elements must be mechanically stable and should have a low temperature coefficient overall. Often two or more capacitors are connected in parallel to form the required capacitance, but with their respective temperature coefficients proportioned between negative and positive values to provide a substantially overall zero temperature coefficient. Inductances, where used, are not so easily controlled, for a very small expansion can alter the inductance appreciably. In cases where the stability of oscillation and the frequency have to be controlled closely, the circuit is placed in a temperature-controlled 'oven' which is maintained a few degrees above ambient.

Loading problems arise when an output is taken from the oscillator, as of course it must. Commonly, for oscillators using *LC* tuned circuits, a coupling coil, loosely coupled to the main winding, is used as a take-off. For *RC* oscillators, such as the phase-shift version, the use of an emitter-follower allows a low impedance output point to be obtained with negligible loading on the oscillator

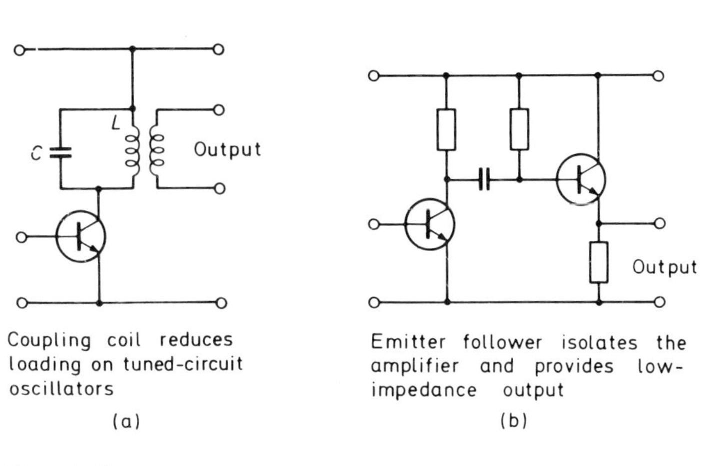

Coupling coil reduces loading on tuned-circuit oscillators

(a)

Emitter follower isolates the amplifier and provides low-impedance output

(b)

Figure 4.12

device itself. The best solution is this last case, where a 'buffer' amplifier is used between the oscillator proper and the required output terminals. This method can be applied to any form of oscillator. *Figure 4.12* shows two typical forms of coupling an oscillator to the output terminals so that the loading effect is reduced or eliminated.

C FURTHER PROBLEMS ON OSCILLATORS

(a) SHORT ANSWER PROBLEMS (answers on page 152)

1 State the conditions which must be fulfilled in order that a feedback amplifier shall oscillate.

2 Explain why an oscillator can be regarded as an amplifier with positive feedback.
3 What components may be used as frequency-determining elements in an oscillator?
4 Sketch the circuit diagram of a tuned-collector oscillator. Could a FET be used in place of the bipolar transistor? What changes to component values might be necessary?
5 What is the greatest phase shift possible with a combination of a capacitor and a resistor?
6 Sketch the circuit diagram of a phase-shift oscillator using a three-stage RC network. Why are three groups of R and C necessary?
7 What purpose does a buffer amplifier serve?
8 A tuned-collector oscillator circuit (which has no faulty components whatsoever) fails to oscillate when switched on. Where would you first look for the possible cause of trouble?
9 The tuned circuit of an oscillator has $L = 150$ mH, $C = 500$ pF. At what frequency will this oscillator operate?
10 What form of single-stage amplifier could be used if zero phase shift was required in it.
11 Complete the following statements:
 (a) An oscillator is an . . . that provides its . . . signal.
 (b) A two-stage common-emitter amplifier has an overall . . . phase shift.
 (c) In an oscillator the magnitude of the loop gain must be . . .
 (d) The minimum gain required by the amplifier in a Wien-type oscillator is . . .
 (e) The minimum gain required by the amplifier in a phase-shift oscillator is . . .
 (f) The feedback network is that part of an oscillator system that determines the . . .
 (g) An oscillator producing a square wave would be a . . . oscillator.
 (h) A multivibrator oscillator has a feedback fraction equal to . . .
12 Are the following statements true or false:
 (a) A relaxation oscillator always produces rectangular waves.
 (b) A multivibrator has two quasi-stable states.
 (c) If $\beta A_v = 1$, the gain with positive feedback is zero.
 (d) The frequency stability of an oscillator is defined by the amount by which its frequency drifts from the desired value.

(b) MULTI-CHOICE PROBLEMS (answers on page 152)

1 The criterion for oscillations to occur is (a) $A_v = 1/\beta$, (b) $\beta A_v < 1$, (c) $\beta A_v = 0$, (d) $1/\beta = \infty$.
2 In the astable multivibrator, (a) $\beta < 1$, (b) $\beta = 1$, (c) $\beta > 1$, (d) $\beta A_v = 1$.
3 A relaxation oscillator produces (a) sinusoidal waveforms, (b) non-sinusoidal waveforms, (c) triangular waves, (d) rectangular waves.
4 If the time-constants of a multivibrator's coupling components are both halved, the frequency will be (a) halved, (b) doubled, (c) quadrupled, (d) unchanged.
5 The phase shift in a feedback network is 90°. Assuming positive feedback, the phase shift in the amplifier if oscillation is to occur will be (a) 0°, (b) −90°, (c) 180°, (d) −270°.
6 One of the chief advantages of the Wien oscillator is that it (a) produces very high frequencies, (b) is easily tuned with twin ganged resistors, (c) easily produces very low frequency square waves, (d) requires only one transistor amplifier.

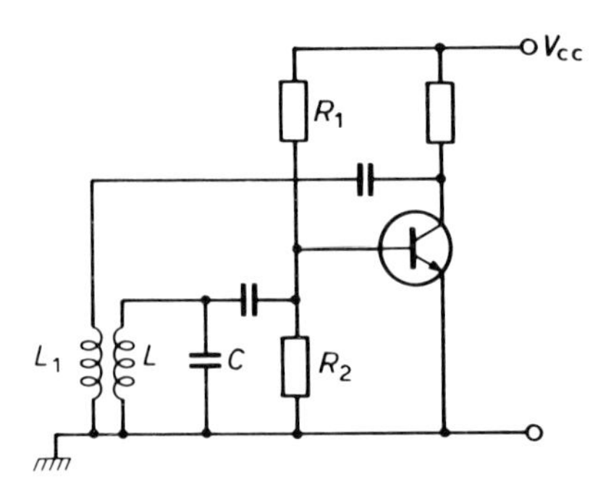

Figure 4.13

7 A disadvantage of the phase-shift oscillator is that (a) it produces only sinusoidal waves, (b) it is difficult to adjust the frequency, (c) it requires a lot of components in the feedback path, (d) it requires a two-transistor amplifier.

8 A tuned-collector type oscillator is not used for low-frequency oscillation production because (a) it is impossible to get sufficient feedback, (b) the feedback becomes excessive and the output distorts, (c) the inductor becomes bulky, (d) a large enough capacitance is difficult to obtain.

9 A buffer amplifier is often used after an oscillator stage in order to (a) stabilise the output level of the oscillator, (b) isolate the oscillator from the effects of external loading, (c) make the oscillator easier to tune, (d) remove any distortion from the oscillator output.

10 In an astable multivibrator, the feedback fraction (a) can be altered by altering the feedback time-constants, (b) can be altered by using higher-gain transistors, (c) is constant irrespective of the time-constants, (d) depends entirely on the amplifier gain.

(c) CONVENTIONAL PROBLEMS

1 In the phase shift oscillator shown earlier in *Figure 4.4* capacitors C are each $0.05\ \mu F$ and resistors R are each $22\ k\Omega$. If the effect of the bias resistors is negligible, what will be the frequency of oscillation? [59 Hz]

2 The phase between the input and output terminals of a certain feedback network is given by

$$\phi = \frac{270^\circ}{1 + f/500}$$

where f is in Hz. This network is used in an oscillator circuit where a single common-emitter amplifier is employed. At what frequency will this circuit oscillate? [250 Hz]

3 In the Wien network shown in *Figure 4.6*, $C_1 = C_2 = 0.1\ \mu F$, $R_1 = R_2 = 25\ k\Omega$. At what frequency will the phase shift between input and output be zero? [63.7 Hz]

4 In the astable multivibrator of *Figure 4.9*, $C_1 = C_2 = 0.001\ \mu F$, $R_3 = R_4 = 100\ k\Omega$. What kind of wave will this circuit generate and what will be (a) its frequency, (b) its mark-space ratio [7.24 kHz, 1:1]

5 Suppose in the previous problem that C_1 is changed to 500 pF and R_4 is changed to 20 kΩ, the other components being unchanged. What will be (a) the new frequency, (b) the new mark-space ratio? [18.96 kHz, 10:1]

6 *Figure 4.13* shows a tuned *LC* oscillator known as the tuned-base oscillator. Keeping the tuned-collector in mind, deduce how this circuit operates. Can you see a disadvantage of this circuit when it is compared with the tuned-collector oscillator.

7 Draw the circuit diagram of a Wien bridge oscillator using either discrete transistors or an integrated circuit amplifier. Identify the components which (a) determine the frequency, (b) determine the amplitude of the output waveform.

8 An amplifier having a voltage gain A_v without feedback has a fraction of its output voltage fed back in series with its input. Derive a general expression for the amplifier gain A'_v when the feedback loop is closed. Hence describe the requirements for oscillations to occur when the feedback is positive.

9 A multivibrator of the form shown in *Figure 4.9* is required to generate a square wave of frequency 800 Hz. If the coupling capacitors are each $0.05\ \mu F$ in value, what should be the values of the base bias resistors R_3 and R_4? [18 kΩ]

10 The multivibrator waveforms shown in *Figure 4.11* show the base waveform of transistor T_1 and collector waveform of transistor T_2. Sketch the base and collector waveforms of transistors T_2 and T_1 respectively with reference to the same time axis.

5 Operational amplifiers

A MAIN POINTS CONCERNED WITH OPERATIONAL AMPLIFIERS

1 Not so many years ago the operational amplifier or **opamp** was a valve-operated, bulky and expensive piece of specialised equipment found exclusively in research laboratories. Its original role was, and still is in many applications, to perform the mathematical operations of multiplication and division, differentiation and integration, particularly in regard to analogue computing. With the advent of miniaturisation, brought about by the arrival of the semiconductor, the opamp soon lost its upper-class position in the rarefied atmosphere of the research establishment and became available even for the use of the vulgar kitchen table experimenter. The basic opamp now available in integrated circuit form is a cheap, high-performance, directly coupled amplifier system capable of high gain and stable operation over a wide range of frequencies including d.c.

2 The integrated opamp is provided with a single-ended output but there are two terminals at the input. This enables the user to select either single- or double-ended input mode. When both inputs are employed, the unit behaves as a **differential** amplifier; for single-ended input, one or other of the differential inputs is earthed.

An ideal model of an opamp would have the following characteristics: (a) an infinite voltage gain, (b) an infinite input impedance, (c) an infinite bandwidth, and (d) zero output impedance. No practical design can ever achieve this ideal but very good approximations can be made. Those opamps based on MOSFET technology approach the ideal conditions most closely, but even the types using bipolar transistors, such as the standard 741, have quite impressive figures. Typically, these are

Bipolar 741	**MOSFET CA3140**
Voltage gain > 100 dB	Voltage gain > 100 dB
Input resistance > 1 MΩ	Input resistance > 10^{12} Ω
Unity gain bandwidth 0–1 MHz	Unity gain bandwidth 0–4.5 MHz
Output resistance < 75 Ω	Output resistance < 50 Ω

Opamps are usually symbolised in the way shown in *Figure 5.1(a)* and *(b)*, which show the differential and single-ended input modes respectively. The input terminals are marked + and −; the positively marked terminal is known as the **non-inverting** input and the negatively marked terminal is known as the **inverting** input. These terms are self-explanatory; a signal applied to the inverting terminal with the other terminal earthed comes out antiphase to the input. A signal applied to the non-inverting terminal comes out in phase with the input. Clearly, if a signal is applied simultaneously to both inputs, the output will (ideally) be

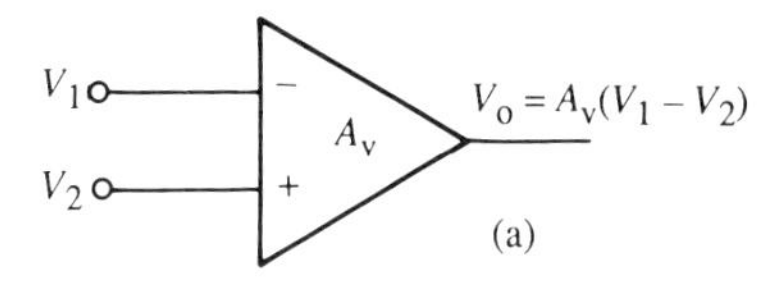

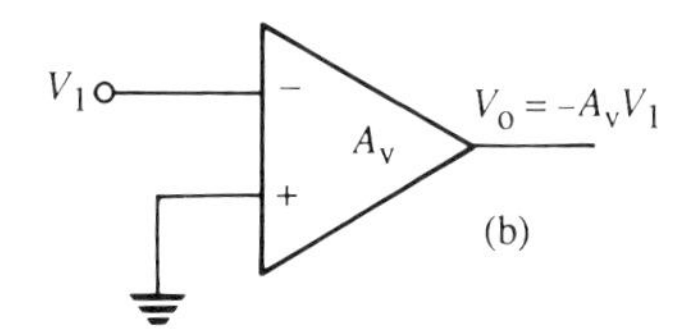

Figure 5.1

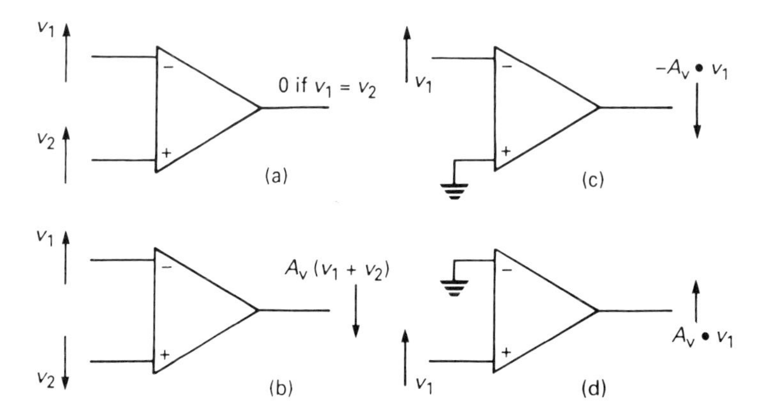

Figure 5.2

zero. When feedback is applied, it must always be applied to the inverting terminal for negative feedback, but the signal may be applied to either or both terminals depending upon the application.

3 *Figure 5.2* summarises the input and output arrangements and conditions for the opamp. At (*a*) two in phase input signals are applied to the differential terminals and the output is the amplifier gain A_v times the *difference* between the inputs; this is **common-mode** working. If $v_1 = v_2$, the output will be zero. At (*b*), two anti-phase inputs are applied and the output is the amplifier gain A_v times the *sum* of the inputs; this is **differential mode** working. For single input conditions, we turn to diagrams (*c*) and (*d*). At (*c*) the input is at the inverting terminal and the output is $-A_v$ times the input. At (*d*) the input is at the non-inverting terminal

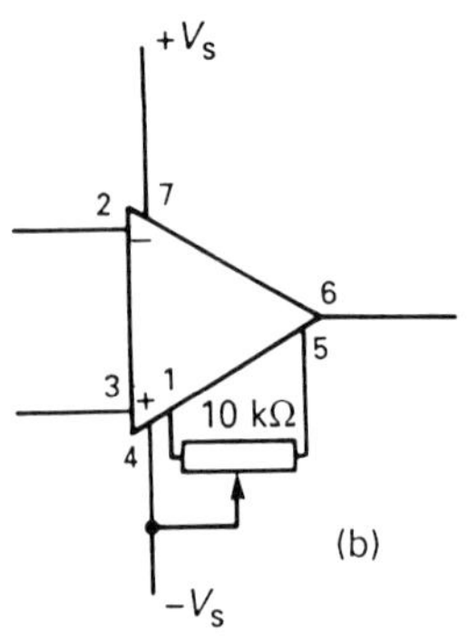

Figure 5.3

and the output is $+A_v$ times the input. Notice that the opamp amplifies differential mode signals but rejects common-mode signals. This has some importance in practical circuits.

4 Ideally, the output of an opamp should be zero when the inputs are zero. This would be so if everything was perfectly matched inside the amplifier, but in real amplifiers this is impossible to achieve. Terminals are provided on opamp packages so that this **offset** voltage which appears at the output even when the inputs are held at zero potential can be neutralised. This is done in practice by connecting a potentiometer between the two **offset-null** pins and taking the wiper to the negative supply rail. This potentiometer (which is a preset control) is then adjusted until the output is brought to zero with both inputs also at zero. *Figure 5.3* shows the connections.

5 It is customary to power opamps by the use of both a positive and a negative supply rail, with the circuit zero, or earth, line being at the electrical centre of the supply. In this way the circuit output is enabled to swing both above and below earth potential to almost the limits of the supply, though this facility is not always required. Most opamps are available in 8-pin d.i.l. integrated packages, though amplifier pairs can be obtained in single 14 or 16-pin units. The connections are standardised to those shown in *Figure 5.4*. The power supply is connected between pins 7 (positive) and 4 (negative); for unbalanced feeding, this negative pin can go directly to the earth line. The output is at pin 6 and the input pins are 2 for the inverting (−) and 3 for the non-inverting (+) configuration. In addition there are two offset-null inputs at pins 1 and 5 which are used in the way illustrated in *Figure 5.3*. Pin 8 may be unused in some types, but it is available for the addition of an external capacitor to modify the frequency characteristics of

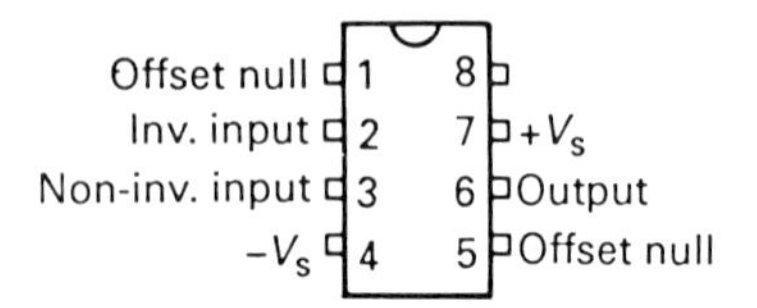

Figure 5.4

the amplifier in others. The working voltage limits and whether or not frequency compensation is necessary must always be looked up in the manufacturer's literature.

6 As we have seen, there is a high gain in the differential mode of operation and zero gain in the common mode. Because of the imperfections already mentioned, the output in common-mode operation is small but *not* zero. When signals are applied differentially to the amplifier, a high gain results, but unwanted signals such as noise voltages are present at both inputs and present a common-mode input; these signals therefore receive only a very small gain which is something to be desired. However, the amplifier can only separate signal and noise in this way if its common-mode output is negligible, and a measure of the opamp's ability to do this is expressed by a factor known as the **Common Mode Rejection Ratio** or **CMRR**. CMRR is calculated in dB and is given by

$$\text{CMRR} = 20 \log \left[\frac{\text{Differential-mode gain}}{\text{Common-mode gain}} \right] \text{dB}$$

90 to 100 dB are typical figures for most commonly available opamps in both bipolar and MOSFET form.

7 Before any external feedback is applied, the inherent (open-loop) gain of an opamp is extremely large, 100 dB being typical. As it stands, this kind of amplification is of no value because the smallest potential difference – a fraction of a microvolt – between the input pins will send the output into saturation and drive the output potential to one or other of the supply voltage levels. The object of having gain values of these high levels is, as we noted earlier, to enable heavy negative feedback to be introduced, so bringing all the advantages that n.f.b. confers and still providing as much useful gain as needed. The application of n.f.b. does not of course guarantee that the circuit system will be stable. An opamp is just as vulnerable as a circuit built up from discrete components. At high frequencies the open-loop gain falls as it does in any amplifier, and the internal capacitances also introduce phase shifts which modify the phase relationships of 0° and 180° which are tacitly assumed for the non-inverting and inverting configurations respectively at low frequencies. In many cases, therefore, some sort of frequency compensation has to be applied to maintain stability over the desired operating range. The manufacturer's recommendations must be consulted for this, but usually a small capacitor within the range 10–100 pF connected between the appropriate pins (usually 1 and 8) on the opamp takes care of things up to the frequency limits of some 250 kHz, irrespective of the amount of feedback applied.

B WORKED PROBLEMS ON OPERATIONAL AMPLIFIERS

Problem 1 The differential gain of an amplifier is 20 000 and the common-mode gain is 5. What is the CMRR of this amplifier? Is this a good figure?

$$\text{CMRR} = 20 \log \frac{20\,000}{5} \text{dB}$$

$$= 20 \log 4000 = \mathbf{72\ dB}.$$

This is not a particularly good CMRR, though it would be acceptable in all but very stringent circumstances.

Problem 2 Describe the circuit arrangement for an opamp to be used as an inverting amplifier and derive an expression for the closed-loop gain of this circuit.

The circuit arrangement of *Figure 5.5* belongs to the general category of inverting circuits. The common feature of these circuits is that the non-inverting input is connected to the earth line. Both the input signal *and* the feedback are applied to the inverting input. The most important thing about this circuit is that whenever some of the output is fed back to the inverting input (negative feedback) the voltages measured on both the inverting and non-inverting input points *will be the*

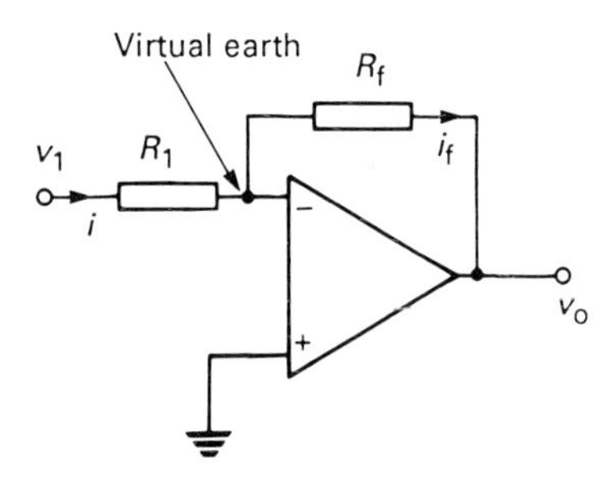

Figure 5.5

same. The output of the opamp will adjust itself, either producing a current or absorbing a current, to keep these voltages identical.

When the non-inverting input is earthed, therefore, the inverting terminal will be held at 0 V. Since this input is at earth potential but not actually connected to earth, it is referred to as a **virtual earth**. Because of this, the full input voltage v_1 will develop across the input resistor R_1 and the signal current in R_1 will be $i_1 = v_1/R_1$. Now considering the feedback resistor R_f which is connected between the virtual earth point and the output terminal, the current i_f in R_f will be $-v_o/R_f$. Since negligible current will flow *into* the opamp because of its high input resistance, we may think of the input circuit as a source of current which must flow into R_f. Hence $i_1 = i_f$ and so $v_1/R_1 = -v_o/R_f$. The closed-loop gain, therefore, is

$$A_v' = \frac{v_o}{v_1} = -\frac{R_f}{R_1}$$

Notice that the gain expression includes the sign inversion and that its magnitude is determined solely by the ratio of the external resistors. This expression is strictly true only if the open-loop gain is infinite; for opamps in which the gains are of the order of 10^5 or more, the expression is accurate enough for most practical purposes.

Problem 3 How might the basic inverting amplifier be modified for use as a summing amplifier?

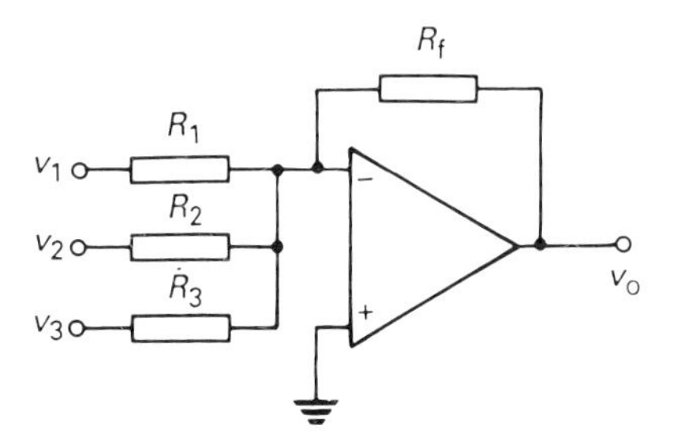

Figure 5.6

If the basic inverting amplifier is modified to accept additional signal inputs and summing resistors are added, the circuit performs the mathematical operation of addition. *Figure 5.6* shows such a circuit system, and because of its ability to add, the virtual earth point is often referred to as the **summing point**. Let the input voltages be v_1, v_2 and v_3. Then the currents in the input resistors R_1, R_2 and R_3 are respectively

$$i_1 = \frac{v_1}{R_1} \qquad i_2 = \frac{v_2}{R_2} \qquad i_3 = \frac{v_3}{R_3}$$

All these input currents flow into R_f, generating an output voltage given by

$$v_o = -(i_1 + i_2 + i_3)R_f$$

$$= -R_f \left[\frac{v_1}{R_1} + \frac{v_2}{R_2} + \frac{v_3}{R_3}\right]$$

If the three input resistors are equal in value and *also* equal to R_f, the output voltage will equal, literally, the sum of the input voltages. For unequal values, the output will depend upon the ratios of the various input resistors to the feedback resistor, and v_o will be a scaled or 'weighted' average of the inputs. So the circuit behaves as an *analogue* adder. The output voltage total cannot, of course, exceed the supply potential.

Problem 4 Derive a general expression for the gain of a non-inverting amplifier configuration. In what form is this circuit known as a voltage follower?

The basic non-inverting amplifier is shown in *Figure 5.7*. Input is applied to the non-inverting terminal and feedback from the output is applied to the inverting terminal. A fraction of the output, defined by the ratio of the resistors R_2 and R_3, is fed back and from basic feedback theory we have

$$\beta = \frac{R_3}{R_2 + R_3}$$

and for A_v very large

$$A_v' \rightarrow \frac{1}{\beta} \rightarrow 1 + \frac{R_2}{R_3}$$

The overall gain can now range from a minimum value of unity to an upper limit determined again by the ratio of two resistances. It is useful to remember that in the non-inverting circuit the closed-loop gain is given very closely by $1/\beta$ and in the inverting circuit by $(1 - 1/\beta)$.

The minimum gain occurs when R_2 is zero; if at the same time R_3 is made infinite (that is, removed entirely), the circuit becomes that shown in *Figure 5.8*. Here the whole of the output is fed back to the input, but this is a specialised case known as the **voltage follower**. The gain of this circuit is clearly unity, and the phase shift is zero. It is, therefore, a sophisticated version of the emitter- or source-follower, and is a useful matching device for coupling together points of high and low impedance.

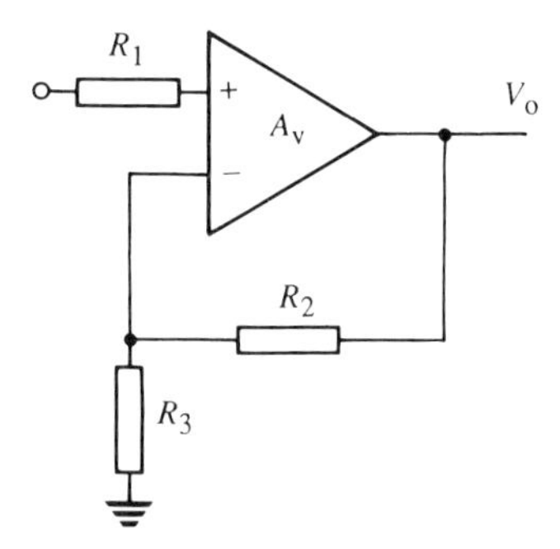

Figure 5.7

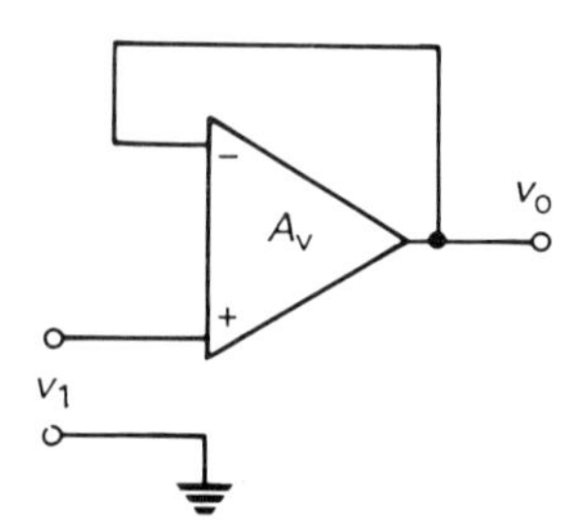

Figure 5.8

Problem 5 In the operational summer of *Figure 5.6* earlier, suppose the input voltage levels to be $v_1 = 2$ V, $v_2 = 3$ V, $v_3 = 4$ V. Also suppose $R_1 = R_f = 10$ kΩ, $R_2 = 18$ kΩ, $R_3 = 20$ kΩ. What will be the output voltage? Why is this output *not* equal to 9 V?

Using the expression derived for the output of the opamp summer and substituting values we have, working in kilohms:

$$v_o = -15\left[\frac{2}{10} + \frac{3}{18} + \frac{4}{20}\right]$$

$$= -15\left[\frac{36 + 30 + 36}{180}\right] = 8.5 \text{ V}$$

As we saw above, the input voltages to a summing amplifier will only be added 'correctly' if the input resistors and the feedback resistor are equal in value, all the ratios then being equal to 1. In this problem, the resistor values were unequal, so the output voltage was a scaled summation of the inputs.

Problem 6 *Figure 5.9* shows a differential amplifier configuration. Discuss briefly the advantages and disadvantages of this mode of operation.

This configuration is widely used in instrumentation and has the advantage that in-phase or common-mode signals at the inputs produce no variation at the output. Such in-phase signals include thermally generated noise, d.c. fluctuations

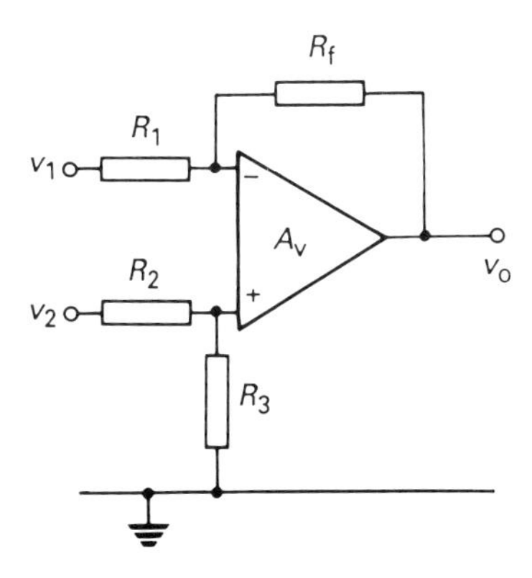

Figure 5.9

and drift. The differential amplifier discriminates against these unwanted intruders and responds only to significant (wanted) signal variations.

If we assume an infinite open-loop gain and infinite input resistance, then the inverting and non-inverting input points will be at the same common-mode input voltage, hence,

$$v_0 = -\frac{R_f}{R_1}(v_2 - v_1)$$

For optimum performance, the resistances to earth should be matched with

$$\frac{R_f R_1}{R_f + R_1} = \frac{R_2 R_3}{R_2 + R_3}$$

It is of some importance to note that there is a marked difference between the input resistance of the two input points as seen by the signals. At the inverting input we see the virtual earth point, hence the input resistance is simply R_1. At the non-inverting input we see R_2 in series with R_3 since no current flows into the non-inverting terminal itself. Where high gain is concerned, so that R_f is considerably greater than R_1 (and the same ratio applies to R_3 and R_2), this discrepancy between the effective input resistances can be very large. This can lead to excessive noise pickup on the high resistance input point (the non-inverting input) relative to that on the low resistance input point. One way of overcoming this problem is to precede the differential amplifier with two unity gain voltage followers as seen in *Figure 5.10*. This maintains a very high input resistance on both signal inputs.

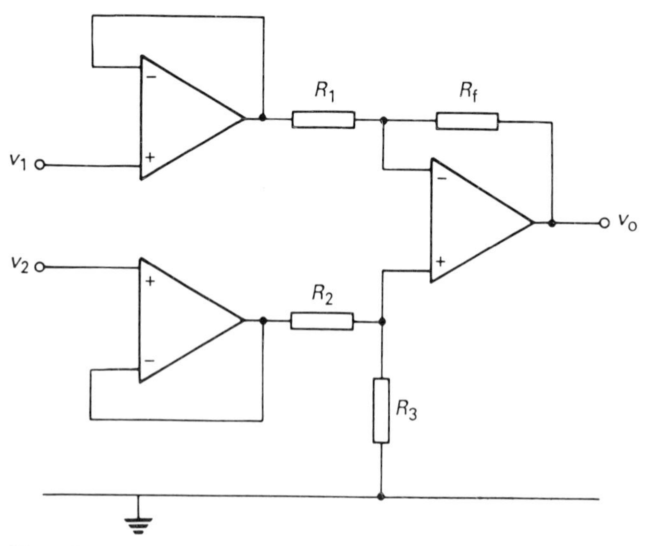

Figure 5.10

Problem 7 In the circuit of *Figure 5.9*, $R_f = R_3 = 470\ k\Omega$, $R_1 = R_2 = 10\ k\Omega$. What is the amplifier gain and what is the resistance seen at each input?

Gain $A'_v = R_f/R_1 = 47$

At the v_1 input point we see only the effect of R_1, hence the input resistance here is 10 kΩ.

At the v_2 input point we see the effect of R_2 and R_3 in series, hence the input resistance here is 480 kΩ.

Problem 8 The three input resistors to the summing circuit of *Figure 5.6* are each 100 kΩ. What value of R_f is required so that the output is the true sum of the three input voltages? What assumption is made in this calculation? If the gain of the amplifier is actually only −100, what value of feedback resistor would be needed such that the output stayed unchanged?

As we have seen, for $v_o = v_1 + v_2 + v_3$, all the resistance ratios must be equal to unity, hence the value of the feedback resistor must also be 100 kΩ.

The assumption made in this calculation is that the amplifier has infinite open-loop gain and an infinite input resistance. Under these conditions, the total current flowing in the input resistors passes into the feedback resistor, and from this the expression for the output voltage is obtained.

When the amplifier has a finite gain, the equation for v_o becomes

$$E - v_o = -R_f\left[\frac{v_1 - E}{R_1} + \frac{v_2 - E}{R_2} + \frac{v_3 - E}{R_3}\right]$$

where E is the potential at the (ideally) virtual earth point. Let the amplifier gain be A, then substituting $E = -v_o/A$ in the previous equation, we get

$$\frac{v_1}{R_1} + \frac{v_2}{R_2} + \frac{v_3}{R_3} = -\left[\frac{v_o}{AR_f} + \frac{v_o}{R_f} + \frac{v_o}{AR_1} + \frac{v_o}{AR_2} + \frac{v_o}{AR_3}\right]$$

But $R_1 = R_2 = R_3 = 100\ k\Omega$

$$\therefore \quad \frac{1}{100}[v_1 + v_2 + v_3] = -\left[\frac{v_o}{AR_f} + \frac{v_o}{R_f} + \frac{3v_o}{100A}\right]$$

$$v_1 + v_2 + v_3 = -v_o\left[\frac{100}{AR_f} + \frac{100}{R_f} + \frac{3}{A}\right]$$

For $v_1 + v_2 + v_3 = -v_o$ the bracketed term must equal 1. Therefore, substituting $A = 100$, we have

$$\frac{1}{R_f} + \frac{100}{R_f} + \frac{3}{100} = 1$$

$$\therefore \qquad R_f = 104\ \text{k}\Omega$$

Problem 9 An opamp is connected as shown in *Figure 5.11*. The non-inverting input is held at a steady d.c. level by a 1.5 V battery supply (which we might call the reference voltage), while the inverting input can be adjusted between 0 and 3 V by potentiometer R_1. Explain what happens at the output as R_1 is adjusted from 0 to 3 V.

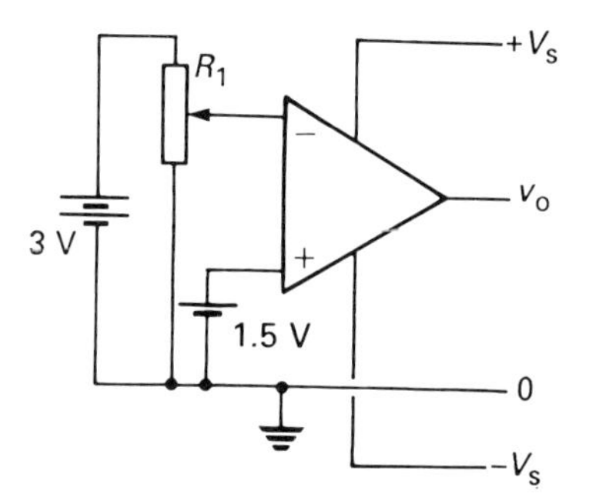

Figure 5.11

As the inverting input is increased from zero towards the mid-point of the potentiometer setting (corresponding to a potential setting of 1.5 V), v_1 is less than the reference level, hence the opamp output will be positive. Since the amplifier is operating at its very large open-loop gain value, any voltage difference at the inputs will drive it into saturation; hence the output will be at its maximum positive saturation value, just below the positive supply rail level. As soon as the slider of R_1 passes the mid-point, v_1 will be greater than the reference level, hence the output will switch immediately to its maximum negative saturation level. The curve of output voltage against input differential voltage is the transfer characteristic and is illustrated in *Figure 5.12*.

An opamp used in this way is known as a zero-crossing comparator. It compares one input with a reference level (or another signal) and switches the polarity of the output when the input levels change relative sign. A simple square wave generator can be made by applying a small amplitude sine wave to the v_1 input position, the reference level simply being earth.

Although the opamp cannot be damaged by driving it into saturation, provided

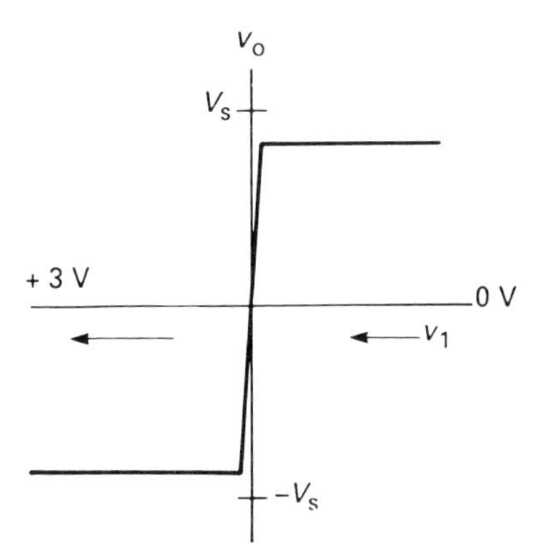

Figure 5.12

the supply limits have been observed, an important point arises regarding the time of recovery from a saturated condition. When very rapid switching from one level to the other is required, a long recovery time can affect the switching operation.

Problem 10 How might operational amplifiers be used for the mathematical operations of integration and differentiation?

The basic **integrator** circuit is shown in *Figure 5.13(a)*. As before, for an ideal amplifier, $i_1 = i_f = v_1/R_1$. But now i_f flows into a capacitor C and hence represents the rate of change of charge. Hence $i_f = -\,\mathrm{d}q/\mathrm{d}t$ and so

$$\frac{v_1}{R_1} = -\frac{\mathrm{d}q}{\mathrm{d}t} = -C\frac{\mathrm{d}v_o}{\mathrm{d}t}$$

since $\mathrm{d}q = C\,.\,\mathrm{d}v$. Integrating, we get

$$v_o = -\frac{1}{CR_1}\int v_1\,\mathrm{d}t$$

Hence the output is the integral of the input and the gain of the circuit is $-1/CR_1$. The output will only equal the 'true' integral of the input if $CR_1 = 1$, that is the amplifier will have unity gain. This is difficult to achieve in practice because of the inconveniently large values necessary for both R_1 and C. In analogue computer applications, R_1 is usually taken as 100 kΩ and C as 1 μF, giving $CR_1 = 0.1$ and an integrator gain of -10. This is allowed for in setting the various levels within the computer.

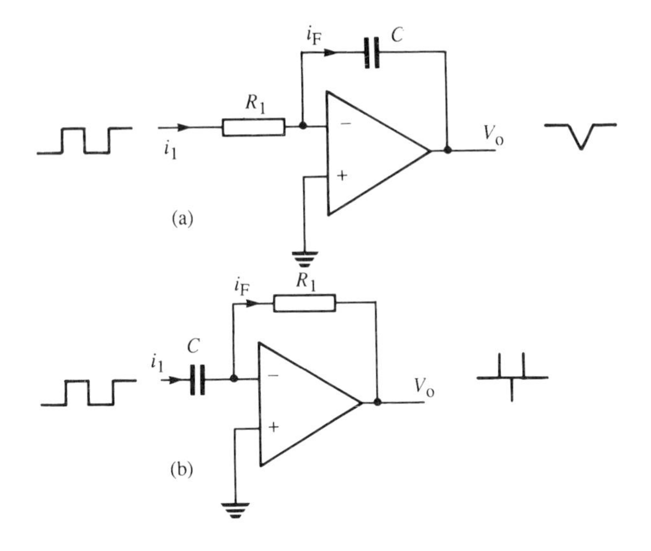

Figure 5.13

One of the main applications of the integrator opamp is as a ramp generator. If a *constant* d.c. voltage $-V$ is applied to the input of the circuit, the output will be given by Vt/CR_1 and this represents a linear ramp of gradient V/CR_1. *Figure 5.13(a)* shows the output waveform for a full cycle of square-wave input. The output ramps correspond to the leading and trailing edges of the square wave input. If the input is a continuous square wave, the output will be a continuous triangular wave.

The basic **differentiator** circuit is shown in *Figure 5.13(b)*. Here resistor R_1 and capacitor C are interchanged. Again, we assume that $i_1 = i_f = -v_o/R_f$, so that i_1 is now the rate of change of charge and therefore

$$-\frac{v_o}{R_f} = C\,\frac{dv_1}{dt}$$

and

$$v_o = -CR_f \cdot \frac{dv_1}{dt}$$

Thus the output is directly proportional to the differential coefficient of the input voltage with respect to time. For a continuous square-wave input this time, the output is a series of very sharp pulses (of theoretically infinite amplitude) corresponding to the input voltage transitions where change is taking place.

Differentiators are not so easily adapted for use in analogue computers as are integrators because any noise on the input signal appears as a much larger proportion of the output signal, the actual amplitude being a function of the gradient (or rate of change) of the input voltage.

Problem 11 For the integrator circuit of *Figure 5.13(a)* suppose $R_1 = 100\ \text{k}\Omega$, $C = 0.47\ \mu\text{F}$. What slope will the output ramp have for an input of 2 V? How long will it take for the output to rise to 1 V after the application of the input?

If $R_1 = 10^5\ \Omega$ and $C = 0.47 \times 10^{-6}$ F, the integrator gain will be 10/0.47 = 21.3 V/s per volt of input. For a 2 V input, therefore, the slope of the ramp will equal 0.0213 V/ms × 2 = 0.0426 V/ms. At this rate, a level of 1 V will be reached in 23.5 ms.

Problem 12 What is the **slew rate** of an operational amplifier?

There is a limit to the rate at which the output voltage of an amplifier can change. This lagging effect comes about because of the presence of internal capacitance and the movement of charge carriers by diffusion in the base region of bipolar transistors. The greatest possible rate of change of the output voltage is known as the *slew rate(s)*.

Slew rate is usually measured in volts per microsecond, hence the maximum frequency at which an amplifier can usefully operate is not only a function of the bandwidth but also of the slew rate. However, with a sinusoidal input waveform, whereas the bandwidth limits the *gain* at the higher frequencies, the slew rate leads to distortion. When the frequency and amplitude of the output waveform are such that the maximum rate of change exceeds the stated slew rate, distortion must follow. A sinewave is changing most rapidly as it passes through zero; hence the frequency at which the zero-crossing slope of a sinewave equals the slew rate is the greatest frequency for distortionless full rated output. To put this into a mathematical form, let the waveform be $v = \hat{V} \,.\, \sin \omega t$, then the time rate of change is $\mathrm{d}v/\mathrm{d}t = \omega\hat{V} \,.\, \cos \omega t$ and this has a maximum value when $t = 0$, i.e. the zero crossing points on the time axis. Then $\mathrm{d}v/\mathrm{d}t(\max) = \omega\hat{V}$ and for the amplifier to operate without distortion, the slew rate must not exceed this figure.

Usually the slew rate limit occurs in the early stages of an amplifier. For any given amplifier the slew rate is constant, so the designer can go either for a large output *or* a good high frequency response, but he cannot have both at once. The 741 opamp has a slew rate of about 0.5 V/μs, but the 741S version is much better at about 20 V/μs.

Problem 13. The 741 opamp has a slew rate of 0.5 V/μs. What will be the maximum frequency at which a 741 will operate without distortion if an output amplitude of 1 $\hat{V}$ is provided? How does the 741S compare with the figure you obtain?

The slew rate must equal (or exceed) the value of $\omega\hat{V}$, hence

$$f = \frac{\text{slew rate}}{2\pi\hat{V}} = \frac{0.5}{2\pi \times 1 \times 10^{-6}}\text{Hz}$$

$$= \mathbf{79.6\ kHz}$$

In the case of the 741S where the slew rate is given as 20 V/μs we have

$$f = \frac{20}{2\pi \times 1 \times 10^{-6}}\text{Hz}$$

$$= \mathbf{3.18\ MHz}$$

C FURTHER PROBLEMS ON OPERATIONAL AMPLIFIERS

(a) SHORT ANSWER PROBLEMS (answers on page 152)

1 Opamps are so called because they were designed originally to perform mathematical operations. True or false?
2 Negative feedback can be obtained by feeding all or part of the opamp's output back to the non-inverting input. True or false?
3 A summing amplifier does not necessarily give the arithmetic total of the input voltages. True or false?
4 Make a list of the ideal properties that an opamp should have.
5 A voltage follower has unity gain and is non-inverting. What is the purpose of such an amplifier?
6 Define the terms: common-mode signals, differential-mode signals, common-mode rejection ratio, slew rate.
7 One of the terminals on an opamp is marked 'Offset Null'. What is the purpose of this terminal?
8 Distinguish between a differential- and a common-mode signal.
9 What voltage would you measure at the inverting input of an opamp with negative feedback if the non-inverting input was taken to −5 V?
10 If, in *Figure 5.5* $R_1 = 10\ \text{k}\Omega$, $R_f = 330\ \text{k}\Omega$, what is the closed-loop gain of the amplifier?

(b) MULTI-CHOICE PROBLEMS (answers on page 152)

1 An amplifier circuit in which feedback is applied in a loop from output to input is (a) a closed-loop amplifier, (b) an open-loop amplifier, (c) an oscillator, (d) none of these.
2 The terminal marked '+' on a circuit diagram is (a) the inverting input, (b) the non-inverting input, (c) the supply input, (d) none of these.
3 A virtual earth point is one (a) which while not connected to earth is at earth potential, (b) which is connected to earth, (c) which can be earthed to avoid possibility of instability.

4 When an opamp is used as a differential amplifier (a) it will perform the operation of differentiation, (b) its output will be the stage gain times the sum of the input signals, (c) its output will be the stage gain times the difference of the input signals.
5 The output from a comparator will be positive if (a) input + is greater than input −, (b) input − is greater than input +, (c) the two inputs are equal, (d) inputs are different irrespective of which is greater.
6 An opamp with a capacitor in the feedback loop performs as (a) differentiator, (b) oscillator, (c) integrator, (d) comparator.
7 A voltage follower (a) has unity gain and is inverting, (b) has unity gain and is non-inverting, (c) has a gain dependent upon the feedback resistor, (d) has a gain equal to the open-loop figure.
8 A summing amplifier (a) gives the algebraic sum of the input voltages, (b) gives the arithmetic sum of the input voltages, (c) gives a weighted sum of the input voltages.
9 If a continuous square wave is applied as input to a differentiator, the output will be (a) a continuous triangular wave, (b) a sine wave, (c) a continuous series of very sharp pulses, (d) a slowly rising ramp.
10 Slew rate determines (a) the bandwidth of an amplifier, (b) the gain of an amplifier, (c) the maximum distortionless frequency limit of an amplifier, (d) the amount of permissible feedback in an amplifier.

(c) CONVENTIONAL PROBLEMS

1 How can an operational amplifier be used as (a) an inverting, (b) a non-inverting amplifier? State the amplifier gain expression for each type. What assumption(s) is made in using these expressions?
2 A voltage follower has unity gain and is non-inverting. What advantages does such an amplifier have? Draw a circuit diagram and explain briefly how the circuit works.
3 Explain why the gain of an amplifier is independent of the open-loop gain of the opamp employed.
4 Why is a differential amplifier used in the input stage of an operational amplifier?
5 Design a simple amplifier with a voltage gain of −100 and an input resistance of 1 kΩ (see *Figure 5.14*).

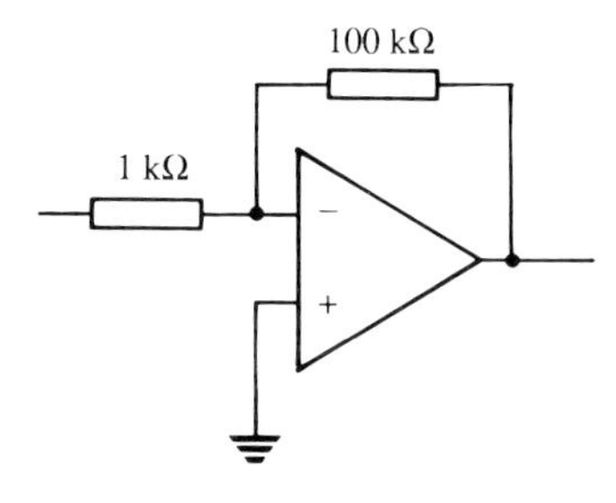

Figure 5.14

6 Three voltages are to be added by using the opamp circuit of *Figure 5.6*. If each resistor at the input is 100 kΩ, what must be the value of R_f such that $v_o = v_1 + v_2 + v_3$? If the amplifier gain is actually −200, what value should be given to R_f to maintain the output at the same value?

[100 kΩ, 102 kΩ]

7 How might an operational amplifier be used to compare two voltage levels and distinguish which is the greater?

8 A transducer element generates an e.m.f. of 1 V in series with a resistance of 3 kΩ. It is first connected to an indicating voltmeter which has a resistance of 500 Ω and then to the same voltmeter by way of an ideal voltage follower opamp. Compare the voltage and power delivered to the voltmeter in the two cases.

[0, 143 V, 41 μW; 1 V, 2 mW]

9 In an inverting amplifier circuit, R_1 is made equal to R_f so that a nominal gain of unity is obtained. If the open-loop gain of the amplifier is A, what should be the value of A for an error of 1% in the inversion?

[$A = 200$]

10 In an integrator circuit (see *Figure 5.13(a)*), suppose $R_1 = 100$ kΩ, $C = 100$ nF. What is the integrator gain? If the input is a 50 Hz squarewave of 2 V amplitude, what will be the slope of the output ramp? What will be the amplitude of the output ramp?

[0.1 V/ms per volt, 0.2 V/ms, 2 V]

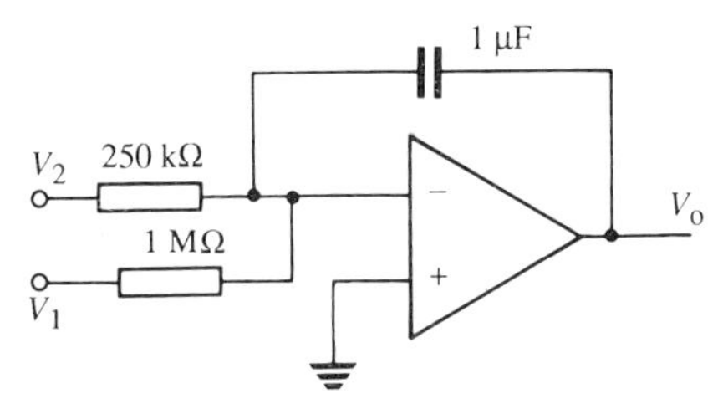

Figure 5.15

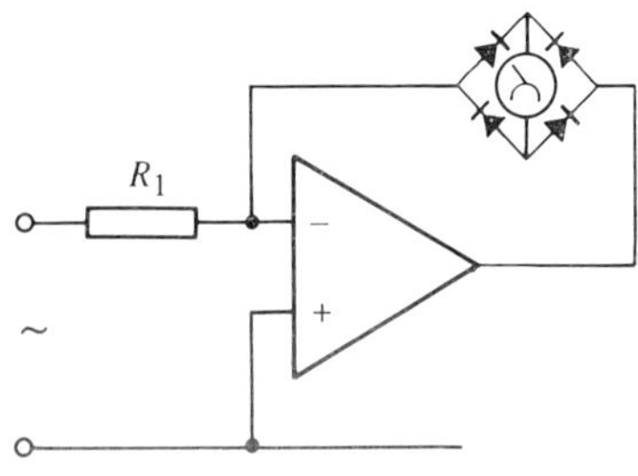

Figure 5.16

11 Obtain an expression for the output voltage of the circuit shown in *Figure 5.15*.

[$-\int(v_1 + 4v_2)\,dt$]

12 AC voltmeters using a bridge rectifier are inaccurate at low voltages because of the bend in the characteristics of the diodes used in the bridge. Explain how the circuit of *Figure 5.16* might overcome this problem.

6 DA and AD converters

A MAIN POINTS CONCERNED WITH DA AND AD CONVERTERS

1 Nature goes about her business in an analogue way; events happen 'smoothly' and 'continuously' in the sense that we are not aware of any sudden jumps in the nature of things. Any changes in physical happenings, even though they may sometimes seem very abrupt, instantaneous even, always take place through a continuous transitional state. Time itself is continuous; we can divide it up into minutes, seconds, microseconds, but it is there 'all the time'. There are no gaps in time, so there can be no gaps in events which exist in time. So time is an **analogue** quantity, just as things like velocity, pressure and temperature are analogue quantities. Such quantities, in other words, can take up *any* value or position within a range of values or positions. An analogue meter, for example, has a pointer which can take up *any* position along its scale length; whether we can actually read it to a particular degree of accuracy is beside the point. The pointer can settle on any one of an infinity of positions, as can the mercury in a thermometer or the sea level between low and high tides. Walking up a sloping plank is an analogue way of changing your level; going up a ladder is a digital method!

2 In electronics, many devices such as computers are not analogue but **digital** systems. When we want to store the output from an analogue device such as, say, a varying voltage or current, in a computer memory, the data has to be converted from its analogue (continuous) form into a digital (discontinuous) form expressed in a binary code which will be accepted by the computer.

To facilitate the transition, an **analogue-to-digital** converter (**ADC**) is required. An ADC produces a pulsed form of output in which each analogue input voltage or current variation is converted into a corresponding binary word, this action taking place at a rate dependent upon the rate of variation of the analogue signal. A digital voltmeter is an everyday example of an ADC in action; the applied voltage is an analogue quantity which is converted inside the instrument to a train of pulses which are counted over a given time period, then decoded to drive the familiar LED or liquid crystal display. Analogue voice signals are often converted to digital form for transmission over long-distance telephone and radio links. At the destination they are converted back to analogue – listening to a digital signal would not be very informative.

3 So it is necessary to be able to convert a digital signal into an analogue form. And for this a **digital-to-analogue** converter (**DAC**) is necessary. This circuit produces an output voltage or current which is proportional to the magnitude of the binary number applied at the input. *Figure 6.1* gives an idea of this sort of conversion,

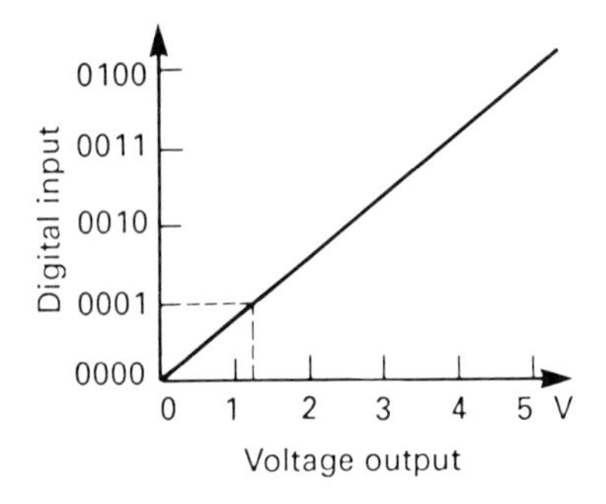

Figure 6.1

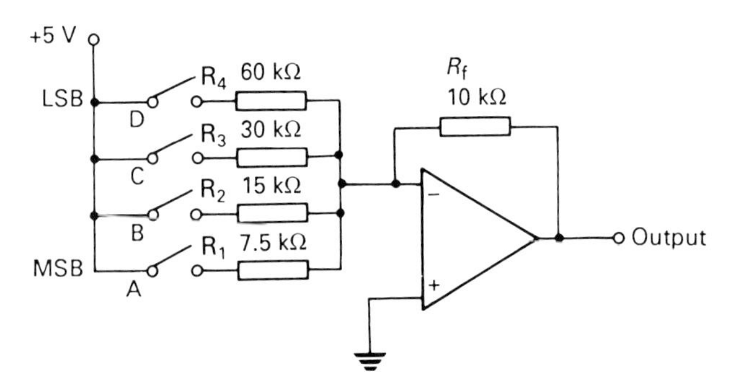

Figure 6.2

though it must be kept in mind that the vertical scale represents a discontinuous quantity – there are no intermediate values between the binary number positions. DACs are usually easier to understand in their working principles than ADCs.

4 A simple form of DAC based on the summing principle of an operational amplifier is shown in *Figure 6.2*. This is known as a binary-weighted DAC. We saw earlier how current flowing through resistors R_1 to R_4 are summed at the virtual earth point (the inverting input terminal) and converted to a weighted or proportional output voltage by the action of the opamp and the feedback resistor R_f. If a 4-bit binary word (or number) ABCD is applied to the input resistor network by way of matching switches, the resistors produce binary weighted currents according to the value of the word. This is done by giving each resistor in the array a value *inversely* proportional to the significance of the input bit. Thus a resistor connected to the most significant bit (MSB), may have a value R; the next resistor will then be $2R$, the next $4R$, and so on. Suppose the input is the binary number 0101, or decimal 5. Then switches B and D will be closed, switches A and C open. The information is being fed into the converter in parallel form, 'side by side', not in sequence. With the resistor values shown, the current through R_2 will

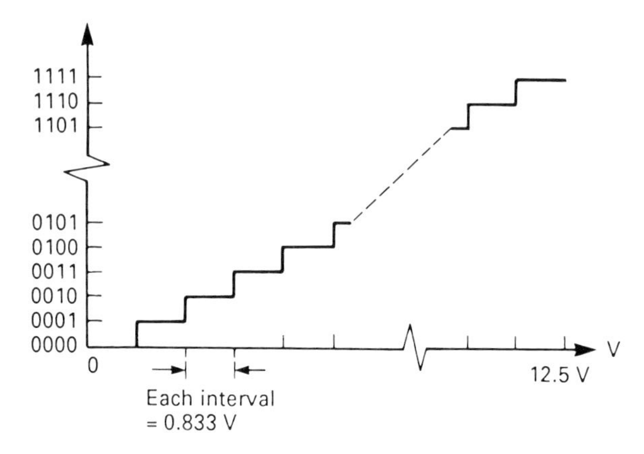

Figure 6.3

be $5/(15 \times 10^3)$ A = 334 μA, and through R_4 will be $5/(60 \times 10^3)$ A = 83.3 μA. The total current at the summing point (and hence flowing into R_f) will be 417.3 μA, so that the output from the opamp will be $(-10^4 \times 417.3)$ μV or −4.17 V. Now suppose the input to change to 1000 or decimal 8, then only switch A will be closed. The input current to the summing point will now be $5/(7.5 \times 10^3)$ A = 667 μA and the output voltage will change from −4.17 V to −6.67 V.

Similar analyses will show that the output increases in 15 steps (for a 4-bit input), each step being of amplitude 0.833 V, as the input ranges from 0000 to 1111 (decimal 15). By feeding the inputs from a 4-bit binary counter and displaying the output on an oscilloscope, a **staircase** waveform will be obtained showing each discrete step in the progression, rather as shown in *Figure 6.3*. The negative sign of the output voltages has been ignored here as the analysis is unaffected by its omission.

The voltage output (analogue) is therefore proportional to the input (digital) magnitude.

5 An ADC is the inverse of the DAC, but the circuitry is much more complex than that required for DACs, and in fact many ADC systems use DACs as part of their makeup. The principle behind ADCs is that the conversion from a continually changing analogue signal to a discrete digital signal requires that the analogue signal is regularly 'sampled' and that each sample is then converted into a corresponding digital value. *Figure 6.4* shows an analogue waveform which is being sampled at the rate of four samples per cycle. The amplitudes of these samples can then be translated into a digital signal which can later be recreated into the original wave. Do not confuse the actual sampling pulses with digital equivalents; a conversion has to be made where the digital *value* is proportional to the amplitude of the samples.

It might seem that it would be necessary to make the sampling rate as high as possible, but this is not so and an analogue signal can be reconstructed adequately if the sampling rate is $2f$ samples per second where f is the highest frequency component of the signal. A speech signal having a bandwidth of 5 kHz, for

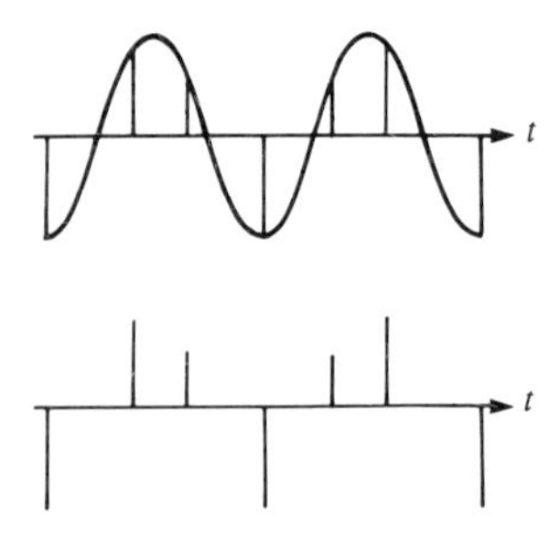

Figure 6.4

example, could be reconstructed from a set of samples taken 10 000 times a second. The actual form that an ADC takes depends upon the bandwidth of the input signal.

6 A technique known as **sample and hold** eliminates error which comes about from trying to digitise a rapidly changing analogue signal such that the signal level has altered before the conversion of the sample is completed. *Figure 6.5* shows this effect. The error voltage, δ V, depends upon the rate of change of the input signal level; if this is appreciable, the converter output will tend to represent the input level at the end of the sample period rather than at the beginning. In the sample and hold technique an electronic switch takes a very short duration sample of the input waveform and stores or holds this sample in a capacitor. The ADC then digitises the stored voltage. Only when the conversion is complete is the next sample switched through and held for the following conversion. The time for which the switch is closed and the capacitor is charging up must be extremely short; this is known as the **acquisition time**. *Figure 6.6* shows the principle.

7 A basic form of ADC known as a **simultaneous** or **flash** converter, is shown in *Figure 6.7*. The opamps are used as comparators, threshold voltages being applied to the inverting inputs through a potential divider chain made up, in this example, of resistors R_1 to R_4. The reference voltage (V_{ref}) applied to this chain represents the full scale or maximum permissible input signal level. In operation

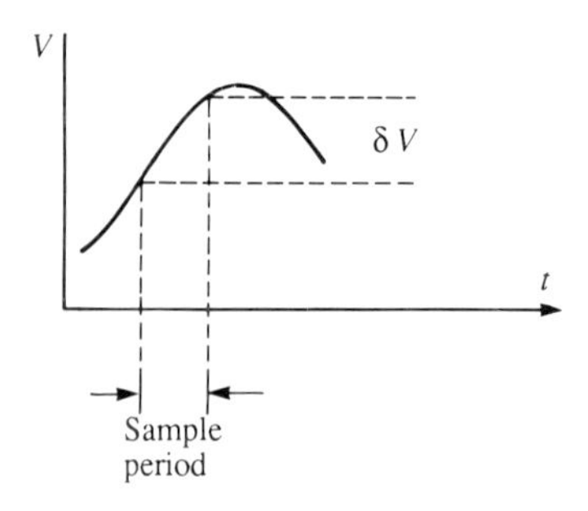

Figure 6.5

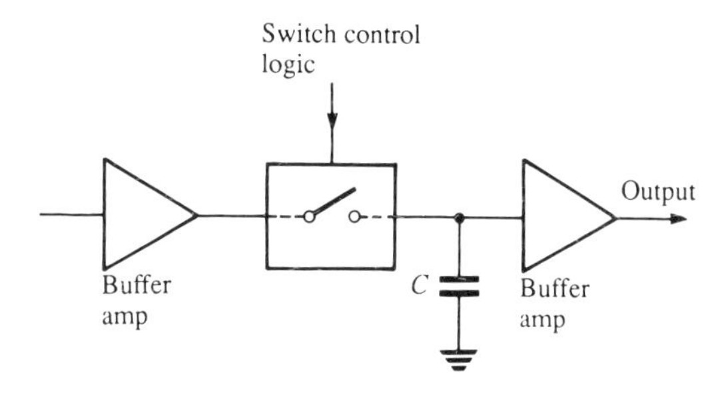

Figure 6.6

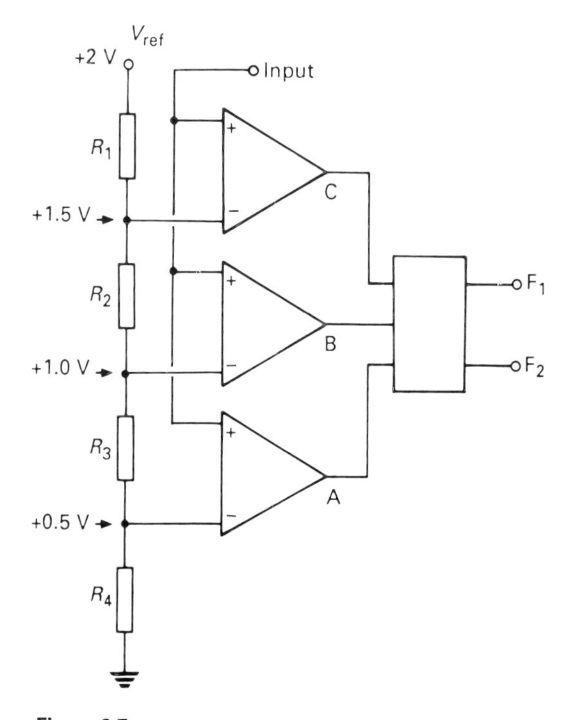

Figure 6.7

each comparator compares the analogue input voltage with its reference voltage. By making the divider resistors of equal value, the voltages at each of the inverting inputs increase (in this case) by one-quarter of the total reference voltage, that is, increments of 0.5 V from 0 to 2 V. The analogue signal to be digitised is then applied to all three non-inverting inputs in parallel. The output

Table 6.1

Input (V)	Comparator output A	B	C	Binary F_1	F_2
0–0.5	0	0	0	0	0
0.5–1.0	1	0	0	0	1
1.0–1.5	1	1	0	1	0
1.5–2.0	1	1	1	1	1

of each comparator will therefore go high if the input voltage is greater than its own particular reference level. So if, for instance, the input voltage is 1.3 V, comparators A and B will have high outputs (logic 1) and comparator C will have a low output (logic 0). Any input less than 0.5 V will not operate any of the comparators, and an input greater than 1.5 V will trip all three. As the output code is not binary, an encoder has to be used to follow the comparators so that the various output combinations of high and low can be converted into true binary. *Table 6.1* shows a truth table for the system.

Three comparators will resolve an input voltage to one only of four input levels. As *Table 6.1* shows, this is equivalent to 2 binary bits of resolution. To extend the coverage, a greater number of both comparators and reference levels are needed. The accuracy of this form of ADC is completely dependent upon the reference levels and hence upon the accuracy of the resistors used in the divider chain. The operating speed is very high and these converters are therefore used for the digitisation of analogue voltages from high frequency sources. A typical 8-bit converter can produce a digitised output within 50 ns of the input being applied.

8 There are a number of characteristics associated with both DACs and ADCs. The first of these is the **resolution** of the system. In the example discussed in *Figure 6.2* there were 4-bits in the input data and the converter had 2^4 or 16 output or **quantization** (discrete) levels; the resolution was therefore 1 part in 16 or 6.25 per cent. The accuracy or **quantization error** of a converter is defined in terms of the **quantization intervals** or the number of steps to be found in the staircase output. The number of steps for an N-bit converter is clearly $(2^N - 1)$; there are 16 numbers between 0 and 15 but only 15 intervals. Hence the analogue output can be no more accurate than one part in this number. If the total output excursion of the converter is known, then the quantisation error is expressed as $\pm$ half the quantisation interval. So, referring to *Figure 6.3* as an example, where the total voltage range is 12.5 V, the converter has an error or $\pm 12.5/(2 \times 15) = \pm 0.417$ V.

For an 8-bit converter which would have 255 intervals, over the same voltage range the error would be $\pm 12.5/(2 \times 255) = \pm 0.0245$ V. So the greater the number of bits, the smaller the quantisation error, as we might have guessed.

Linearity error is a measure of the extent by which the actual output differs from a mean straight line drawn along the staircase and marked (a) in *Figure 6.8*. A linearity error curve is illustrated by the broken line marked (b), while an error resulting from scaling (or gain) inaccuracy is shown in line (c). Finally, line (d) which runs parallel to the wanted output line (a) comes about because of offset error, the output not being zero when all the inputs are zero.

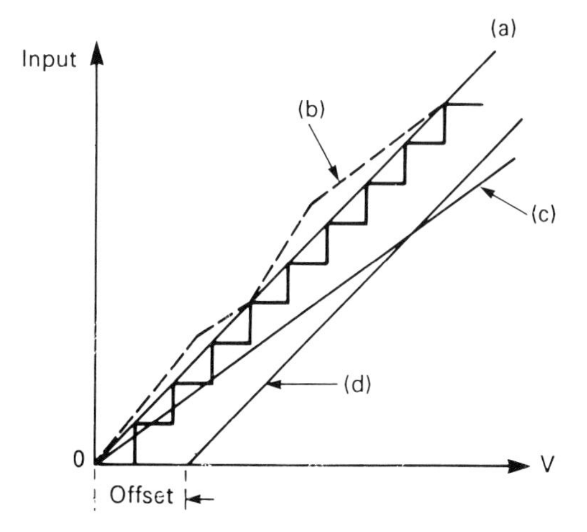

Figure 6.8

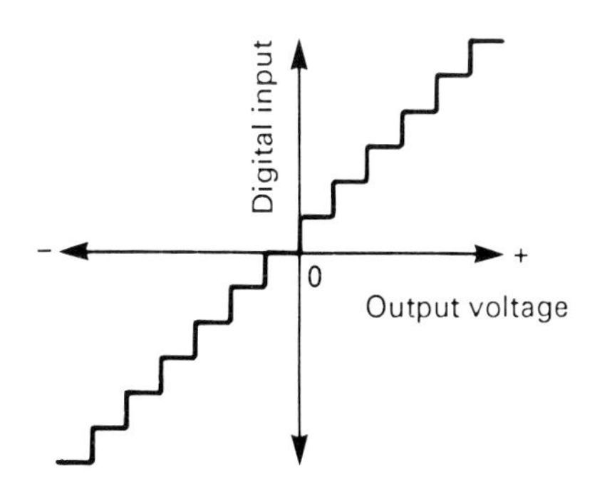

Figure 6.9

9 The DAC discussed under *Figure 6.3* provides a single polarity output; such converters are termed **unipolar**. Converters are often required which give an output running across both positive and negative values and these are termed **bipolar** converters. It is customary in these systems to have both positive and negative reference lines instead of the positive reference and earth line so far discussed. A typical bipolar staircase output is shown in *Figure 6.9*.

B WORKED PROBLEMS ON AD AND DA CONVERTERS

Problem 1 What is the main disadvantage of the binary-weighted converter shown in *Figure 6.2*? Discuss an alternative form of ADC which overcomes this disadvantage.

The binary-weighted converter has the disadvantage that a separate different-valued resistor is required for each input bit. Anything over six bits leads to a range of resistance values which is too great for proper operation. A 12-bit device, for example, assuming that we begin with $R_1 = 10\ \text{k}\Omega$ corresponding to the MSB, would require a 20.48 MΩ resistor for the LSB. Values such as this are not easy to manufacture with the degree of accuracy and stability which is necessary.

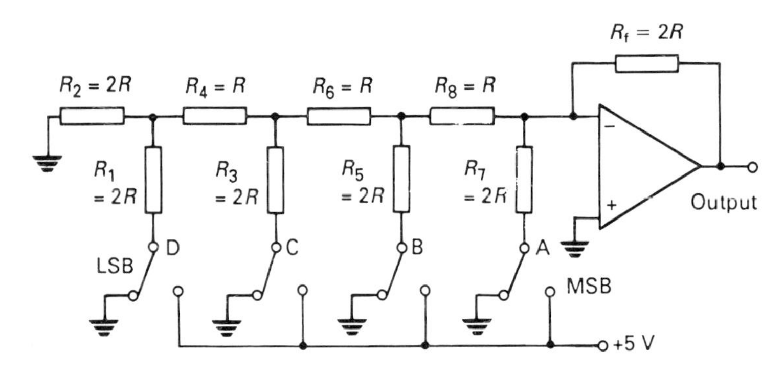

Figure 6.10

An alternative system which produces binary-weighted currents but which uses only two values of resistance is shown in *Figure 6.10*. As in the previous case, the weighted currents are converted to proportional voltages by the opamp and feedback resistor R_f. The 5 V reference line is illustrative only; commercially available converters using the ***R*–2*R* ladder** principle as this method is known, operate with other levels. The switching is of course electronically controlled and operated (in this example) by the 4-bit parallel binary input.

In theory the actual values chosen for the resistors are unimportant provided that all the Rs have the same value and the $2R$s are double this. In practice the values selected should be low enough to provide a low output impedance but not so low as to put an appreciable loading on the reference line. Typical values are $R = 10\ \text{k}\Omega$, $2R = 20\ \text{k}\Omega$. To follow the operation, suppose that switch A (representing the MSB) is connected to the reference line and the other switches to earth. Then R_1 $(=2R)$ and R_2 $(=2R)$ are in parallel to earth, so their equivalent resistance is simply equal to R. This equivalent R adds to R_4 $(=R)$ to form another $2R$ in parallel with R_3 $(=2R)$ and earth; see *Figure 6.11(a)*. The combination of R_3

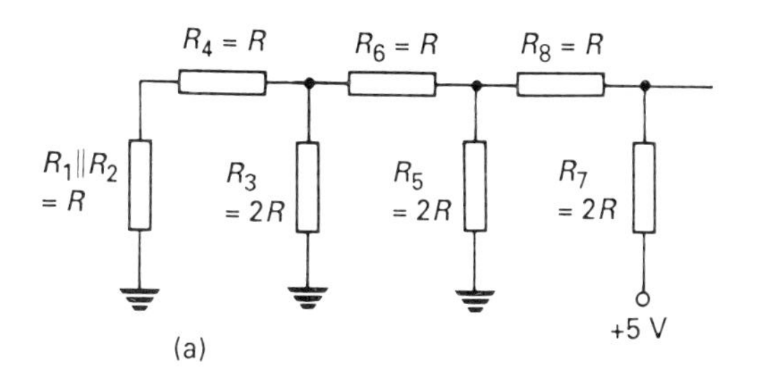

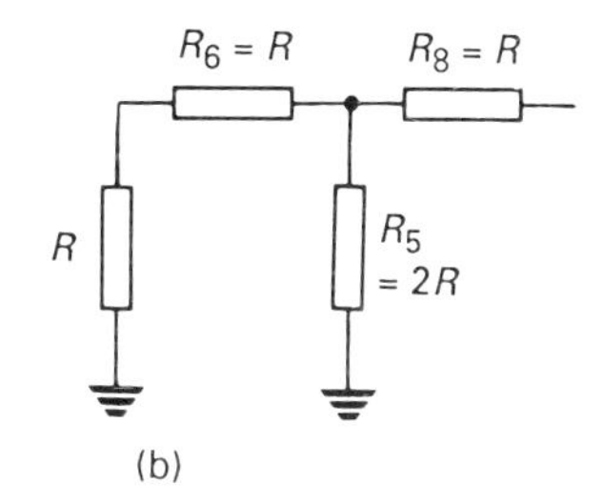

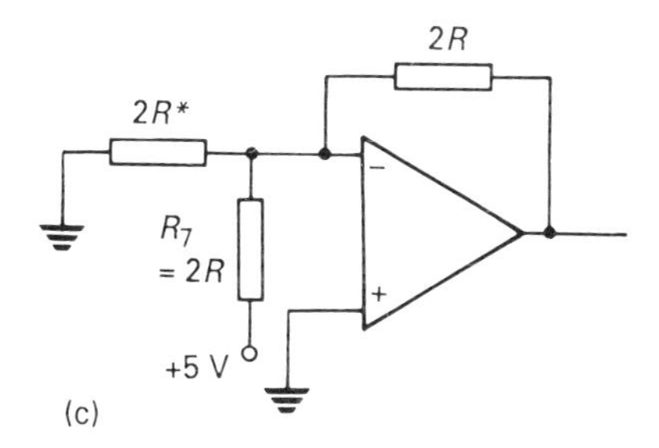

Figure 6.11

($=2R$) in parallel with the previous two equivalent Rs (in series) reduces to a single R in series with R_6; see *Figure 6.11(b)*. Carrying on in this way (try it for yourself) leads to the circuit shown in *Figure 6.11(c)*. A study of this last diagram reveals that no current will flow through the final equivalent resistor (designated as $2R^*$) because one end goes to earth and the other end to the virtual earth terminal of the opamp. This resistor, therefore, plays no part in the further analysis. The other resistor, R_7, connects to the 5 V reference line; hence, taking its value as 20 kΩ, a current of $5/(20 \times 10^3)$ A or 0.25 mA flows through R_7 and the feedback resistor R_f which is also equal to $2R$ or 20 kΩ. The output voltage is therefore 5 V, representing the analogue of the MSB at the binary input.

You will find it a useful exercise to work through the ladder network with other binary input combinations.

Problem 2 What is meant by output settling time?

Output settling time is another converter parameter which has importance. It is defined as the time the converter needs to settle within ±0.5 LSB of the final value after the input data changes. It is usually specified with all bits switched on or off at room temperature. A typical figure is 0.1 μs. The converter is limited, therefore, to the handling of the inputs whose period is less than the settling time.

Problem 3 Describe an experiment using an integrated DAC which illustrates much of the foregoing text.

Practical DACs available in integrated circuit form, often have reference levels which are built in, and the values are chosen such that a direct correlation exists between the output voltage and the decimal equivalent of the binary input. An 8-bit system having a reference of 2.55 V, for example, will, for an input range of 0 to 11111111 (or 0–255 decimal) have an output of 0.01 V per digit input. An input of 187 (decimal equivalent) will therefore produce an output of 18.7 V.

An integrated DAC suitable for experiments is the ZN428, a systems diagram of which is given in *Figure 6.12*. This is a typical monolithic converter which, in

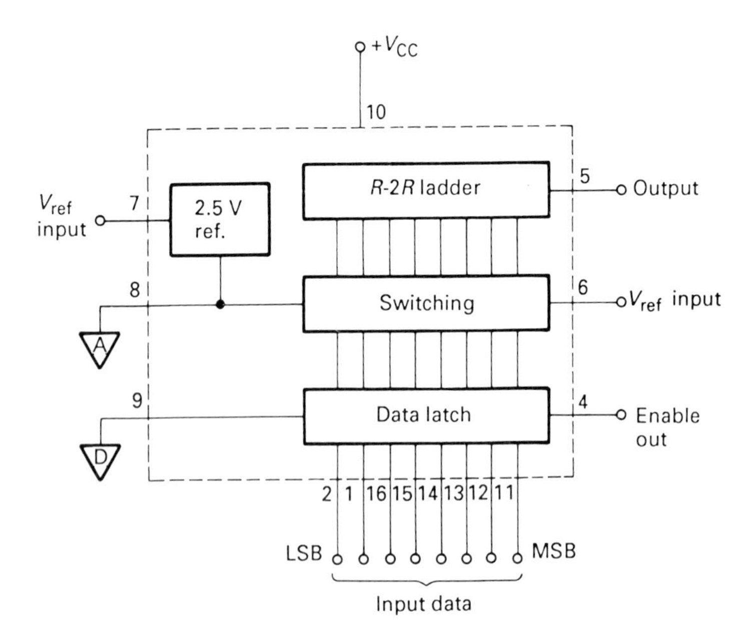

Figure 6.12

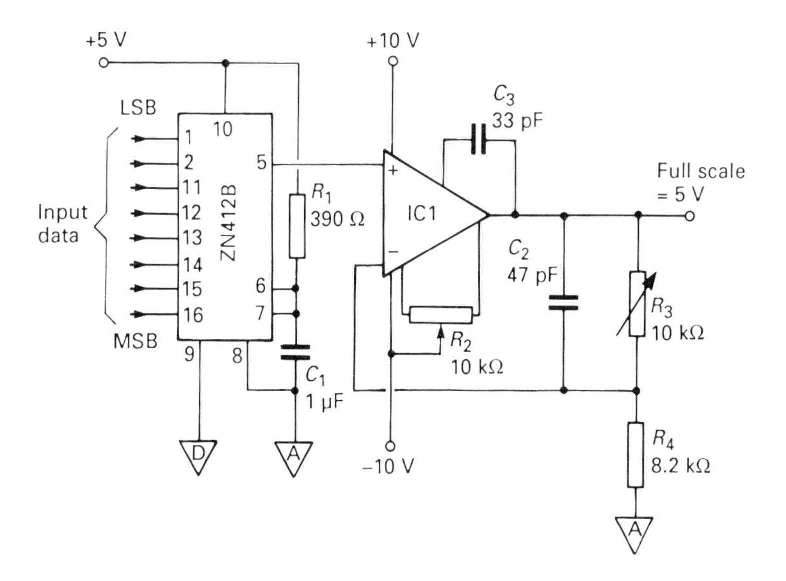

Figure 6.13

addition to the R-$2R$ ladder network and an optional 2.5 V internal reference, has input latches which will hold the input data when enabled. This facilitates up-dating from a computer bus line, for example. The input is TTL and 5 V CMOS compatible. The switching system is transistor operated.

A practical circuit design is given in *Figure 6.13* and from this a number of experiments can be derived. The 8-bit input can be driven from TTL logic outputs. The included reference source requires an external load resistor R_1 and a decoupling capacitor C_1. In addition, an opamp for which the 531 is suitable, is needed to provide both buffering and gain. An offset-null trimmer control is also necessary to achieve optimum accuracy at low output voltage levels and the potentiometer R_3 is adjusted for the required full-scale output with all inputs held high. The enable input should normally be set low. The system provides a unipolar output and if the given voltage levels are used, the output will have a full-scale range of +5 V.

Setting up is quite easy: put all input bits to 0 and adjust the offset control to give zero output. Then put all input bits to 1 and adjust R_3 to give an output of 4.98 V, that is, full scale less 1 LSB.

Problem 4 Describe a form of ADC which depends upon a binary counter and a DAC in its assembly.

Possibly one of the simplest types of ADC, both in economy of components and understanding, is the **single-counter converter** which includes in its makeup a

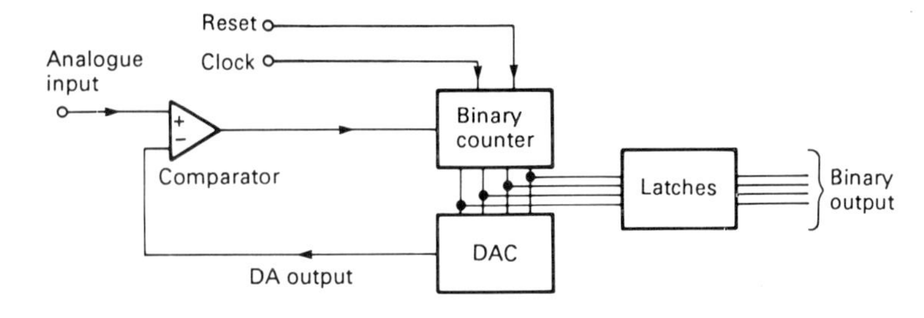

Figure 6.14

DAC of the kind already discussed. *Figure 6.14* shows a block diagram of the system. In addition to the ADC there is a binary counter and a comparator. There may also be a latching system so that the binary output can be held in the latch until required.

At the start of the conversion sequence we assume that the analogue input is 0, that a reset signal has been applied to the counter and that the output of the ADC is 0. The inverting input to the comparator is therefore also at 0. When some voltage is applied to the non-inverting terminal, the comparator output goes high and enables the counter, which begins to count the clock pulses. Each clock pulse advances the counter one step and the input to the DAC consequently increases. The output from the DAC increases also, by one step for each clock pulse. This output feeds into the comparator and, as long as this voltage is below the analogue input, the output from the comparator remains high. As soon as the DAC output exceeds the analogue input level, however, the comparator output will trip low and shut off any further clock pulses to the counter. The digital output will then be held in the latches at a value corresponding to the analogue voltage level at the input. The counter is then reset and the sequence repeats.

Problem 5 What are the disadvantages of the single-counter type of ADC? How can they be overcome?

The single-counter type of converter is a relatively slow operator, being dependent upon the speed of the counter and the DAC. A very precise DAC is also essential. Since the count has to start at 0 for each conversion cycle, the time for each conversion depends upon the amplitude of the analogue signal. Further, during a conversion, the DAC output may correspond exactly to, or be very slightly below, the analogue level at the comparator; the comparator will then fail to trip until the counter has moved through a further clock cycle. The greatest quantisation error is, therefore, one complete interval.

An improvement can be made by replacing the up-only counter by an up-down counter. This results in a system known as a **tracking** or **follower converter**. Here the counter counts up when the comparator output is high but counts down when the comparator output is low. This routes the clock pulses to the count-up or

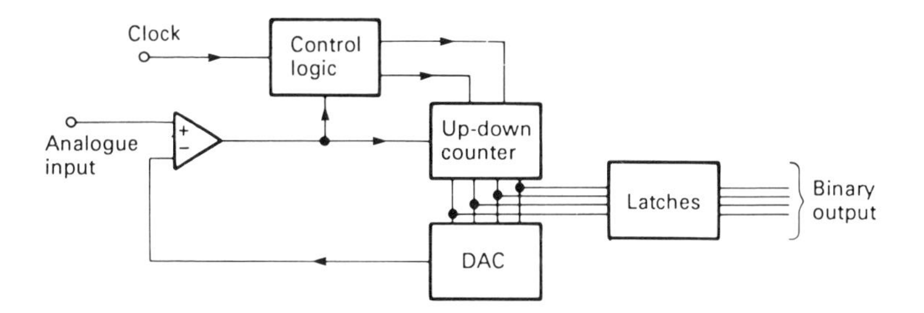

Figure 6.15

count-down inputs by means of gate switches on the comparator output. *Figure 6.15* shows a block diagram of the system.

At the start of a conversion, assume that the DAC output is at 0, then with some analogue input voltage, the output of the comparator will be high. This routes the clock pulses to the count-up input of the counter and the output of the DAC increases until it exceeds the analogue level. The comparator then trips low, the clock is transferred to the count-down input of the counter, and the counter reverses from the previous figure until the correct level is reached.

Problem 6 The tracking converter has itself a disadvantage. Explain what this is.

If the input to the tracking converter remains unchanged during a conversion cycle, the first one-down count will drop the output from the DAC immediately below the analogue level, so tripping the comparator output high and initiating the reversal of the count. The following one-up count will again be sufficient to trip the comparator, and the one-up, one-down sequence will go on for as long as the analogue input remains constant. This is a disadvantage of this type of converter; the output oscillates about the least significant bit for an unchanging input.

Problem 7 Describe a converter based on the principle of successive approximation.

The **successive approximation converter** is very popular in practical designs. It has an advantage over the counter systems in that N bits of resolution can be secured with only N clock pulses. In essence, the method follows on from the counter-type converters except that the counter is replaced by a register and some additional control circuitry is included between this register and the comparator. A simplified block diagram is shown in *Figure 6.16*.

The operation is as follows: a reset signal (the **start-conversion** pulse) sets all bits in the register to 0. The DAC output is thus 0, and any input signal at the analogue terminal causes the comparator output to go high. On the application of the first clock pulse, the register turns on its MSB to the DAC; all other bits stay at 0. This produces a voltage from the converter (V_1). This voltage is compared with the input voltage and a decision is made on the next clock pulse edge to set the

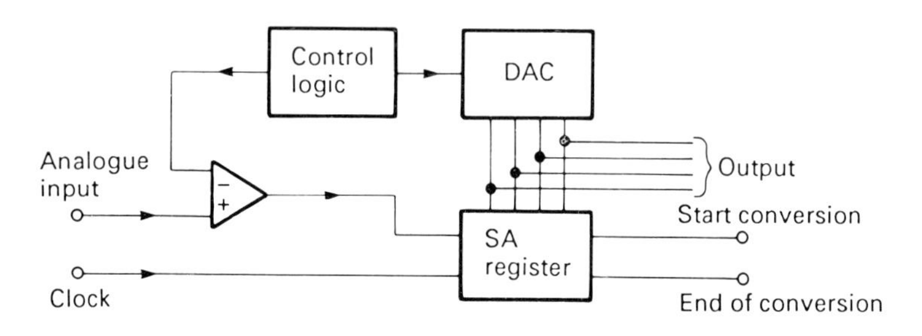

Figure 6.16

MSB to 0 if V_1 is greater than the input or retain a 1 if V_1 is less than the input. In either case, on the next clock edge the register will set the next MSB. In turn it will keep or reset this second bit, again by comparing the DAC output with the input. This process is repeated until the LSB is reached. A bit is retained if V_1 is less than the input or reset if V_1 is greater than the input. When all the bits have been tried, the register **end conversion** output goes high; the digital output from the converter is then a valid representation of the input voltage. This output is latched until the next start-conversion pulse.

The total conversion time is equal to N clock pulses where N is the number of bits in the register. This time does not depend upon the amplitude of the analogue input as it did with the counter circuits. *Figure 6.17* shows a conversion cycle for an 8-bit successive approximation converter.

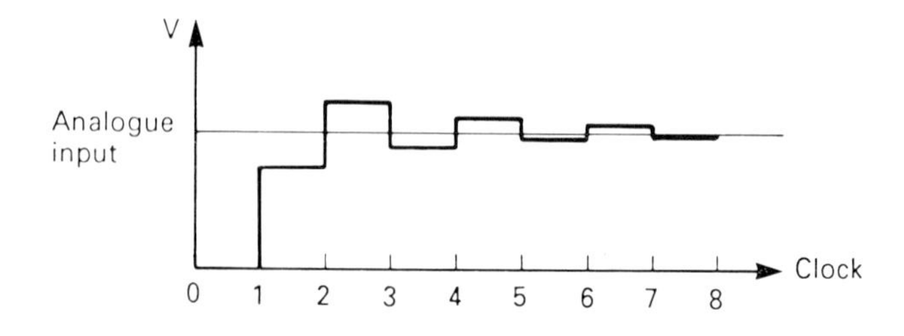

Figure 6.17

Problem 8 What is the principle used in digital voltmeters for conversion from an analogue input to a digital signal appropriate to energise a numerical display?

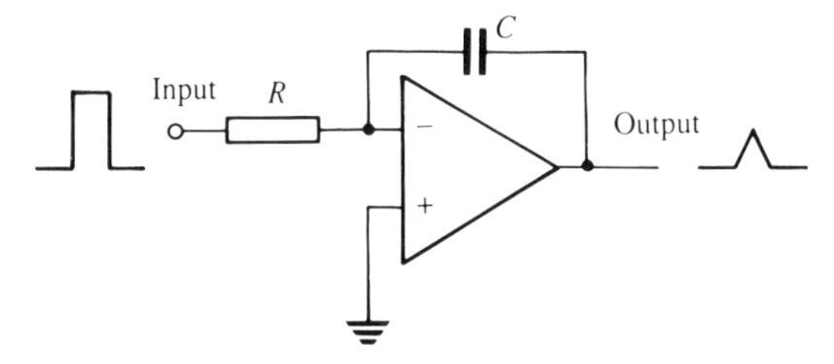

Figure 6.18

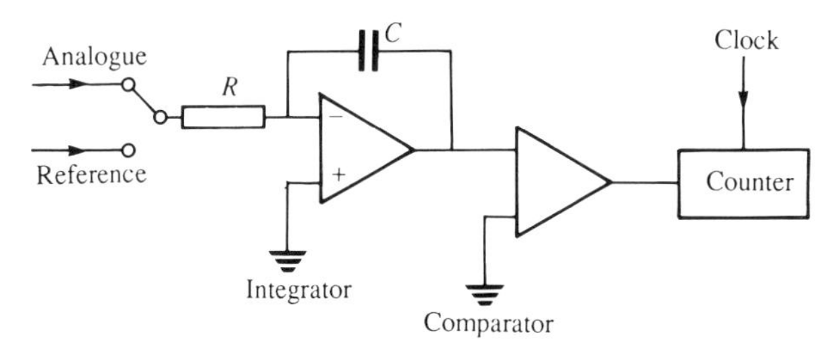

Figure 6.19

This form of ADC generally makes use of the opamp **integrator**.* We have seen earlier that the output of an integrator, for a pulsed input, ramps up and down on the leading and trailing edges of the pulse respectively, see *Figure 6.18*. The principle of operation is that the ADC of *Figure 6.19* uses an electronic switch to change the input voltage of the integrator between a reference voltage and the analogue input. By integrating the analogue voltage for a fixed time t_1 then counting the clock pulses for the time t_2 which the integrator takes for its output voltage to rise to the reference (or in other systems to fall back to zero), the analogue to digital conversion is made. Integration-type converters are slow acting but have very high resolutions.

Problem 9 What particular advantages does the transmission of information over lines or radio links in digital form have over analogue signals?

Digital signals are less susceptible than analogue signals to deterioration by stray pickup, noise and attenuation in long lines and radio links. Each binary digit or

* A very full description of this form of conversion will be found in *Electrical and Electronic Principles 3* (2nd edition).

bit of information exists in only one of two possible states; changes in wave shape or amplitude can be tolerated to quite a high degree before loss of information occurs. Another advantage is that digital transmission makes the data acceptable to computers.

Problem 10 What is a shaft position encoder?

This is a direct method of converting analogue to digital signals. A flat disc which can rotate on a shaft is divided into a number of annular rings, one ring for each digit in the number. Brushes make contact with each of these rings which are divided into areas of positive contact (representing logic 1) and insulating areas (representing logic 0). The conducting pattern makes contact with the brushes as the shaft revolves and a binary output is obtained which corresponds with the analogue position of the disc. Up to 1000 positions per revolution are possible by this method. The system is useful in that direct mechanical connection can be made to machinery.

C FURTHER PROBLEMS ON AD AND DA CONVERTERS

(a) SHORT ANSWER PROBLEMS (answers on page 152)

1 What is the purpose of a digital to analogue converter?
2 In a 4-bit binary weighted resistor type of DAC, the resistor representing the least significant bit is 20 kΩ. What value represents the most significant bit?
3 What advantage does the *R*-2*R* converter have over the weighted resistor type?
4 How many comparators are needed to construct a 6-bit parallel ADC?
5 Why is a 'flash' converter so called?
6 Define the terms: (a) resolution, (b) linearity, (c) quantization steps, as applied to AD and DA converters.
7 A DAC has a 6-bit input data signal. What is the percentage resolution of this converter?
8 What is the principle behind the sample-and-hold technique? When is it used?
9 How many quantization levels does a 4-bit signal range have? How many quantization intervals?
10 Which has the greatest operating speed, an integrator type comparator or a successive approximation type of comparator?

(b) MULTI-CHOICE PROBLEMS (answers on page 152)

1 An ADC (a) converts a constant voltage to a digital signal, (b) converts a continually changing voltage to a digital signal, (c) converts a sinewave to a digital signal.
2 A 6-bit digital signal has (a) 15, (b) 31, (c) 63, (d) 127 quantisation intervals.
3 The largest value resistor in a binary weighted network of a 6-bit DAC is 64 kΩ. The lowest value resistor is (a) 1 kΩ, (b) 2 kΩ, (c) 4 kΩ, (d) 16 kΩ.
4 A 6-bit unipolar converter has a total output range of 10 V. If the input digital signal is 101101, the output will be (a) 3.07 V, (b) 7.14 V, (c) 8.5 V, (d) none of these.

5 The quantisation error of the DAC in the previous example is (a) ±0.01 V, (b) ±0.08 V, (c) ±0.16 V, (d) ±0.32 V.

6 A sinewave of frequency 5 kHz will be recreated successfully from a sample rate of (a) 2500 per second, (b) 5000 per second, (c) 10 000 per second, (d) 20 000 per second.

7 Unipolar DACs produce (a) a positive output, (b) a negative output, (c) both a positive and a negative output, (d) a positive or a negative output.

8 In a flash converter having a 3-bit digital output, the number of comparators needed is (a) 3, (b) 5, (c) 7, (d) 9.

9 The accuracy of a flash comparator depends upon (a) the resistor values used in the divider chain, (b) the reference voltage, (c) the gains of the comparators, (d) the accuracy of the resistors used in the divider chain.

10 An *R*-2*R* ladder network, regardless of the number of applied bits *N*, has (a) only two resistor values, (b) only four resistor values, (c) the same number of resistors as there are bits, (d) $(2^N - 1)$ resistors.

11 The digital output of a shaft position encoder depends upon (a) the speed of rotation of the shaft, (b) the angular position of the shaft, (c) the diameter of the shaft.

(c) CONVENTIONAL PROBLEMS

1 The inverting input of an opamp is earthed and negative feedback is applied from output to non-inverting input. What voltage would you measure at the non-inverting terminal?

[0 V]

2 What voltage would you find at the inverting input of an opamp with negative feedback if the non-inverting input was taken to −5 V?

[−5 V]

3 Define the following terms: (a) binary-weighted network, (b) bipolar DAC, (c) quantization interval, (d) follower converter, (e) parallel converter.

4 A 4-bit weighted resistor DAC has a LSB resistor value of 100 kΩ. What are the other resistor values? How many output levels does this converter have?

[50 kΩ, 25 kΩ, 12.5 kΩ; 15]

5 For a DAC define the following: (a) quantization error, (b) linearity error, (c) offset error.

6 What is a major disadvantage of a weighted resistor type of DAC? How is this overcome in the *R*-2*R* ladder network?

7 Draw a diagram of a 4-bit *R*-2*R* DAC. In such a circuit, the *next* to MSB switch is closed and the other three switches are open. Draw an equivalent circuit for this condition (refer to *Figure 6.10* if you wish).

8 Draw a block diagram of a successive-approximation type AD converter and briefly describe its function.

9 Why is it necessary to use a sample-and-hold technique when rapidly changing analogue signals are being digitised? How often should a 50 Hz sinewave be sampled? What is meant by acquisition time?

10 Give example types of converter that are (a) slow speed, high resolution, (b) high speed, medium resolution, (c) very high speed.

[(a) integrator, (b) successive-approximation, (c) parallel]

11 The circuit shown in *Figure 6.20* is designed to produce a binary output count at terminals A, B, C, D which is proportional to the magnitude of the analogue input voltage. The counter is 4 D-type flip-flops and the bistable is a conventional

twin-nand gate type. Explain the action of the entire circuit, assuming that the counter has just been reset to zero and commences counting immediately. Assume also that there is a small steady d.c. voltage applied to the analogue input terminal.

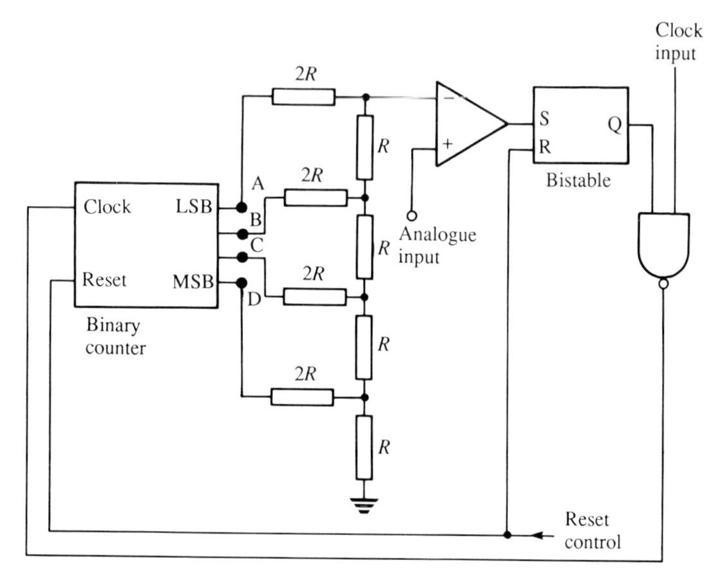

Figure 6.20

7 Controlled rectifiers

A MAIN POINTS CONCERNED WITH CONTROLLED RECTIFIERS

1 The power delivered to a load, such as a lamp or motor, can be controlled by nothing more elaborate than a rheostat. In *Figure 7.1(a)*, a rheostat is connected in series with a power source (a battery) and a load of some kind. When the resistance setting of the rheostat is zero, full power is delivered to the load. When the resistance setting of the rheostat is at a maximum, only a fraction of the full power is delivered to the load. This method is, of course, very wasteful, since the rheostat converts the surplus energy into unwanted heat. *Figure 7.1(b)* shows an alternative which is even less elaborate than the rheostat. Here a switch is connected in series with the battery and the load. When the switch is open, no power is dissipated in the load or in the switch since the current flowing in the circuit is zero. When the switch is closed, maximum power is dissipated in the load but again no power is dissipated in the switch since the voltage developed across it is zero. This argument assumes that the switch is ideal, has either zero (closed) or infinite (open) resistance. No real-life switch would have these properties, but the point being made is that power is only being wasted during the periods when the switch is actually opening or closing. There are then conditions in which a finite resistance appears, however momentary, so power is dissipated within the switch itself and not within the load.

2 To control the power delivered to a load with maximum efficiency we require a switch with as near ideal characteristics as we can get and then the flow of power to the load is varied by altering the open-to-closed ratio or **duty cycle** of the switch operation. In this, the speed of switching is of first importance. In practice, it is usual to vary the off period while keeping the *sum* of the successive on and off

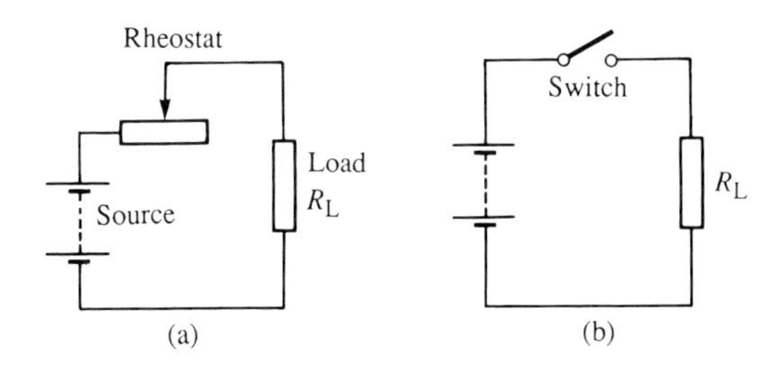

Figure 7.1

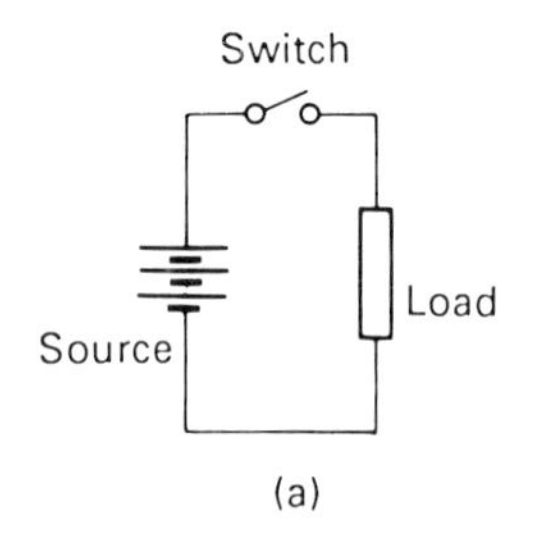

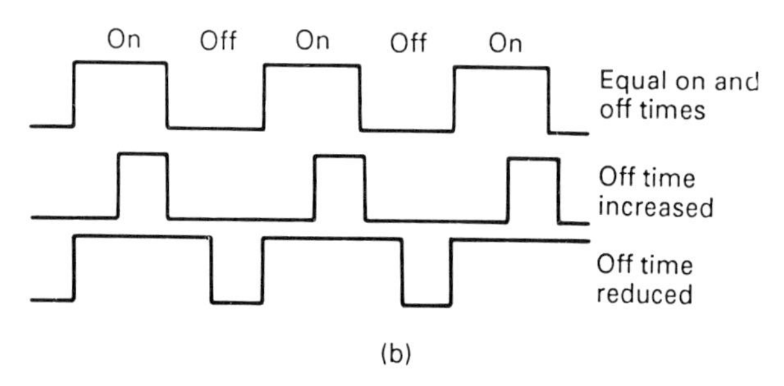

Figure 7.2

periods constant; this is illustrated in *Figure 7.2*. The point to notice is that the pulse repetition frequency is unchanged. The choice of this frequency is dependent upon the application but because it is only the average power which is being varied the frequency has to be high in relation to the most rapid rate at which the load power is required to vary. Since the control system usually operates on a.c. mains supplies, the repetition frequency is either 50 or 100 Hz, depending upon the circuit configuration.

3 The name **controlled rectifier** is applied to those systems in which the on period of the load circuit (which may involve very heavy currents) is controlled by an auxiliary light-duty circuit in which the current is very small. We have seen in an earlier part of the course (*Electronics 2*) that the transistor can be used as a switching device. Although widely used in digital systems where only low-level power is controlled, the transistor has the disadvantage that there is an appreciable voltage drop across the device when it is switched on, and leakage current when it is switched off. Further, a continuous base current is required to keep a transistor switched on. So, although the transistor is fast acting and has easy-to-drive characteristics, other devices have been designed specifically for the control of large power-consuming circuits.

4 The **silicon controlled rectifier** (SCR) or **thyristor** is the basic device in a number of such semiconductor assemblies designed for power control. It is a four-layer device made up of alternate wafers of *n*- and *p*-type material arranged as shown in *Figure 7.3*. We might treat this arrangement as three *p-n* junctions in series, as indeed it is. By bringing out three connecting leads, A, G and K, the device becomes the standard form of thyristor. The terminals are known respectively as

the **anode**, **gate** and **cathode** and the circuit symbol for this device (which is a *p*-gate controller) is shown in *Figure 7.4*. The property of the thyristor which makes it such a versatile power controller is the fact that it does not conduct until a current pulse flows into the gate connection. Once it has been switched on in this way, a form of positive-feedback regeneration takes place which keeps the thyristor switched on until the voltage applied between anode and cathode is removed. *Figure 7.5* shows a simple circuit in which the a.c. mains supply is applied to the load resistor R_L by way of a thyristor; if a train of short duration pulses are applied to the gate, the thyristor conducts for the remainder of the (otherwise normal) half-wave rectified output, this remainder being dependent upon the phase of the gate pulses relative to the a.c. input waveform. The fraction

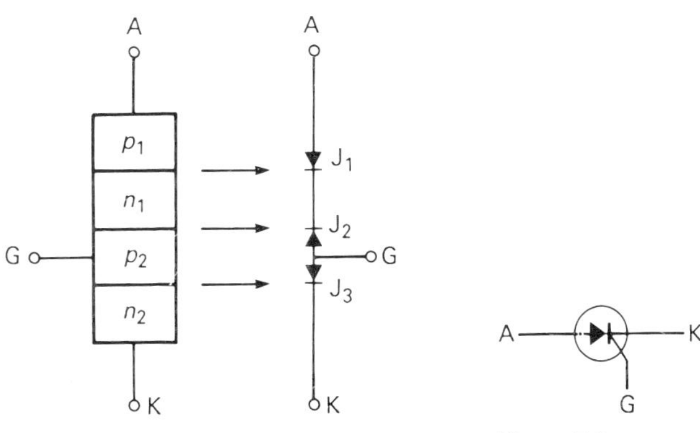

Figure 7.3

Figure 7.4

of the waveform passed to the load by the thyristor, therefore, determines the average power dissipated in the load. The negative half-cycle of input is not transmitted because the thyristor switches off as soon as its anode–cathode voltage falls to zero. Thyristors can be used to control very high powers since reverse-biased junctions can be made which will withstand many hundreds of volts. They are available to switch many thousands of amperes with a voltage drop of only a volt or so. This is where thyristors score over the power transistor switch.

5 Thyristors are made in which the gate lead is brought out not from a *p*-type layer as shown in *Figure 7.3* earlier, but from an *n*-type layer. These forms of thyristor are known as *n*-gate anode-controlled devices to distinguish them from the *p*-gate cathode-controlled types. Anode-controlled thyristors, symbolised as in *Figure 7.6(a)*, can be switched by using negative-going gate pulses. In the so-called silicon controlled switch (SCS) a gate lead is brought out from both the *n*- and *p*-type layers; the symbol for this is shown in *Figure 7.6(b)*.

6 The standard thyristor is a half-wave device; the negative half-cycle of the input must be rejected irrespective of what the gate signal happens to be doing at the time the anode is negative with respect to the cathode. For full-wave operation, two thyristors are enclosed within one package, the anode of each being connected internally to the cathode of the other. A single-gate input terminal is

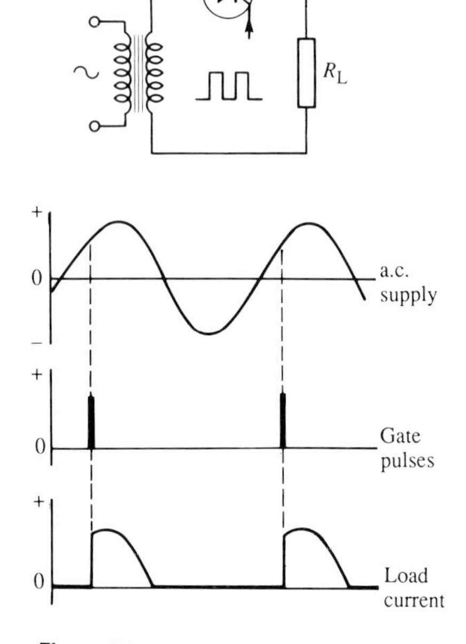

Figure 7.5

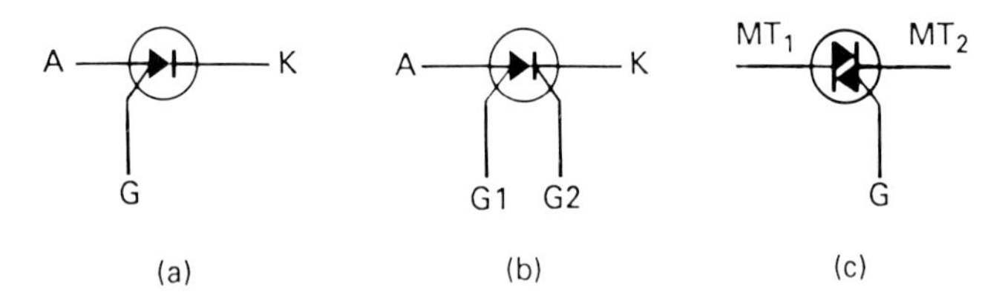

Figure 7.6

then connected internally to the two gate layers. These devices are called **triacs** or bi-directional thyristors and the circuit symbol is shown in *Figure 7.6(c)*. Triacs can be switched from an off to an on state for either polarity of the applied a.c. voltage with positive or negative gate pulses. The words 'anode' and 'cathode' are not strictly applicable as such to the triac and these terminals are referred to as just that: **main terminal MT_1** and **main terminal MT_2**. The gate input is always referred to MT_1 in the same way as it is referred to the cathode in the ordinary uni-directional thyristor. A gate current of 20 mA is usually sufficient to trigger triacs rated to control currents up to some 30 A.

B WORKED PROBLEMS ON CONTROLLED RECTIFIERS

Problem 1 A d.c. supply is switched so that the load current follows the waveform shown in *Figure 7.7*. What percentage of full power is dissipated in the load?

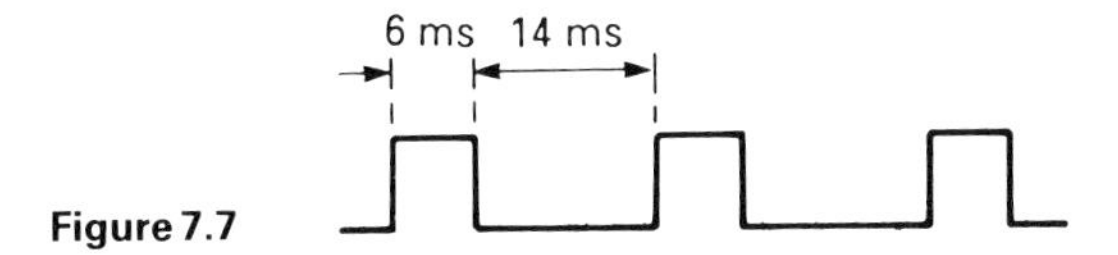

Figure 7.7

From *Figure 7.7* the load current flows for 6 ms and is then off for 14 ms. The fraction of full power (which would be represented by a continually on waveform) is clearly 6/(14 + 6) or 3/10. The percentage of full power is therefore **30 per cent**.

Problem 2 *Figure 7.8* shows the load current waveform for a simple full-wave rectified output in which the conduction is delayed by an angle ϕ radians in each half cycle. Derive an expression giving the average current in the load for all values of ϕ between 0 and π radians. What is the average current when $\phi = 0$, $\pi/3$, $\pi/2$?

When $\phi = 0$, the average current will be that of a fully rectified sinewave and we should be able to recall that this is $2\hat{I}/\pi$ or $0.637\hat{I}$. When the start of each half cycle is delayed by an angle ϕ, then

$$I_{av} = \frac{\hat{I}}{\pi}\int_{\phi}^{\pi} \sin \omega t \, d(\omega t)$$

$$= \frac{\hat{I}}{\pi}[-\cos \omega t]_{\phi}^{\pi} = \frac{\hat{I}}{\pi}[-\cos \pi + \cos \phi]$$

$$= \frac{\hat{I}}{\pi}[1 + \cos \phi]$$

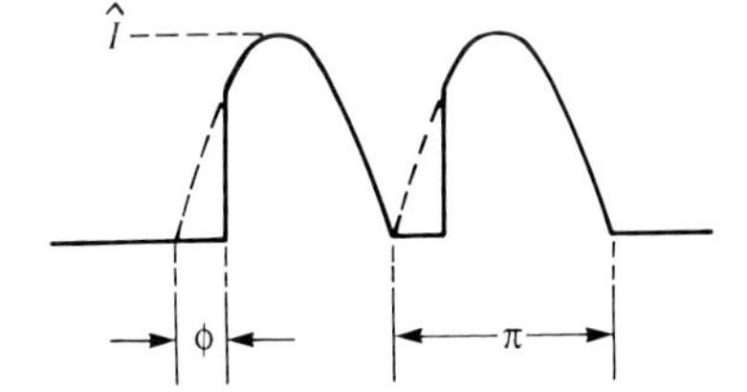

Figure 7.8

You can now check the case when $\phi = 0$. For $\phi = \pi/3$

$$I_{av} = \frac{\hat{I}}{\pi}\left[1 + \cos\frac{\pi}{3}\right] = \mathbf{0.478\hat{I}}$$

and for $\phi = \pi/2$

$$I_{av} = \frac{\hat{I}}{\pi}\left[1 + \cos\frac{\pi}{2}\right] = \mathbf{0.318\hat{I}}$$

From these results we see that the average load current is progressively reduced as the starting delay on each half-cycle is increased, becoming zero when $\phi = \pi$.

Problem 3 By considering the layered form of construction of a thyristor as shown in *Figure 7.3*, show that this might be construed as a pair of interconnected transistors. Hence deduce how the thyristor action takes place.

Glancing back at *Figure 7.3* we can see that what we have in effect are two transistors, one *n-p-n*, one *p-n-p*, derived by imagining the four layers of the thyristor to be split down the centre and displaced one layer width, as *Figure 7.9* shows. The transistor equivalents are then interconnected in the manner depicted. Assume now that the gate terminal is at zero volts or slightly negative. The *n-p-n* transistor of the pair (T_2) will be cut off and its collector current will be zero. The collector current of T_2 provides base current for the *p-n-p* transistor T_1 (and conversely), hence if one transistor is turned off, the other must also be turned off. In this condition the thyristor as a whole is switched off or said to be in a **blocking state**.

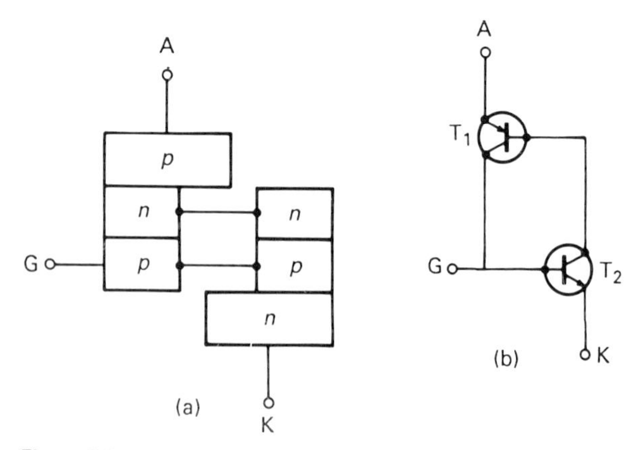

Figure 7.9

If now the gate terminal is made sufficiently positive, collector current will begin to flow in T_2. This provides a base current for T_1 which also begins to conduct. As the collector of T_1 now provides additional base current for T_2, it takes over from the gate input and a cumulative action is established which in a very short space of time drives both transistors into saturation. In this condition the thyristor as a whole is switched on. Thus the gate input acts as a trigger which switches the device from its blocking state to the conducting state in a time typically of the order of 1 μs.

Once the thyristor is switched on in this manner, the gate input has no further control over events and the thyristor will remain switched on. The minimum forward current necessary to *maintain* conduction after the removal of the gate current is the **latching current**, I_L. Generally the collector current of T_1 is very much greater than the gate current and for this reason the gate input is ineffectual in switching the thyristor off. The only way this can be done is by reducing the anode–cathode voltage so that the current falls below a critical level known as the **holding current**, I_H. This occurs when the anode voltage is reduced to a value below that necessary to maintain a current at the latching level.

We can now draw a characteristic curve for the action of the thyristor, and this is shown in *Figure 7.10*. In terms now of the four layer construction (*Figure 7.3*) this characteristic is explained in the following way: for the characteristic to the left of the origin, suppose the gate to be held at zero or a small negative potential and the anode voltage to be increased negatively. Then avalanche breakdown of

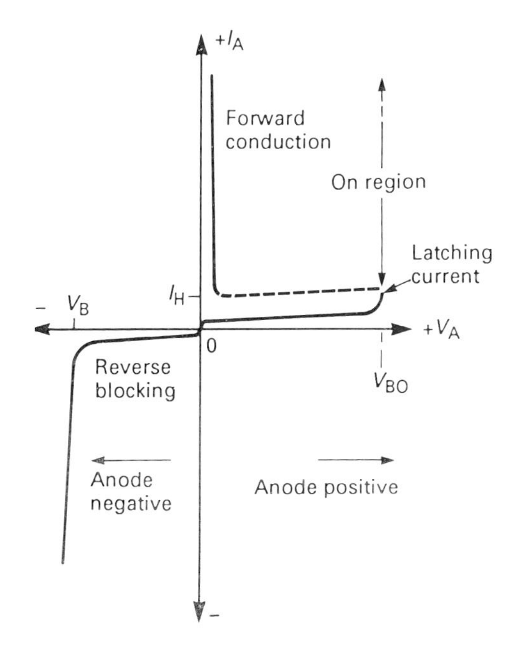

Figure 7.10

the two outer junctions, J_1 and J_3, will occur at some voltage ($-V_B$) and a large reverse current will flow through the assembly. Because the breakdown of two junctions is necessary, this reverse voltage can be very large. If now the anode is increased positively, the centre junction J_2 will approach breakdown and this avalanche current will have the same polarity as a positive gate current. Hence, as breakdown nears, the thyristor will 'snap' itself into the conducting state. The anode potential which will switch the thyristor on, even though the gate current is zero, is called the **breakover voltage**, V_{BO}. Once conduction takes place, a heavy forward current flows and the anode–cathode voltage drop falls to a very low value, typically 1 V.

Problem 4 What effect does a positive base current have on the basic characteristic of *Figure 7.10*? Sketch a family of curves showing the effect of a range of positive gate currents.

The effect of gate current over the $I_G = 0$ condition is to *reduce* the anode potential at which breakover occurs. *Figure 7.11* shows a family of curves for a series of gate current levels from 0 to 50 mA. The actual breakover points depends upon the fabrication of the thyristor but these can range from 100 V to 2000 V with current capacities up to as much as 1000 A. The holding current I_H in the illustration is taken to be about 50 mA which is typical for small power thyristors. For $I_G > 50$ mA, the thyristor is permanently on and behaves as a conventional rectifier diode.

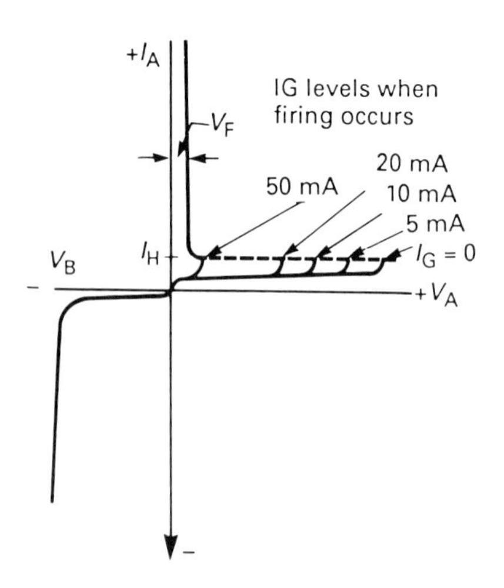

Figure 7.11

Problem 5 *Figure 7.12* shows a thyristor used to control a unidirectional load current. The input is 240 V. 50 Hz mains supply and the gate is forward biased by a current which makes the breakover voltage V_{BO} to be 200 V. Sketch the output waveform of voltage across the load R_L and discuss the limitations of this method of control.

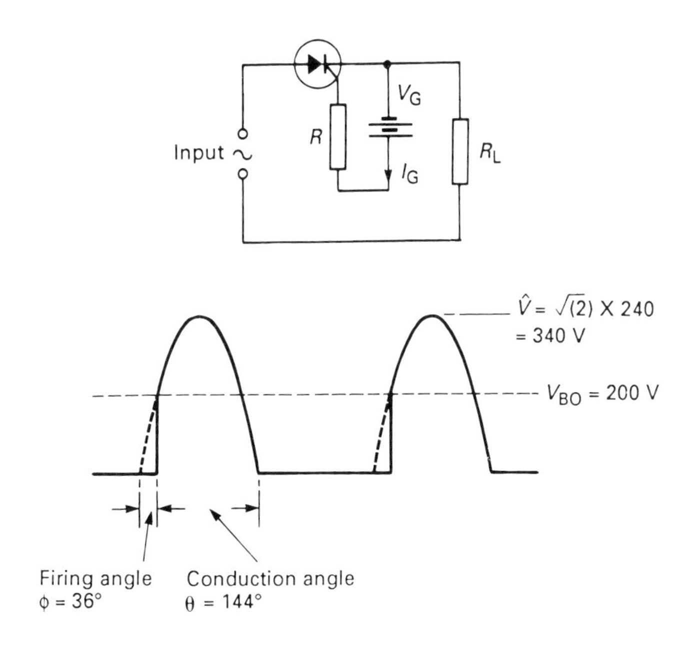

Figure 7.12

On all negative half-cycles of the input waveform the thyristor will be switched off and the output voltage will be zero. On the positive half-cycles, the thyristor will switch on as soon as the anode voltage reaches 200 V and will remain switched on until the input falls to zero. It will then remain off until the succeeding positive half-cycle reaches V_{BO} again.

The output waveform will consequently follow the form illustrated in *Figure 7.12*. The angle represented by each full half-cycle is 180°. The thyristor therefore conducts over 144° or, as we say, the **conduction angle θ** is 144°. The angle of 36° is known as the **firing angle** or the delay angle, ϕ.

This sort of circuit has the merit of simplicity. Its main disadvantage is that the thyristor can be fired only within the first 90° of the positive half-cycle, so we cannot obtain a conduction angle which is less than 90°, that is, one which occurs in the 'second half' of the half-wave.

Problem 6 Return for a moment to the discrete transistor equivalent of a four-layer thyristor which was shown in *Figure 7.9*. From a consideration of this diagram, derive an expression for the thyristor current in terms of the separate transistor common-base current gains (α_1 and α_2), and so explain the thyristor action in terms of this expression.

Let the common-base current gains of transistors T_1 and T_2 be respectively α_1 and α_2. Consider the current I crossing the collector junction of T_1: this current is made up of $\alpha_1 I$ (the hole component of T_1), $\alpha_2 I$ (the electron component of T_2), and the total leakage current I_{CBO}. Hence

$$I = \frac{I_{CBO}}{1 - (\alpha_1 + \alpha_2)}$$

From this expression, if the *sum* of the two current gains is unity, the current will (in theory) be infinite, though limited in practice by the external circuit resistances. You should recall that the actual value of the α gain of a transistor depends upon the current flowing, being quite low at small values of emitter current but rising towards unity as the current increases. By proper design, it is possible to arrange that the sum of the αs at a very low forward bias will be just less than unity; the current will then be, perhaps, ten times I_{CBO}. This is still an extremely small current and the thyristor as a whole is an effective block to the passage of current. If the forward bias is now increased, the breakover voltage is eventually reached and avalanche breakdown occurs in the centre junction, augmenting the gains α_1 and α_2. As $(\alpha_1 + \alpha_2)$ tends to unity, I becomes very large; in fact, the increase in current itself tends to increase $(\alpha_1 + \alpha_2)$ so that the avalanche multiplication need not be very great to sustain the condition $(\alpha_1 + \alpha_2) = 1$. The thyristor is now switched on.

The locus of the characteristic between the breakover voltage V_{BO} and the holding current I_H (see *Figure 7.10*) is given by $(\alpha_1 + \alpha_2)$. By reducing the supply voltage, the current is reduced and a point is reached (I_H) where the current no longer sustains the α sum at unity; switch off then occurs. The breakover voltage is reduced on injecting current into the gate since the current flowing, and hence $(\alpha_1 + \alpha_2)$, is increased. Values of α less than 0.5 would not, of course, be found in ordinary discrete transistors where figures greater than 0.995 are the general norm.

Problem 7 Suppose that the α dependence on the voltage and current conditions of the equivalent transistors built into a thyristor is expressed as

$$\alpha = 0.5 + \frac{V}{1000} + \frac{I}{50}$$

where I is in mA. Estimate the values of V_{BO} and I_H for this thyristor. The figures used are hypothetical, but which of the results, V_{BO} or I_H, is likely to be the more accurate?

At the breakover point the voltage is high but the current is low, so by ignoring the effect of current on the gain, we get

$$0.5 = 0.35 + \frac{V_{BO}}{1000}$$

since the gains α_2 for the two sections of the thyristor must each be 0.5. Hence $V_{BO} = 150$ V.

At the holding level the voltage will be low and most of the extra gain will be due to I_H. Hence

$$0.5 = 0.35 + \frac{I_H}{50}$$

from which $I_H = 7.5$ mA.

I_H is likely to be the more accurate out of these two solutions. The current value at V_{BO} will be only slightly less than I_H while the on voltage is practically the same as that of I_H – look at the characteristic curve. The current contribution to the gain values of the equivalent transistors is therefore dominant.

Problem 8 What is meant by phase control? What are its advantages over control by rheostat?

Phase control is the process of rapid switching which connects a supply to a load for a controlled fraction of each half (or full) cycle of the supply, the fraction being determined by phase angle. It has the advantage that virtually no power is dissipated as heat because the power is not being current regulated as it is in rheostat control but by switching. A thyristor is either fully on or fully off and in either case the internal power wastage is negligible.

Problem 9 Sketch a circuit diagram of a simple thyristor control system operating from a.c. mains supplies that will illustrate the process of phase control. What are the disadvantages of this circuit and how might they be overcome?

A simple circuit arrangement which might be used to control the motor speed of an electrical drill or act as a lamp dimmer is shown in *Figure 7.13*. The timing of the effective gate input is adjustable, the components performing this duty being the network R_1, R_2 and C.

When R_2 is at its minimum value, the gate voltage will have practically the same magnitude and phase as the a.c. input V, R_1 being a small value safety resistance. Hence the gate voltage exceeds the firing potential almost at the start of the positive half-cycle and the thyristor will conduct over 180°, simulating the effect of an ordinary rectifier. *Figure 7.14(a)* shows this situation. When R_2 is increased so that $(R_1 + R_2) = X_C$, the gate voltage will lag 45° on V; the thyristor will now fire only when the anode voltage has reached an angle of 45° into the positive

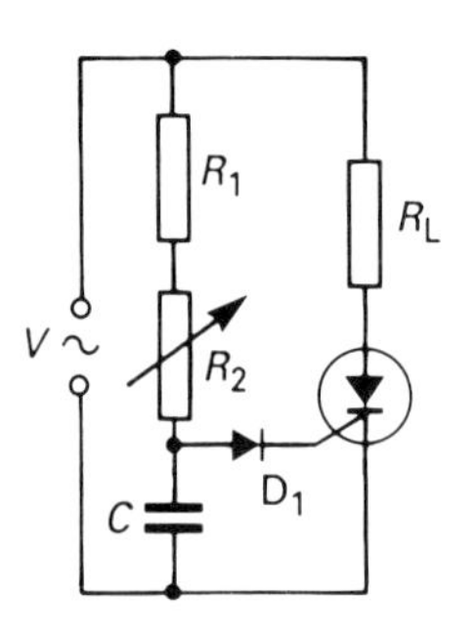

Figure 7.13

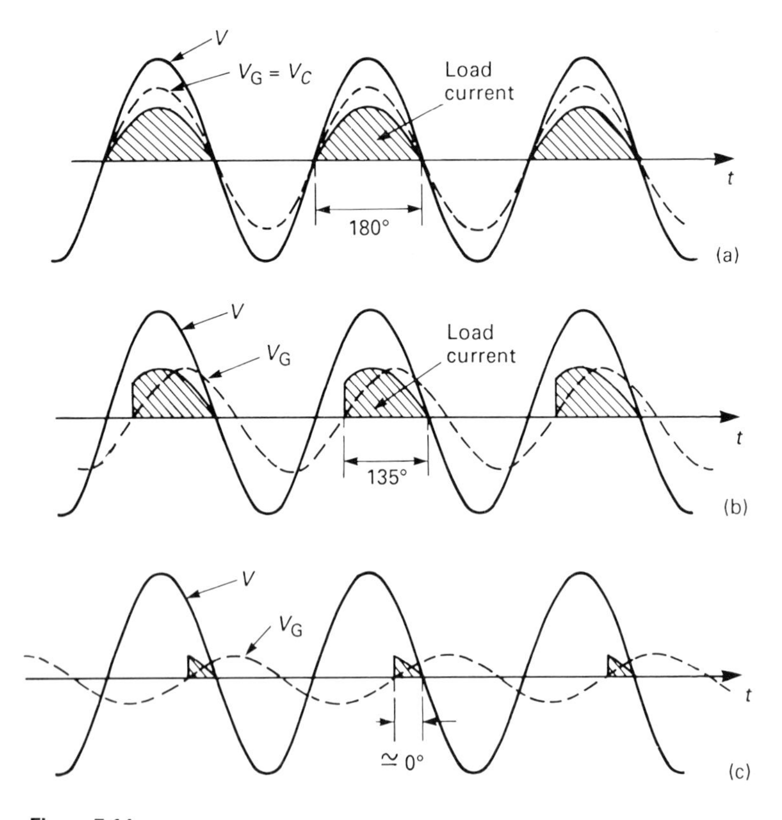

Figure 7.14

half-cycle, giving a conduction angle of 135°; see *Figure 7.14(b)*. If R_2 is now increased to its maximum value so that $(R_1 + R_2)$ is very much greater than X_C, the gate potential will lag the input by an angle approaching 90°; further, the gate voltage will be very small and the thyristor will not fire until the gate is almost at its maximum voltage. The conduction angle is then approaching zero. *Figure 7.14(c)* illustrates this last case. Hence the conduction angle can be varied from almost 180° to 0°.

When the input cycle is negative, the thyristor remains switched off, and the purpose of diode D_1 is to prevent the gate being driven to a large negative level which could damage the thyristor or cause spurious operation. The disadvantages of this circuit is that the maximum output is limited to half-wave operation (this can be overcome by using a triac), and that the *RC* network cannot give more than a 90° phase shift, so the control is poor at low power levels, tending to jump from off to a high output which leads to motor 'shuddering' or severe lamp flickering. Replacing diode D_1 with a diac (see Chapter 8) gives the advantage that the gate is triggered much more precisely. A double *RC* circuit can also be introduced to allow a phase shift approaching 180°.

Problem 10 A thyristor has a latching current of 40 mA and is fired by a pulse of 100 μs duration. It feeds an inductive load made up of a coil of inductance 1 H having a resistance of 40 Ω from a d.c. source of 120 V. It is found that the thyristor does not remain on when the firing pulse ends. Explain why this is so, and suggest a method of overcoming this problem.

We can solve things with a little mathematics here. As soon as the voltage is applied to the load, the current will rise exponentially towards its maximum possible value of V/R or $120/40 = 3$ A; see *Figure 7.15*. Since the time-constant of L/R is long compared with the pulse length, 0.025 s to 0.0001 s, we can assume that the rise in current is linear over the 100 μs period of the gate pulse. Hence the rate of rise of current is V/L or $120/1 = 120$ A/s. Therefore, after 100 μs the current will be 12 mA.

But this level is well below the latching level of 40 mA. Hence the thyristor will not remain switched on when the gate pulse ends. A solution is to connect a

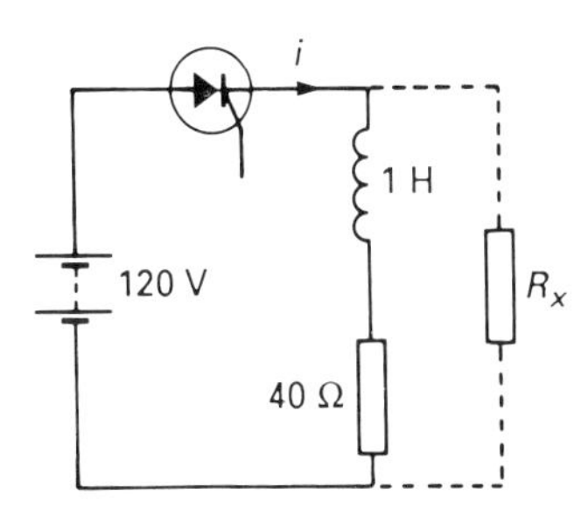

Figure 7.15

resistor in parallel with the load so that a current of 40 mA can be forced to flow immediately at switch on. If this resistance is R_x then

$$R_x = \frac{120}{(40 - 12) \times 10^{-3}}\ \Omega$$

$$= \mathbf{4.28\ k\Omega}$$

A preferred value of 3.9 kΩ could be used here. The addition of this resistor, of course, introduces a waste of power.

Problem 11 What is meant by integral cycle control?

This is an alternative to phase control which has certain advantages for particular applications. It also avoids the generation of radio-frequency interference which can be a problem in high-power thyristor control because of the rapidity of switching and the heavy current interruptions involved. Integral control methods (or **burst-firing**) overcome the problem by having the thyristor switched on for an *integral* number of cycles (or half-cycles), then turning it off for an approximately similar period, both switching actions taking place at the moment the input is passing through the zero voltage point. *Figure 7.16* shows the principle of the method; the gate firing pulses are electronically generated and controlled so that the zero voltage crossings on the input wave coincide with the switching points.

The method is used where heavy heating loads are concerned, like furnace control; switching every cycle is unnecessary in such situations because of the usually long thermal time constants of such loads. Little variation of the heater temperature will occur if the input to the heating elements consists of a number of cycles-on and a number of cycles-off as opposed to switching every cycle.

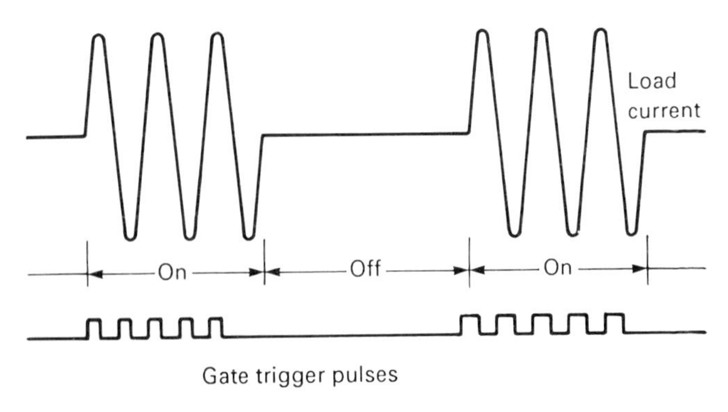

Figure 7.16

Problem 12 A thyristor circuit shown in *Figure 7.17* has a fixed gate potential such that the device fires when the voltage across C has reached 100 V and switches off when the voltage has fallen to 50 V. What sort of waveform appears across the thyristor load R_L? If $R_L = 120\,\Omega$ resistive and $C = 10\,\mu F$, what power is dissipated in R_L for each cycle of operation?

This circuit produces a sawtooth waveform across the load represented by the discharge curve of C as its voltage falls from 100 V to 50 V. The energy stored in

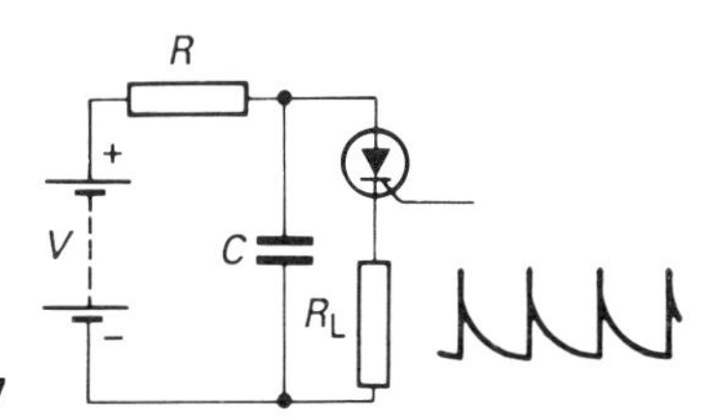

Figure 7.17

the capacitor during the charge (via resistor R) is dissipated in R_L during the discharge cycle.

Ignoring losses in the thyristor, the energy released by C in discharging is $\frac{1}{2}CV^2 = \frac{1}{2}(100^2 - 50^2)\,C$ joules and so

$$\text{Power given up} = \frac{(100^2 - 50^2)}{2t}\,C\ \text{W}$$

where t is the time of discharge. This can be found from the exponential equation for the discharge of a capacitor:

$$50 = 100\exp - \frac{t}{120 \times 10^{-5}}$$

from which we get $\qquad 833t = \log_e 2$

$\therefore \qquad t = 0.83$ ms

Substituting values into the power equation, we get

$$P = \frac{(100^2 - 50^2)}{2 \times 0.83} \times 10^{-2}$$

$$= \mathbf{45.2\ W}$$

B FURTHER PROBLEMS ON CONTROLLED RECTIFIERS

(a) SHORT ANSWER PROBLEMS (answers on page 152)

1 Supply the missing word(s): (a) Thyristors control large amounts of . . . with minimum control . . ., (b) A thyristor turns on when the . . . voltage is exceeded, (c) A thyristor can be turned on by timed . . . pulses

2 What are the advantages of thyristors as switches over (a) *p-n* diodes, (b) transistors?
3 Sketch from memory the general appearance of a thyristor anode characteristic curve, fully labelled.
4 Define the terms (a) latching current, (b) breakover voltage, (c) holding current, (d) breakdown voltage.
5 Draw the circuit symbols for (a) a *p*-gate thyristor, (b) an *n*-gate thyristor, (c) a triac.
6 What is the basic difference between the function of a thyristor and that of a triac?
7 What advantage, if any, does an actual triac have over two standard thyristors wired up in inverse parallel?
8 Why wouldn't integral cycle control be suitable as a lamp dimmer?
9 Why doesn't removal of gate current turn a thyristor off? Under what condition might this in fact happen?
10 Why might it be possible for a thyristor control circuit to malfunction if the load is highly inductive?
11 Make a list (not only those given in the text) of the possible sources of loss in a thyristor.

(b) MULTI-CHOICE PROBLEMS (answers on page 152)

1 Thyristors can be thought of as (a) two diodes connected in series, (b) two diodes connected in parallel, (c) three diodes in series, (d) two transistors in parallel.
2 A triac is two thyristors connected (a) anode to cathode in series, (b) anode to anode in parallel, (c) anode to anode in series, (d) anode to cathode in parallel.
3 A triac gives (a) a d.c. output, (b) an a.c. output, (c) a half-wave rectified output, (d) a full-wave rectified output.
4 Phase control of a thyristor has the advantage that (a) it works on both a.c. and d.c. (b) it rectifies the supply current, (c) it gives control over 360° of the input waveform, (d) it controls the gate current in short pulses.
5 A triac is superior to a thyristor because (a) it is more reliable at high current densities, (b) it amplifies the supply input, (c) it can be triggered into conduction in either direction, (d) the pack contains two thyristors.
6 Holding current is (a) the minimum anode current which will maintain the **on** state, (b) the gate current which will just trigger the thyristor, (c) the gate current below which the thyristor switches off, (d) none of these.
7 Breakover voltage is (a) the reverse avalanche level, (b) the gate voltage which causes the thyristor to switch on, (c) the anode voltage level which allows the thyristor to switch off.
8 Integral cycle control is most suitable for (a) lamp dimming, (b) d.c. series motor control, (c) furnace control, (d) car ignition systems.
9 Burst firing is another name for (a) integral cycle control, (b) phase control, (c) full wave rectification, (d) a train of gate trigger pulses.
10 When a thyristor is switched on the forward voltage drop across it is (a) about 1 V, (b) equal to the supply voltage, (c) equal to the breakover voltage, (d) zero.

(c) CONVENTIONAL PROBLEMS

1 What is the r.m.s. value of the waveform shown in *Figure 7.7* earlier?

[$0.547\hat{I}$]

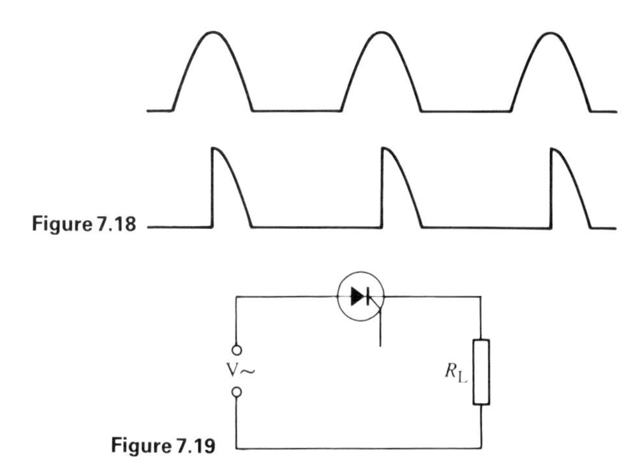

Figure 7.18

Figure 7.19

2 In *Figure 7.18* the load current waveforms in a switched d.c. system are shown. What fraction of full power does each represent?

[Both half power]

3 Describe, using appropriate diagrams, the operation of a thyristor. Sketch the anode characteristic and indicate curves for a range of gate currents starting at zero.

4 With the help of a circuit diagram, explain how a thyristor can be used to control the speed of a d.c. series motor or act as a lamp dimmer.

5 In *Figure 7.19*, $V = 200$ V and $R_L = 10\ \Omega$. For a conduction angle of 120°, find the average load current and power. What are these figures for a conduction angle of 75°? If the voltage across the thyristor is 1.2 V when switched on, what power is lost for the conduction angle of 120°?

[4.77 A, 746 W; 2.36 A, 336 W; 5.76 W]

6 A thyristor is used with fixed bias in a circuit similar to that shown in *Figure 7.17*. Capacitor C charges up by way of resistor R and the thyristor fires when the voltage across C is 150 V, switching off again when the voltage falls to 75 V. What power is dissipated in the load if $R_L = 100\ \Omega$, for each cycle of operation?

[122 W]

7 The average current I_{av} in the full-wave rectified waveform which is delayed by an angle ϕ is given by $I_{av} = \hat{I}(1 + \cos\phi)/\pi$. Draw a curve showing how the average current changes with respect to the delay angle for values of ϕ from o to π radians.

8 A triac is used to regulate the power delivered to a furnace which may be considered resistive. What percentage of the maximum possible power is delivered to the furnace if the firing angle is delayed by 45°?

[91 per cent]

9 Knowing the appearance of the anode characteristic curves of a thyristor, deduce the equivalent characteristic of a triac.

10 The relationship between the breakover voltage V_{BO} and the gate current I_G (mA) of a certain thyristor is $V_{BO} = 450 - 20I_G$. The thyristor is connected to a load resistor across a 240 V a.c. supply. Derive an expression and plot a graph showing how the load current varies with the gate current I_G. What are the limitations of using the thyristor in this circuit?

8 Triggering devices

A MAIN POINTS CONCERNED WITH TRIGGERING DEVICES

1 There are a number of switching and triggering devices which, like the thyristor and the triac, are multilayer devices; among these are the diac, the gate turn-off thyristor (GTO) and the programmable unijunction transistor (PUT). The unijunction transistor (UJT), while not in the multilayer category, is a versatile device which is not only suitable for firing thyristor and other control circuits but has applications in timing and clock pulse generation. We will examine the UJT first of all.

2 Like an ordinary transistor, the **unijunction transistor** has three terminals, but unlike the transistor it has, as its name implies, only one internal semiconductor junction. Its physical construction is shown in *Figure 8.1*, together with its theoretical circuit symbol. A narrow bar of lightly doped *n*-type material has an off-centre single *p-n* junction diffused into the bar. The connection taken from the *p*-region at this junction is known as the emitter. Connections are also taken from each end of the bar and known respectively as Base 1 and Base 2. The way the UJT operates can be followed with the help of an equivalent circuit, and this is given in *Figure 8.2*.

The silicon bar acts as a resistor (as it does in the FET) which is effectively tapped into by a diode where the junction is situated. Suppose the various terminals of the UJT are biased in the manner shown in the diagram, and a load

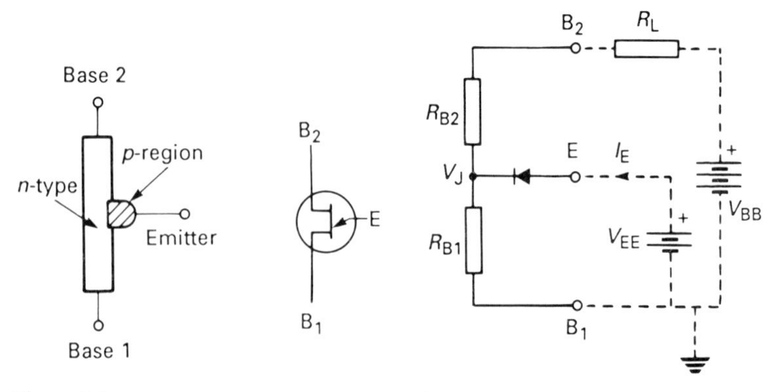

Figure 8.1

Figure 8.2

resistor included in the lead to B_2. Then a current will flow through the bar resistances R_{B1} and R_{B2} and the potential, V_J, at the cathode of the diode will be

$$\frac{R_{B1}}{R_{B1} + R_{B2}} \times V_{BB}$$

which will be some fraction η of the voltage on B_2 with respect to B_1. Hence

$$V_J = \eta V_{BB}$$

where η is the **intrinsic stand-off ratio**.

3 If the emitter bias V_{EE} is less than V_J, the junction will be reverse biased and only the leakage current will flow in the emitter circuit. However, if V_{EE} is increased to $V_J + 0.6$ V, the junction will become forward biased and a current will flow from the emitter to B_1. This happens because the injected carriers are holes; these will be repelled by the positive B_2 end of the bar and attracted to the negative B_1 end. This increased flow through the equivalent resistor R_{B1} leads to an effective reduction in its resistance which in turn reduces the emitter voltage. Thus we have a condition where the emitter *current* is increasing but the emitter *voltage* is decreasing. This is a condition of **negative resistance**; a graph drawn showing current against voltage will have a negative gradient. A typical current-voltage characteristic for a unijunction transistor is shown in *Figure 8.3*. The peak point on the curve, V_p, is that voltage at which the unijunction switches on, and here $V_p \simeq \eta V_{BB} + 0.6$ V. As more holes are injected into the bar, saturation is eventually reached and the emitter voltage begins to rise again with increasing emitter current.

4 Unijunction transistors, because of their operating characteristics and the presence of negative resistance, are used as oscillators, particularly for the

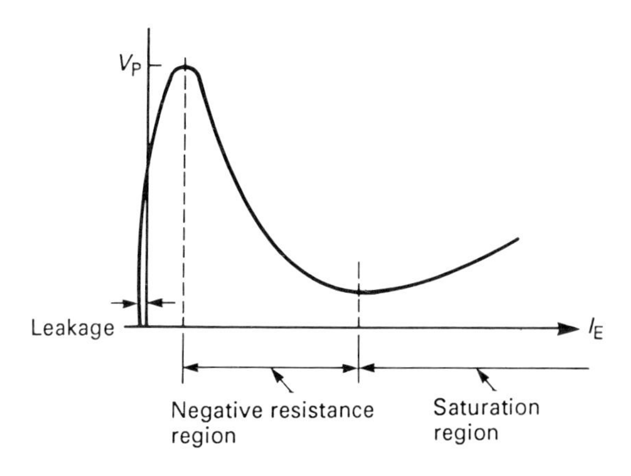

Figure 8.3

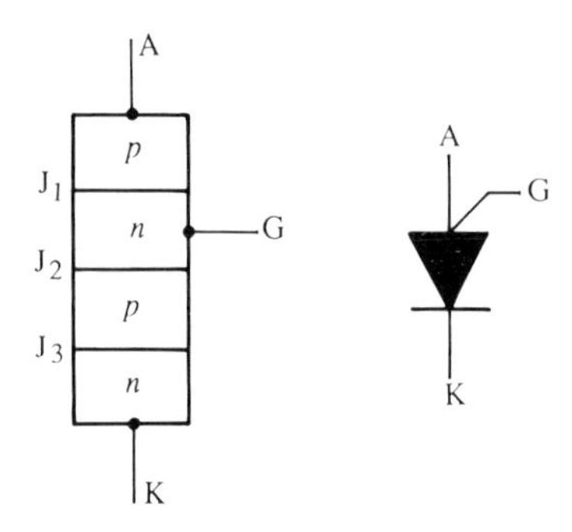

Figure 8.4

production of sawtooth waveforms from which, in turn, sharp triggering pulses can be derived for the control of thyristors and triacs.

5 The construction form of the **programmable unijunction transistor** is quite dissimilar to that of the UJT, so although its name might give the impression that it is a simple variant of the UJT, it is in fact a four-layer device very much like the ordinary thyristor. *Figure 8.4* shows its form; the only difference between it and the thyristor is the gate connection. Here the gate connects to the n-type layer nearest the anode rather than the p-type layer nearest the cathode. In the circuit symbol shown on the right of the diagram, the gate is consequently depicted entering the anode; you should compare this with the standard thyristor symbol where the gate is shown entering the cathode.

The PUT is set up for operation by adding two *external* resistors which simulate the internal resistors RB_1 and RB_2 found in the UJT. Let these two added resistors now be R_1 and R_2, and suppose them to be connected as shown in *Figure 8.5*. We now have in effect a UJT but with the facility of control over the values of R_1 and R_2 so that the stand-off ratio η is adjustable, as is V_p. By increasing the anode–cathode voltage (the anode now acting as a UJT emitter does) a point is

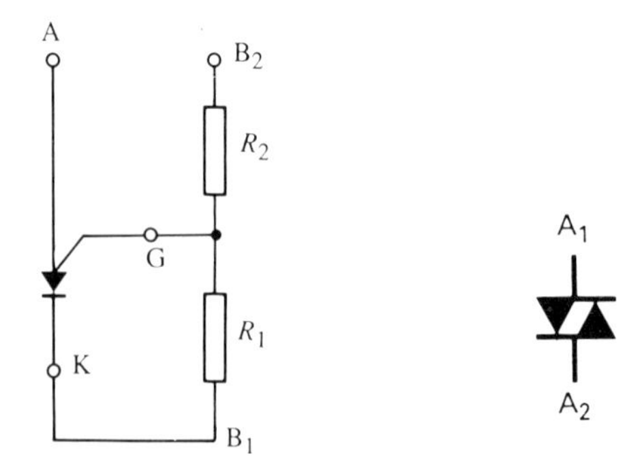

Figure 8.5

Figure 8.6

reached (V_p) where the anode-gate junction J_1 becomes forward-biased and the PUT turns on. Further increases in anode ('emitter') voltage now lead to a region of negative resistance where the anode current falls as the anode voltage increases; the operating characteristic of the PUT is therefore similar in appearance to that of the UJT except that 'emitter' current is replaced by 'anode' current.

6 The **diac** (or **DI**ode **A**lternating **C**urrent) is a silicon bidirectional device suitable for firing thyristors. It is essentially a gateless triac; in fact, its circuit symbol shown in *Figure 8.6* is seen to be the same as that of a triac without the gate terminal. There are two terminals, anode A_1 and anode A_2, but although the device looks rather like two diodes in parallel, its make-up is much more involved. Five layers of semiconductor material are involved, but if we imagine the structure to be split down the middle as depicted in *Figure 8.7*, we may treat it as a pair of electrically separate, but physically connected four-layer devices in parallel, each reversed with respect to the other. Hence the form of the circuit symbol. The diac switches on only when the breakover voltage (strictly zener breakdown here) in *either* direction is exceeded.

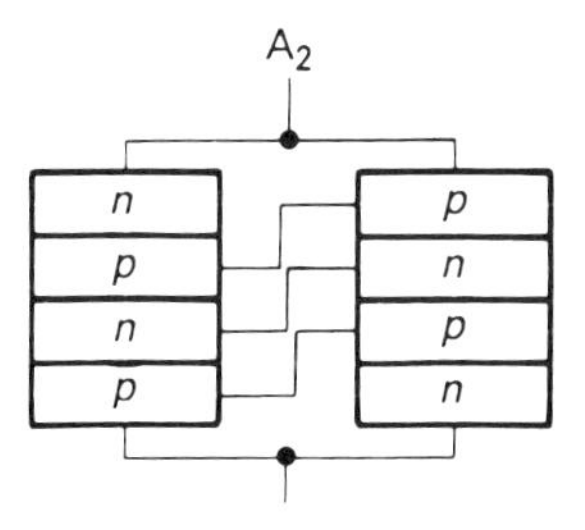

Figure 8.7

Suppose anode A_2 is made positive with respect to anode A_1, and the voltage is increased. It will help to understand the operation of the diac if we visualise *Figure 8.7* as a discrete transistor circuit made up as shown in *Figure 8.8*. What we have in essence is a four-layer arrangement making up two transistors on each side (which is essentially the assembly of a thyristor) with two zener diodes in the gating circuits. These zeners are themselves part of the system, so you must not search for them as discrete entities in the actual diac construction. When A_2 goes positive with respect to A_1, the current will initially be small, but as the voltage is increased, the breakdown point of zener Z_2 will be reached. When this occurs, base current will flow from T_3 through Z_2. T_3 then begins to turn on so that its collector current increases and supplies drive to the base of T_4. T_4 in turn now switches on and applies more base current to T_3. This effect is regenerative and in a very short time T_3 and T_4 will be switched into saturation. The resistance between A_2 and A_1 is then very small. This state of conduction cannot be removed until the applied voltage falls below the holding level. When the polarity between A_2 and A_1 is reversed, T_1 and T_2 form the conducting path in place of T_3 and T_4.

So the diac is a device which, looked at from either direction, remains off until the applied voltage reaches a precise level, when it then switches hard on. These switching characteristics are illustrated in *Figure 8.9*.

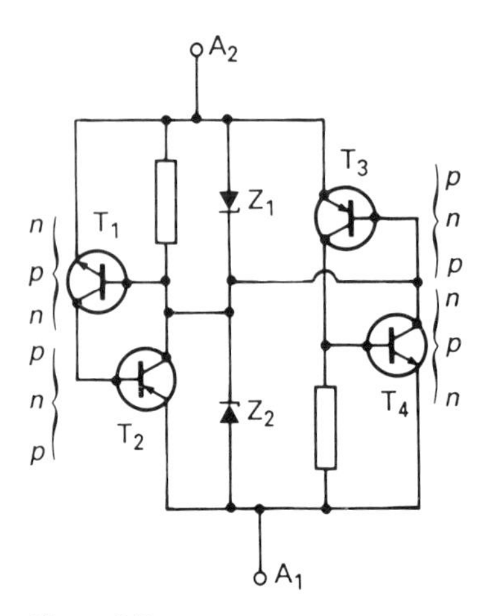

Figure 8.8

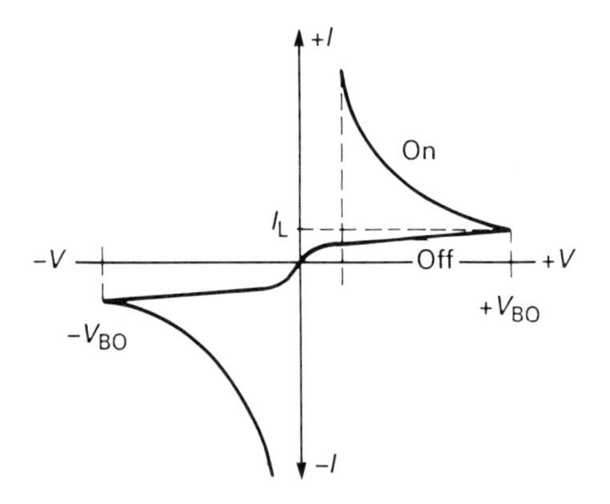

Figure 8.9

7 The **gate turn-off thyristor** (GTO) combines the high-current, high-blocking voltage characteristics of a conventional thyristor with the fast and easy-to-drive characteristics of a bipolar transistor. Like the thyristors already covered, the GTO is a four-layer device and its general structure and circuit symbol are seen in *Figure 8.10(a)*.

The GTO is turned on in the usual way by the application of a positive pulse at the gate, but unlike the conventional thyristor it is turned off when a negative gate

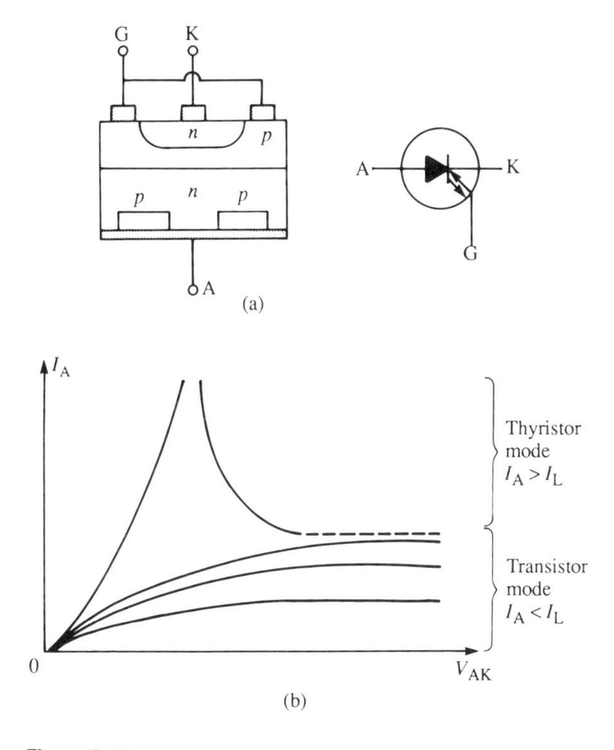

Figure 8.10

pulse is applied. Typically a pulse length of only a few microseconds and of as little as −5 V amplitude is sufficient to effect the turn off.

The forward characteristics of a GTO are of interest. A study of a typical family of characteristics shows that there is a change in the form of the characteristics on either side of the current value which represents the latching level I_L. When the anode current is less than I_L, the device behaves as a high-voltage transistor; anode current rises with increasing anode voltage and then remains substantially constant, its actual level depending upon the gate current. Above the level of I_L the characteristic moves into the thyristor form. If the gate current is less than that required for triggering, I_{GT}, the GTO is off and only a very small leakage current flows between anode and cathode. If the gate current is equal to or greater than I_{GT}, the GTO switches on, a large anode current flows and the voltage drop between anode and cathode is small. As long as the anode current remains below the latching level, however, the GTO will return to the off state if the gate current drops below I_{GT}. See *Figure 8.10(b)*.

B WORKED PROBLEMS ON TRIGGERING DEVICES

Problem 1 A certain unijunction transistor has a total resistance between B_2 and B_1 of 8 kΩ and a stand-off ratio of 0.47. What are the individual values of the internal resistances?

Here we have $R_{B1} + R_{B2} = 8000\ \Omega$

But the ratio $$\frac{R_{B1}}{R_{B1} + R_{B2}} = \eta = 0.47$$

Hence $R_{B1} = 0.47 \times 8000 = \mathbf{3760\ \Omega}$

From this $R_{B2} = \mathbf{4240\ \Omega}$

Problem 2 The 2N2646 unijunction transistor has $R_{B1} + R_{B2} = 7000\ \Omega$ and $\eta = 0.65$ as typical values. It is to be used in the circuit shown in *Figure 8.11*. Show that the output across capacitor C will be sawtooth wave and estimate the period of this wave.

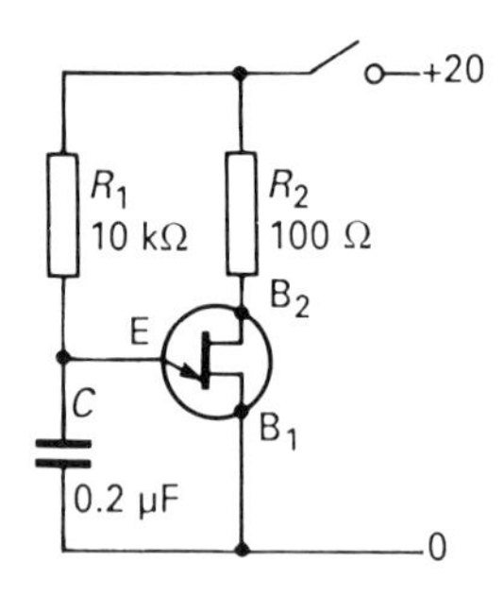

Figure 8.11

For the first part, suppose that at time $t = 0$, the voltage across C is zero and the emitter junction is consequently reverse biased. The UJT is therefore switched off. Capacitor C will charge up through resistor R_1 until at some time (t_1) later the voltage across it will equal the peak voltage V_p. At this point the UJT will switch on and the emitter-to-B_1 voltage will fall, permitting C to discharge through the low resistance path presented by R_{B1}. This turns the transistor off again at time t_2 and after that the charging cycle begins afresh. The voltage waveform developed across C is therefore a sawtooth as shown in *Figure 8.12*.

To establish the period of this waveform we use the diagram of *Figure 8.13*, beginning at time t_0 when C is uncharged and the emitter potential is zero. As

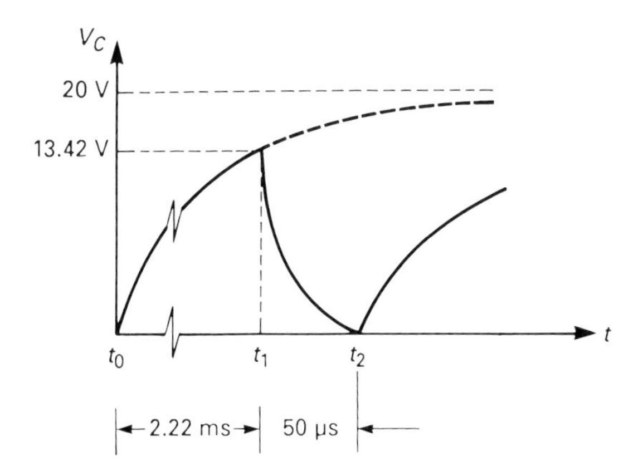

Figure 8.12

soon as the circuit is switched on, current flows through R_2, the unijunction bar B_2 and B_1, and back to the supply. This current will be

$$I_B = \frac{V_{BB}}{R_2 + R_{B1} + R_{B2}}$$

$$= \frac{20}{7100}\,\text{A} = 2.82\,\text{mA}$$

The voltage on B_2 is $V_{BB} - I_B R_2 = 20 - (2.82 \times 10^{-3}) \times 100 = 19.72$ V; call this V_{B2}. Then the voltage at the junction point, V_J, will be

$$\eta \times V_{B2} = 0.65 \times 19.72 = 12.82\,\text{V}$$

The capacitor will charge, therefore, from 0 V towards 20 V with a time constant of $CR = 0.2 \times 10^{-6} \times 10^4$ s or 2 ms. The equation of the charge is then

$$V_c = 20\left[1 - \exp\frac{-t}{2 \times 10^{-3}}\right]$$

This charge will continue until the voltage has risen to 12.28 + 0.6 V when the transistor will switch on. From the diagram of *Figure 8.12*, therefore

$$13.42 = 20\left[1 - \exp\frac{-t}{2 \times 10^{-3}}\right]$$

from which we find t = **2.22 ms**.

This is the period of the wave if we ignore the time of discharge.

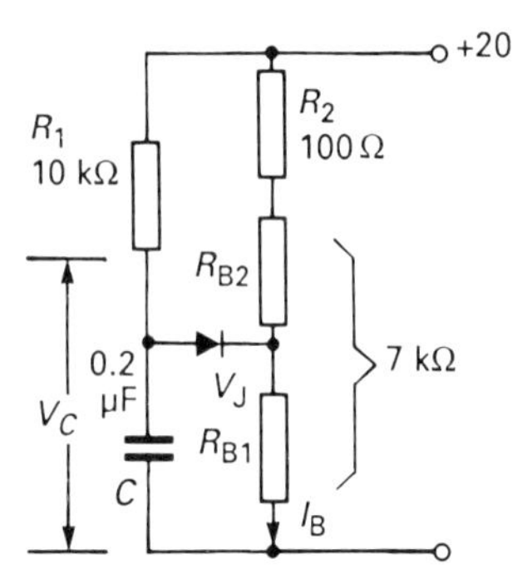

Figure 8.13

Problem 3 Estimate the discharge period of the capacitor in the circuit relating to Problem 2 above. What output is obtained from the B_2 terminal of the UJT?

When the capacitor voltage reaches 13.42 V in the way we have seen, the emitter junction becomes forward biased and hole carriers are injected into the *n*-region; these carriers move towards the negative end of the bar and reduce the effective resistance of R_{B1}. The voltage at the junction point consequently falls and the diode is even more heavily forward biased; this results in an enhanced injection of hole carriers into the bar region, reducing the resistance of R_{B1} even further. This action is self-supportive and the ohmic value of R_{B1} falls very rapidly to its minimum, typically some 20–50 ohms. Hence a very low resistance appears across the capacitor immediately following 2.22 ms of the charging cycle. The time constant of the discharge curve, taking the minimum value of R_{B1} to be 50 Ω will consequently be $0.2 \times 10^{-6} \times 50$ s $= 10$ μs, and the discharge will be completed (in a practical sense) in about 50 μs. *Figure 8.12*, though the scale of the discharge is greatly exaggerated for clarity, illustrates this result.

Since the emitter junction can be looked on as a short circuit during the discharge cycle, the lower end of R_{B2} will be practically at zero volts during this period. So the lower end of R_{B2} will drop from 12.82 V (while C is charging) to, say, 0.1 V while C is discharging. The top end of R_{B2} (terminal B_2) will similarly drop from 19.72 V to a much lower figure – see if you can estimate it. The output at terminal B_2 will consequently be rather as shown in *Figure 8.14* where the

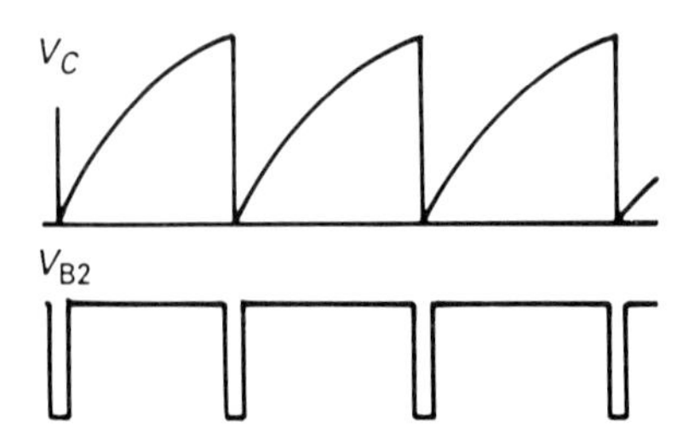

Figure 8.14

sawtooth across C is also drawn for comparison. By placing a resistor in the B_1 lead to earth, a trigger pulse may also be obtained from this point as well.

Problem 4 In the circuit of *Figure 8.11* it is found that if the value of R_1 is too large or too small, the circuit fails to operate. Explain why this happens.

Suppose R_1 to be very large, of the order of a megohm or so. Then the current which would go to charge the capacitor may be comparable only with the normal leakage current of the diode junction. The charge on C would therefore reach only to some low value and remain there, unable to set up enough voltage to initiate the switch-on cycle. If C was an electrolytic having its own small leakage current, the problem would be made worse.

If R_1 is very small, of the order of a few hundred ohms, the current flowing through it and the diode junction after the circuit has fired once might be sufficient to keep the diode in the saturated region of its characteristic. The circuit would then quite clearly fail to oscillate.

Problem 5 Draw a simple circuit showing how a UJT oscillator might be used to trigger a thyristor in a lamp dimming system.

Figure 8.15 shows a suitable circuit. Although it is perfectly possible to trigger the thyristor directly from the B_1 terminal by including a resistor in this lead, the use of an isolating pulse transformer such as T_2 separates the high-voltage mains supply from the trigger circuitry which can then be operated from a low voltage supplied by transformer T_1. This is a quite common technique; often optoisolators (see Chapter 9) are used instead of pulse transformers.

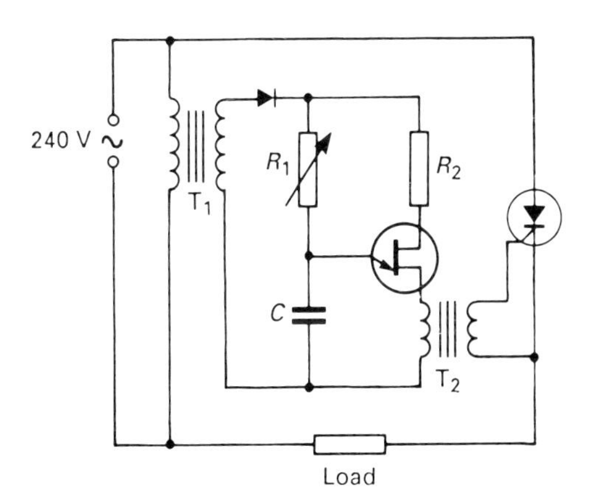

Figure 8.15

Problem 6 Sketch a circuit showing how a diac might be used to generate trigger pulses to a thyristor in a power control system.

A circuit is shown in *Figure 8.16*. The gate trigger pulses are generated by the circuit made up from resistors R_1 and R_2, capacitors C_1 and C_2 and the diac. Assume the instant when the mains input is at zero. As this input increases along the positive half-cycle, point P potential increases as capacitor C_1 charges, but point Q remains at earth potential. At the instant when the voltage at P reaches the diac breakover voltage, the diac turns on and develops a voltage across R_2. A positive pulse is thereby applied to the gate of the thyristor via C_2, which remains switched on for the remainder of the positive half-cycle. Capacitor C_3 is necessary in this circuit to maintain the output at a reasonably constant level while the thyristor is turned off over the negative half-cycle of input.

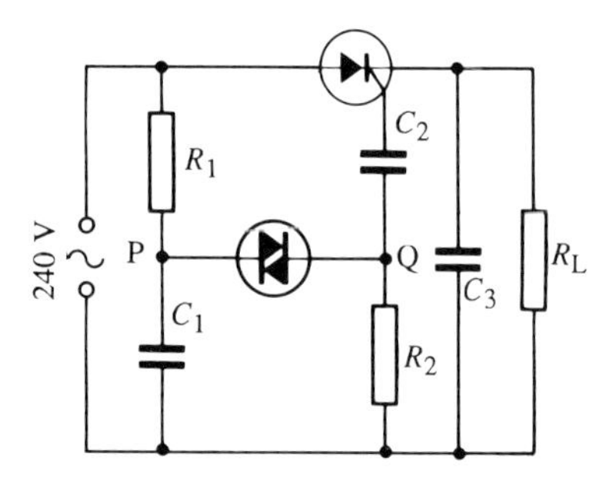

Figure 8.16

Problem 7 What special features do the drive circuits controlling a gate turn-off thyristor require over and above those used with standard thyristors? Describe a suitable form of drive circuit for a GTO power controller.

The GTO operates with very short switching times and requires a polarity change at the gate for switch-on and switch-off respectively. This polarity change is not needed with standard thyristors. To turn the GTO on requires a positive current to be injected into the gate for the necessary turn-on time. To turn the GTO off requires a negative voltage of between -5 and -10 volts typically, applied directly between gate and cathode for a period of several microseconds, thus removing current from the gate electrode. The actual voltage must be less than the gate-cathode reverse breakdown level but high enough to remove the charge necessary to effect the turn-on.

Drive circuits usually operate to provide the turn-off pulse from a capacitor which has been allowed to charge up during the time that the GTO has been conducting, the actual amplitude of the pulse being determined by a regulator diode. *Figure 8.17* shows a basic arrangement. T_1 can be switched on and off

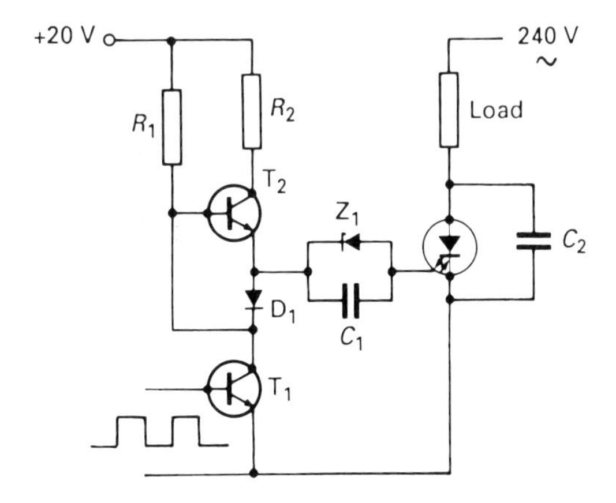

Figure 8.17

alternately by an incoming rectangular waveform. When T_1 is off, a positive drive current flows into the GTO gate by way of R_2 and the emitter-follower T_2, with the resistor setting the initial value. The zener regulator D_2 conducts as soon as its breakdown voltage (typically 10–12 V) is reached, thus holding the charge on C to the level then reached. A small and necessary current then flows into the gate during the conduction period.

When T_1 switches on, T_2 switches off. Capacitor C_1 now discharges via T_1 and D_1, and the negative voltage set by D_2 is applied to the gate of the GTO. For the period in which T_1 remains on, the gate voltage will stay negative because the reverse gate resistance of the switched-off GTO is high and the discharge period of C_1 is consequently long. The capacitor across GTO (C_2) is known as a **snubber**; this limits the rate at which the anode–cathode voltage rises at turn-off. A diode is sometimes used for the same purpose.

Problem 8 Describe how a diac might be used to form an oscillator having a sawtooth output waveform.

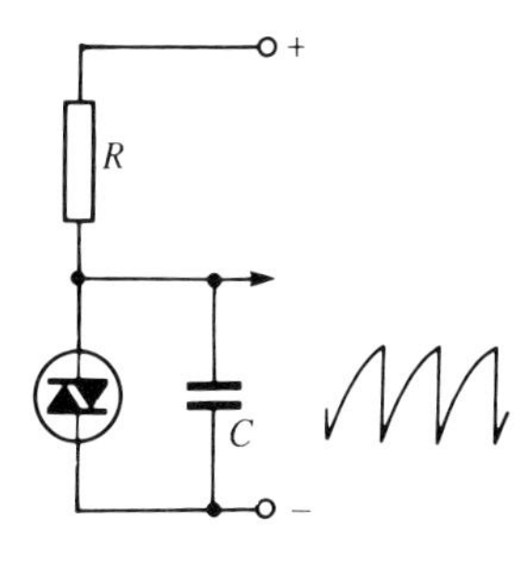

Figure 8.18

Figure 8.18 shows a simple circuit arrangement which will perform the required function. At switch-on the capacitor C charges, as it did for the UJT circuit, by way of resistor R. As soon as the voltage across C exceeds the diac breakover voltage, the diac switches on and discharges C. The consequent drop in the voltage across C, and the diac, then causes the diac to switch off, after which the cycle repeats. The output, as for the UJT, is a sawtooth wave. The polarity of the supply is immaterial, but it must, of course, be greater than the diac's breakover voltage, and it determines the polarity of the output wave.

C FURTHER PROBLEMS ON TRIGGERING DEVICES

(a) SHORT ANSWER PROBLEMS (answers on page 152)

1 The intrinsic stand-off ratio of a UJT depends upon the applied voltage. True or false?
2 The n-channel of a UJT is lightly doped and the p-type emitter is heavily doped. Why is this done?
3 When holes are injected into the n-bar of a UJT from the emitter, why do these move towards the B_1 end of the bar rather than the B_2 end? What is happening to the electron component of the current at this time?
4 Without referring to the text, draw the circuit symbols for (a) a UJT, (b) a diac, (c) a GTO.
5 A diac in its on state can only be turned off by removing the applied voltage. True or false?
6 The emitter junction of a UJT becomes forward biased as soon as V_p exceeds $V_J + 0.6$ V. True or false?
7 A GTO is like an ordinary thyristor except that the direction of the applied anode–cathode voltage is immaterial. True or false?
8 A programmable unijunction transistor has its intrinsic stand-off ratio controlled by two external resistors. True or false?
9 A GTO is a four-layer device which behaves in the same way as a thyristor. True or false?

(b) MULTI-CHOICE PROBLEMS (answers on page 152)

1 The intrinsic stand-off ratio of a UJT depends upon (a) the applied voltage, (b) the bias on the emitter, (c) the internal resistances of the n-type bar, (d) none of these.
2 The stand-off ratio is defined as (a) $(V_p - 0.6)/V_{BB}$, (b) $(V_{BB} - 0.6)/V_J$, (c) $(V_{BB} - V_J)/0.6$, (d) $(V_J + 0.6)/V_p$.
3 The stand-off ratio of a UJT typically has values (a) greater than 1, (b) 0.5 to 0.8, (c) less than 0.3, (d) none of these.
4 In an n-type UJT which is switched on, current flows from (a) B_1 to B_2 only, (b) B_2 to B_1 only, (c) B_1 to emitter only, (d) B_2 to emitter only, (e) none of these.
5 Between the peak voltage point and the valley point, a UJT exhibits (a) a negative resistance region, (b) a high resistance region, (c) a saturated region, (d) none of these.
6 A diac in its off state can only be switched on (a) by applying a positive pulse to the gate, (b) by applying a negative pulse to the gate, (c) by applying an anode–cathode voltage greater than the breakover voltage.

7 A programmable unijunction transistor is switched on by (a) raising the gate voltage, (b) raising the anode voltage, (c) taking the gate to earth.

8 The GTO is a type of thyristor which can be turned off (a) by the application of a negative gate pulse, (b) by introducing a snubber capacitor across its terminals, (c) by earthing the gate momentarily.

9 A diac is (a) unidirectional, (b) bidirectional, (c) neither of these.

(c) CONVENTIONAL PROBLEMS

1 Describe the physical construction of a unijunction transistor and briefly explain its mode of operation.

2 Show how a UJT can be operated as a sawtooth waveform generator. How might a pulse output be obtained?

3 Suppose the circuit of *Figure 8.11* earlier to have the following component values: $R_1 = 100\ \text{k}\Omega$, $R_2 = 2.5\ \text{k}\Omega$, $C = 10\ \text{nF}$. If the same UJT is used, at what frequency will the circuit oscillate?

[About 1.45 kHz]

4 *Figure 8.19* shows a diac used in a circuit which supplies gate pulses to a thyristor. Explain how the circuit operates. Sketch the kind of waveforms you would be likely to see at the points marked P, Q and R.

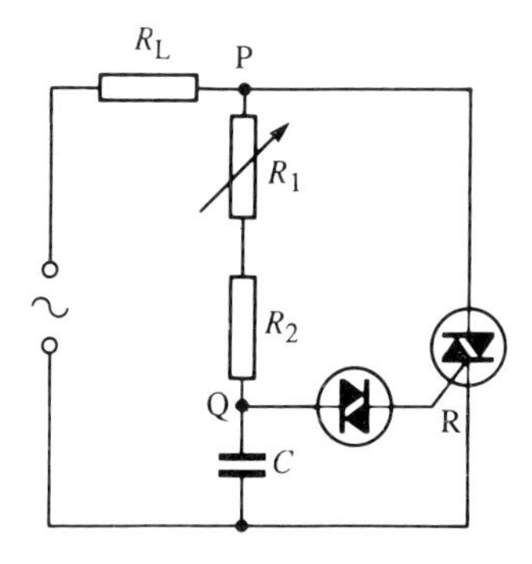

Figure 8.19

5 *Figure 8.20* shows a method of gating a GTO. Using the description given in the text relating to *Figure 8.17* as a guide, explain how this gating control works.

6 The frequency of operation of a UJT oscillating system is given approximately by the equation

$$f = 1/(CR\ \log_e(1/1 - \eta))$$

What value of R will be required if the desired frequency is 10 kHz, and $C = 100\ \text{nF}$, $\eta = 0.5$?

[About 1.44 kΩ]

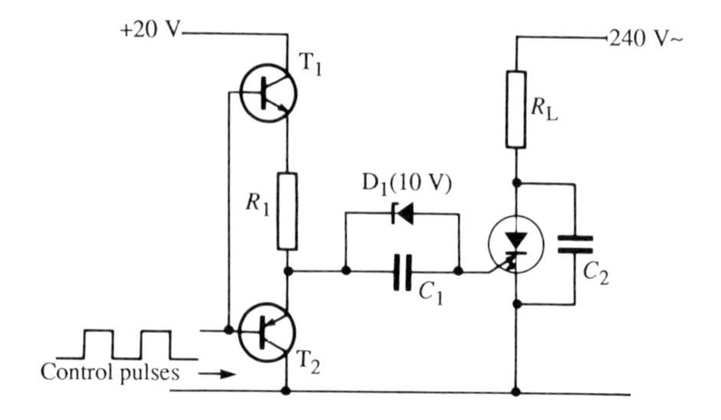

Figure 8.20

7 From a consideration of the operation of a UJT oscillating system, deduce the equation given in the previous example. (This is harder!)
8 Why cannot the properties of a UJT be simulated using two resistors and a discrete diode?
9 It is found during an experiment that when turn-off of a GTO was achieved by discharging a capacitor into the gate and the GTO load was inductive, the triggering became uncertain. Can you explain why this happened?
10 A so-called snubber capacitor (or diode) is often connected in parallel with a GTO to reduce the rate of change of voltage with respect to time acting across the GTO when it switches off. Why is such a protection necessary?

9 Optoelectronic devices

A MAIN POINTS CONCERNED WITH OPTOELECTRONICS

1. Two categories of electronic device come under the heading of optoelectronics: light sources and light sensors. Light sources produce radiant energy which may or may not be located in the visible spectrum, and light sensors respond to this radiant energy by converting it into electrical signals. Among light sources are light-emitting diodes (LEDs), liquid crystal displays (LCDs), hot filament and gas discharge devices. Among light sensors are photoresistors or light-dependent resistors (LDRs), photodiodes and phototransistors. Sources and sensors are usually found together in systems such as fibre optical communications and optoisolator packages.
2. The **light-emitting diode** is derived from two semiconducting materials: gallium arsenide (GaAs) which emits light in the near infra-red region of the spectrum, typically at a wavelength of about 900 nm (1 nm $= 10^{-9}$ m), and gallium arsenide phosphide (GaAsP), an alloy which absorbs the GaAs radiation and emits visible light in the red, yellow and green regions of the spectrum. The last named LEDs find wide application as cheap, low-power-consumption indicator lamps and as the familiar segmented number displays on such equipment as digital clocks and cash registers at supermarket checkout points.
3. **Liquid crystal displays** feature exclusively in most battery-operated equipment such as watches and calculators, where they have a great advantage over the LED display in their very low power consumption. Unlike LEDs, liquid crystals do not emit light but reflect incident light or transmit back light. For this reason they cannot be used in the dark. They are also relatively slow in responding to changes in the energising signal level.
4. **Hot filament** devices are basically incandescent bulbs, though the filaments are operated at much lower temperatures and are disposed in the envelope in such a way that the device indicates numerals or letters of the alphabet when energised. Because of thermal inertia, these devices are slow to respond to signal variations, their power consumption is high and, like incandescent lamps, the failure of the filament renders them useless.
5. **Gas discharge** displays which, like hot filament devices, can often be seen on the pumps at petrol stations, are made on the principle of ordinary neon indicators. Two electrodes, anode and cathode, are enclosed in a gas-filled envelope and when a voltage of sufficient magnitude is impressed between the electrodes, the gas is ionised and the cathode glows a colour characteristic of the gas used. For neon, this is a bright orange. The cathode is usually segmented so that numerals can be formed by energising the appropriate portions of the cathode assembly. These displays are ideal for distant viewing or for high ambient lighting conditions.

The so-called **Nixie tube** operates on the same principle. This is a multiple cathode discharge tube, each cathode being formed into the shape appropriate to the ten digits 0 to 9. By energising the right cathode of the set, the required digits can be displayed in their complete form.

6 Light sensors respond to radiant energy in the form of ultra-violet, visible, infra-red, x-rays, gamma rays and nuclear particles, so the use of the word 'light' might be better replaced by electromagnetic waves. We are interested here only in those devices which respond to the visible and infra-red parts of the spectrum.

Up until the early 1950s, light sensors were made only from selenium, caesium or copper oxide; at the present time sensors are available in which use is made of germanium, silicon, cadmium sulphide, cadmium selenide or lead sulphide as the light-sensitive material. Between them, these cover the whole of the visible, infra-red and heat regions of the spectrum, and selection is made to suit the application. There are two main groupings: those sensors which generate an e.m.f. on receipt of radiation – these include the **photodiode**, the **PIN diode** and the **phototransistor**; and those sensors which do not generate an e.m.f. but instead change their resistance – these are the **light-dependent resistor** types or photoconductive LDRs.

B WORKED PROBLEMS ON OPTOELECTRONIC DEVICES

Problem 1 Describe the principle of operation and a typical form of construction of a light-emitting diode.

The principle of operation of the LED is that in any forward-biased *p-n* junction, recombination of electrons and holes takes place in the vicinity of the junction when forward current flows. When such a combination takes place, the energy

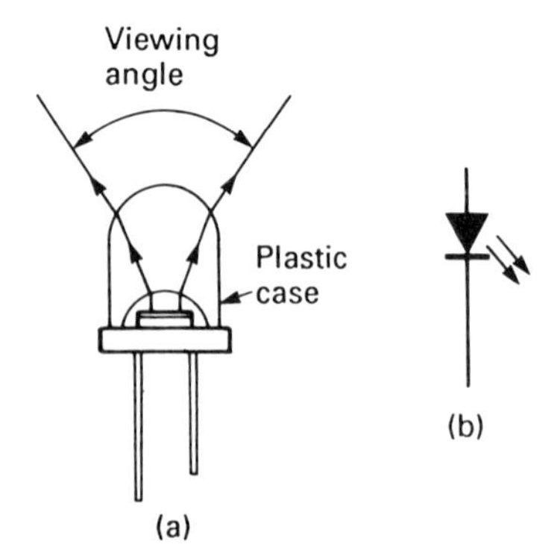

Figure 9.1

released by the electron returning to the valence band appears in silicon crystal as heat but in a gallium arsenide crystal as a photon of light. The intensity of this radiation is proportional to the rate of recombination and so to the magnitude of the diode current. The problem in manufacture is to extract as much of this radiation as possible, keeping in mind that the vital junction is buried between layers of p- and n-type semiconductor material.

To enable light to escape, special doping is necessary to prevent the bulk of the material from reabsorbing the radiation as it passes through, and to ensure that internal reflection is reduced to a minimum. The protective casing also needs careful design to reduce surface reflection losses and to avoid obscuring the junction itself. A typical constructional form of LED indicator is shown in *Figure 9.1*, together with the circuit symbol.

The protective dome is a glass or resin structure, coloured to suit the transmitted radiation and of a refractive index intermediate between that of the n-type semiconductor (which is high) and the surrounding air. The dome also serves to magnify the light-emitting area of the junction. The viewing angle is a function of the design of the package rather than of the actual junction configuration. Whether the casing is domed as illustrated or flat, whether it contains reflectors or diffusers, and so on. The TIL31, for instance, has a domed casing which acts as a lens, but this restricts the viewing angle to about 40° overall. The TL33 on the other hand, has a flat casing which allows a viewing angle of some 120° overall.

LEDs come in a range of sizes, shapes, assemblies and luminosities. Sub-miniatures have lens diameters of 2 mm, miniatures with diameters of 3 mm and standards of 5 mm; these are the round, domed variety. There are also high-intensity types having brightnesses up to 10 times those of the standard range, and they are available in sets having matched luminosities. Flashing varieties are also available in which a built-in integrated circuit switches the diode on and off at a rate of a few hertz. Bicoloured types with outputs of red/green or green/yellow are also available.

Problem 2 How do the electrical characteristics of an LED compare with those of an ordinary diode?

The electrical characteristics of light-emitting diodes are similar to those of conventional diodes except that the forward voltage drop of the LED when illuminated is about 1.5 to 2.0 V against 0.7 V for the ordinary diode. The luminosity of an LED increases as the current through it increases, but excessive current will burn the diode out, as will excessive reverse current. Typically, for indicator LEDs, a current within the range 5 mA to 25 mA is suitable, and in normal use a series resistor is always used to limit the current to a safe level.

Problem 3 An LED is to operate at a current of 10 mA from a 5 V supply. What value of series resistor would be suitable?

We need only an approximate answer to a question of this sort. Assuming that the LED drops 1.5 V when working, this leaves 3.5 V to be dropped across the resistor. At 10 mA, this works out to be 350 Ω. As you are unlikely to have one of these, a 330 Ω preferred 5 per cent resistor would be quite suitable.

Problem 4 How does the LCD differ from the LED? What are the advantages and disadvantages of the LCD compared to the LED?

Liquid crystals are organic substances that are mesomorphic, which means that they lie between the liquid and solid state. A number of such substances are known but the type used in LCDs can be electrically controlled. In the normal state, thin slices of this material appear to be transparent; this is because all the crystals face in the same direction as shown in *Figure 9.2(a)*. If a voltage is impressed across the faces of the slice, the crystals are rearranged and the transparency is affected. For display purposes, a thin layer of liquid is sandwiched between two thin glass plates. A transparent electrically conductive film or back plane is put on the rear glass sheet and transparent sections of conductive film in the shape of the desired segments are coated on to the front sheet. When the crystal is activated by the application of a voltage between the front film segment (or segments) and the back plane, the crystals within the liquid rearrange themselves so that the transmission of light from the back plane is interrupted. The incident light is scattered and absorbed within the liquid slice and those portions of the liquid which are thus affected appear to be dark against a light background. *Figure 9.2(b)* shows this effect. It is not easy to see the segments or characters of an unenergised LCD, though by looking at the surface obliquely the pattern can usually be discerned.

The advantage of LCDs over LED displays is their extremely low power consumption, a matter of microamps per segment rather than milliamps. As such, they are used exclusively in battery-operated devices. They have disadvantages: they cannot be seen in the dark, they are slow in responding to input level changes, and for this reason they cannot be multiplexed. They are not operated by direct voltages as are LEDs but with low-frequency square-wave voltages at a frequency of around 50 to 100 Hz. Direct voltages lead to deterioration of the crystal.

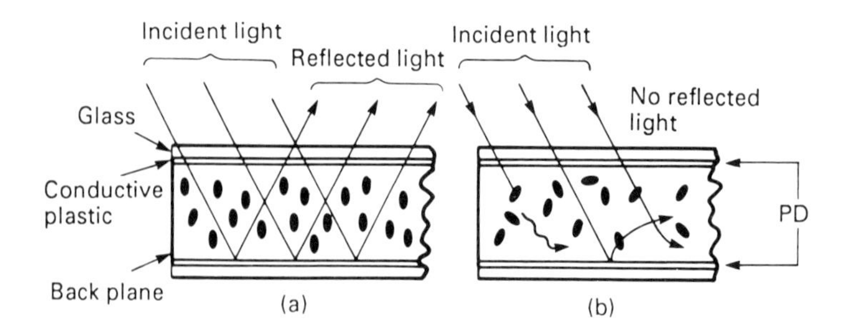

Figure 9.2

Problem 5 Distinguish between seven-segment, hexadecimal and starburst display formats.

Seven-segment displays use LEDs in an arrangement which permits the digits 0 to 9 to be formed. These displays are necessary because they convert, in a visible way, binary signals into decimal notation for the convenience of the observer. A seven-segment module is shown in *Figure 9.3(a)*. Connections from the seven LEDs embedded in the base material are brought out to rear pins which may, if necessary, be plugged into an appropriate d.i.l. holder or soldered directly into a printed circuit board. The common terminal may be connected either to all seven anodes of the LEDs (common-anode) or to all seven cathodes, the choice depending upon the polarity of the applied signals. The other seven connections are lettered *a* to *g* as illustrated, and the individual LED segments may be energised in combinations to form the required digits. A format is also available which provides an indication of polarity in the form ± 1, and this is shown in *Figure 9.3(b)*. Decimal points are also included which may be selected as right-hand or left-hand as required.

These displays are available in a range of character heights: 0.3, 0.43, 0.5, 0.56, 0.8 and 1.0 inch, the usual colours being red or green.

When a number of these display modules have to be used, it is often inconvenient to build up an array from discrete units; it is usual to employ what are known as multiplexed assemblies made up in two- and four-digit format in either common-anode or common-cathode versions. By end stacking such assemblies, an increased number of digits may be built up without the wiring becoming too complicated.

Hexadecimal displays are similar at first glance to the seven-segment types discussed above but they are actually formed from a four by seven arrangement of dots. *Figure 9.4(a)* shows this pattern. Displays of the digits 0 to 9 are similar to those of the seven-segment variety, but the binary inputs for 10 to 15 (not

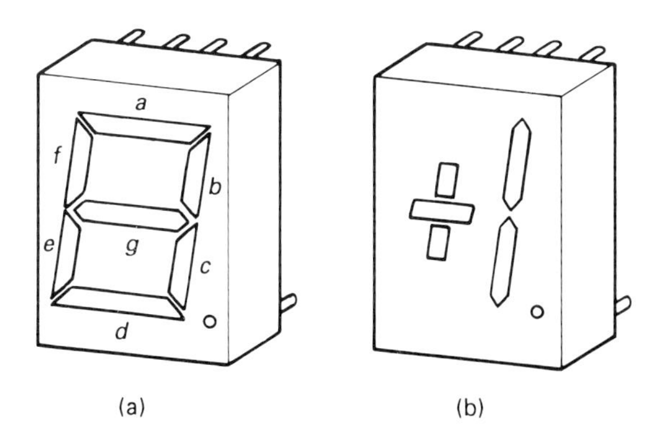

Figure 9.3

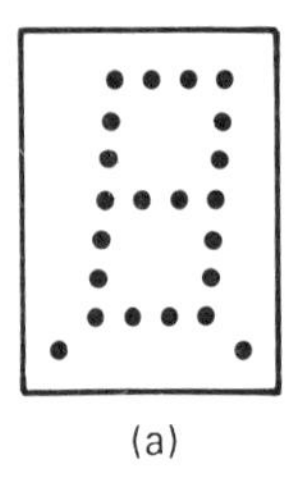

(a)

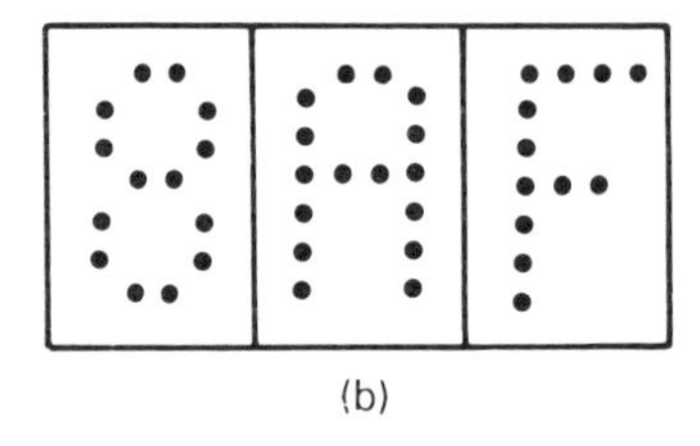

(b)

Figure 9.4

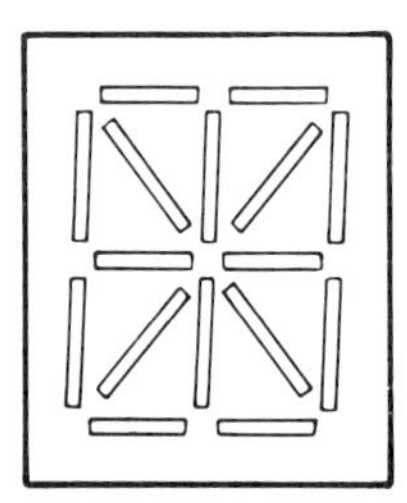

Figure 9.5

meaningful on seven-segment displays) are interpreted as the letters A to F. *Figure 9.4(b)* shows the resultant displays for the figure 8 and the letters A and F.

Starburst or alphanumeric displays consist of 16 bar-segments arranged in the pattern shown in *Figure 9.5*. They are not normally available in single modules but are sold in four digit assemblies with built-in driving circuitry, including a decoder and multiplexer. The digits 0 to 9, the complete alphabet A to Z, and a number of other common symbols such as %, +, −, &, (, and £ to a total of sixty-four presentations, can be produced on these displays.

Most optoelectronic LED displays, when mounted in a finished piece of equipment, are placed behind an acrylic optical filter screen which enhances the display image and reduces spurious reflections.

Problem 6 In wiring up a seven-segment LED display, the connections to segments *a* and *g* were inadvertently reversed. Which of the displayed numerals 0 to 9 will be affected by this?

If we draw up the numerals as they appear normally on a display (see *Figure 9.6(a)*), we see that segments *a* and *g* are concerned, together or individually, in all the numbers except 1. This does not mean, however, that all these numbers will be 'distorted' by the incorrect wiring. When *both a* and *g* are illuminated, the fact that the signals to these segments are changed over does not affect things. Hence the numbers 1, 3, 5, 6, 8 and 9 will still appear to be correct. Where only *a*

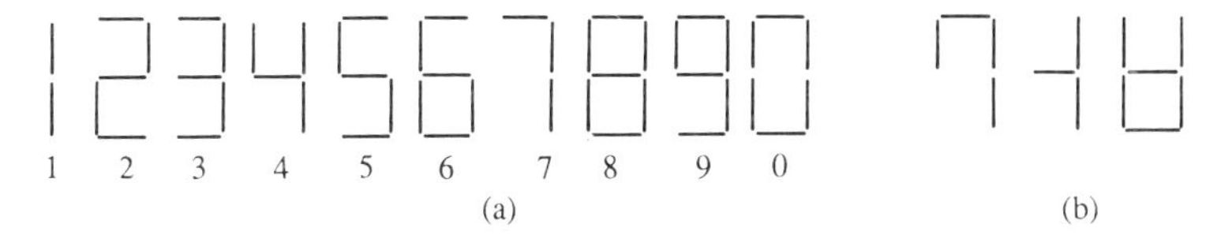

Figure 9.6

or g segments are present, however, the numbers will be affected. These will be 4, 7 and 0, and they will appear as shown in *Figure 9.6(b)*. We have assumed that the figure 6 has a 'top' tail; if it has not, it too will be affected.

Problem 7 Describe the construction, and discuss the characteristics, of a photoconductive light sensor. Illustrate the operation of such a sensor with a simple circuit arrangement.

The photoconductive cell (or photoresistor) is a passive light-sensitive device, the resistance of which decreases as the incident light intensity increases, and conversely. Its action is based on the fact that external energy provided in the form of light photons can act to remove electrons from their parent atoms, so creating mobile charge carriers which reduce the effective resistance of the material.

The most commonly used materials are cadmium sulphide and cadmium selenide, the first of which has a spectral response close to that of the human eye, while the second peaks in the near infra-red. When an external voltage is applied to the cell a current flows in the circuit whose magnitude is a function of the generated carrier density and hence of the intensity of the incident light. These cells are not polarised and may be used either way round.

The semiconductor material is deposited on a ceramic substrate in a zig-zag pattern by a masking process; the whole is then enclosed in a glass or plastic housing with a transparent window area over the sensitive surface. *Figure 9.7*

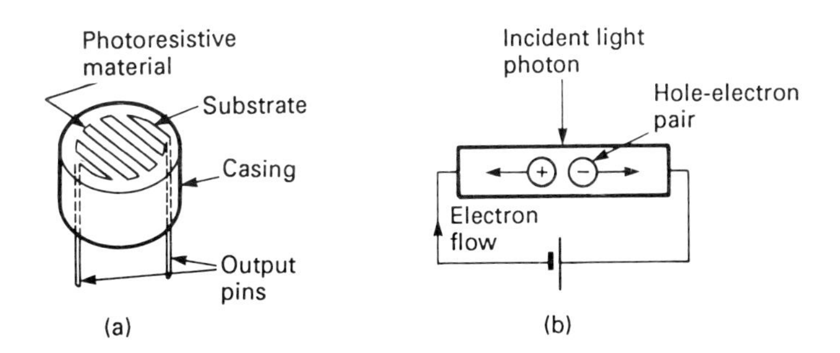

Figure 9.7

shows the general appearance. In some cells the window is coated with a thin film of gold which serves as an electrostatic screen. Leads are brought out from each end of the semiconductor track through the back of the casing; these leads are made up from material which minimises thermocouple offset voltages appearing at the junction of lead and track.

The steady state resistance of a photoresistor in the absence of illumination is called the dark resistance. This may be as much as 10 MΩ or more, falling to a few hundred ohms in bright light – the lit resistance.

Photoresistors have response times that are different for a given change in illumination from dark to light (the resistance fall time) than from light to dark (the resistance rise time). Taking the response time as being between 10 per cent and 90 per cent of the total resistance change, typical figures for these cells are 0.1 s for the rise and 0.35 s for the fall. They are not suitable, therefore, for situations in which the light intensity is changing rapidly, as in chopper amplifiers, though with proper circuit design their use is not totally precluded.

Figure 9.8 shows a simple circuit arrangement in which a photoresistor operates a lamp or relay. The voltage at the inverting input of the opamp is fixed by the potential divider resistors R_1 and R_2, while the non-inverting input is connected to the junction of the photoresistor (an ORP12 is suitable) and a 10 kΩ potentiometer. With the photoresistor illuminated, adjust the potentiometer until the output lamp goes off (or the relay drops out). Shading the photoresistor should now cause the lamp to light (or the relay to operate). Different settings of the potentiometer will affect the luminosity level at which the photocell will trip the circuit. The opamp is, of course, operating as a simple comparator. Reversing the inverting and non-inverting connections will produce the opposite effect; the lamp will go off when the cell is shaded.

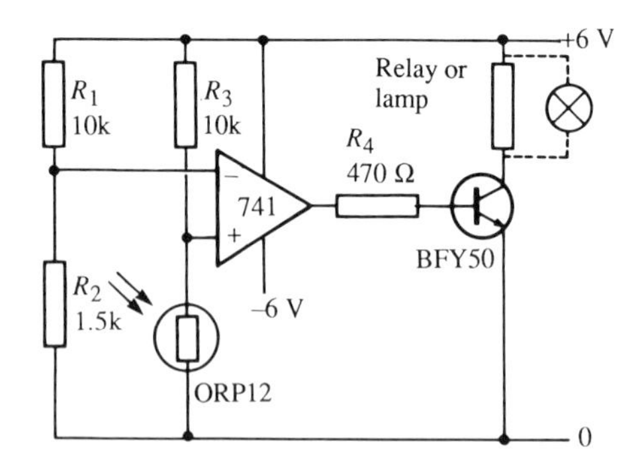

Figure 9.8

Problem 8 What is the principle of operation of a photodiode? Explain what is meant by the photovoltaic mode of operation. Suggest a suitable head amplifier arrangement for use with a photodiode.

A photodiode, which is classed as an active device, is constructed in the same way as an ordinary diode except that the casing has a transparent area so that light can

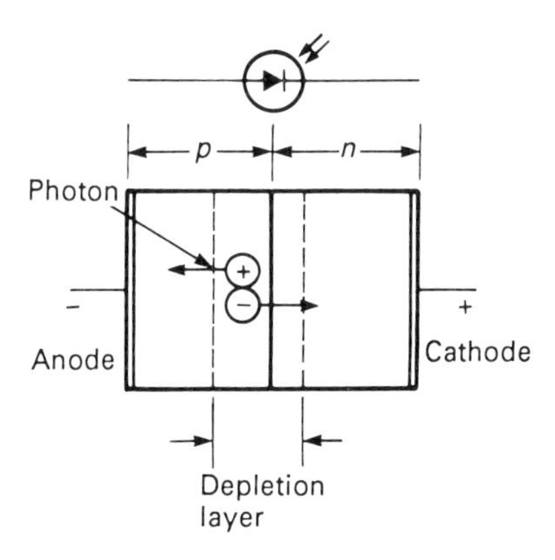

Figure 9.9

fall on the junction. Silicon photodiodes consist of a thin wafer of *n*-type material into which a *p*-type layer is diffused to a depth of about 1 μm. This narrow wafer allows light to penetrate to the *p-n* junction with negligible loss in energy.

Figure 9.9 shows the junction with reverse bias applied and a depletion layer established. This layer, being free of mobile carriers, acts as an insulator between cathode and anode. When a light photon of sufficient energy enters the depletion layer, it is absorbed and its energy released in the form of the generation of a hole-electron pair. Under the influence of the applied field, these pairs separate, the electron moving to the *n*-type cathode, the hole to the *p*-type anode. Hence a current additional to the very small normal leakage current flows in the circuit, even though the diode is reverse biased. *Figure 9.10* shows this effect in terms of the biased diode characteristic. At (a) the diode is shown before the junction is exposed to light; the characteristic is normal. At (b) the diode is shown under the effect of incident light. It is seen that the reverse current level has shifted

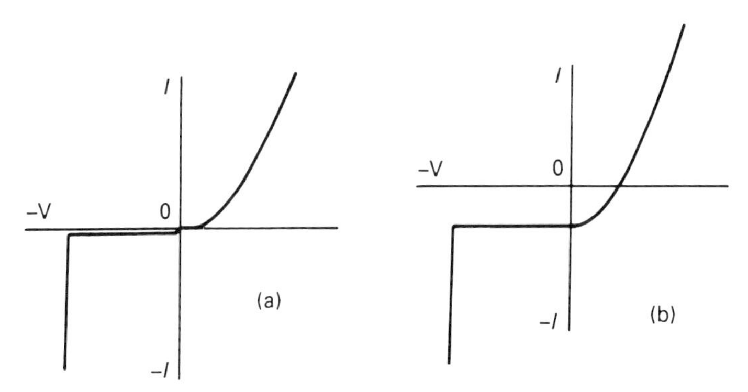

Figure 9.10

downwards but otherwise the curve is virtually unchanged. The significant difference is in the negative portion of the characteristic and here we see that (for a constant reverse bias) a change in the reverse current is obtained which is proportional to the light intensity.

For maximum sensitivity, the depletion layer should be wide. This can be done by increasing the reverse bias voltage, though this increases the leakage current and is undesirable. A layer of pure (undoped) semiconductor material can, however, be introduced between the *p*- and *n*-layers of the junction and this leads to the so-called PIN diode (*p*-intrinsic-*n*). This has two effects: it reduces the junction capacitance, desirable from the point of view of fast response, and it increases the reverse breakdown voltage level. The first of these effects is the more advantageous in the present context.

If a photodiode is used without reverse bias it is said to be operating in the **photovoltaic mode**. As such it is suited for use in optical instrumentation and camera control systems. In the photovoltaic mode, the electron-hole pairs formed when the light falls on the depletion layer diffuse into the enclosing layers, holes into the *p*-type and electrons into the *n*-type material. These extra majority carriers on each side of the junction cause the terminal attached to the *p*-material (the anode) to become positive with respect to the terminal connected to the *n*-material (the cathode). An e.m.f. is consequently generated across the terminals that is proportional to the light intensity. This e.m.f. is typically within the range 50–500 mV on open circuit.

Figure 9.11 shows a suitable amplifier arrangement for a photodiode used in the voltaic mode. Using an FET input opamp, the value of the feedback resistor R_f is chosen to give the required output in terms of the variation in diode current. Typically, R_f will be large, 470 kΩ–1 MΩ, and the output is given closely by the product $I_D R_f$.

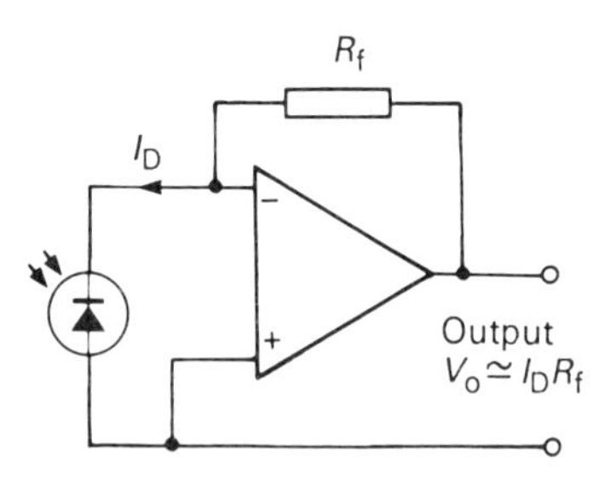

Figure 9.11

Problem 9 In what way does a phototransistor differ from a photodiode?

A **phototransistor** is essentially a photodiode with an integral amplifier. It has the same basic construction as an ordinary transistor but is designed so that light can fall on to the base-emitter junction. When no light falls on this junction, only the usual leakage current flows and this is negligibly small. When light falls on the junction, more hole-electron pairs are created in the manner already described for the photodiode, and the current arising from these is amplified by normal transistor action. Essentially, the base lead is redundant in a phototransistor, but it is often brought out of the casing so that additional collector current control over and above that generated by the light input is available.

Problem 10 What is a bargraph type of indicator? Describe a circuit arrangement which will enable a bargraph to indicate, say, the output volume level of an audio amplifier.

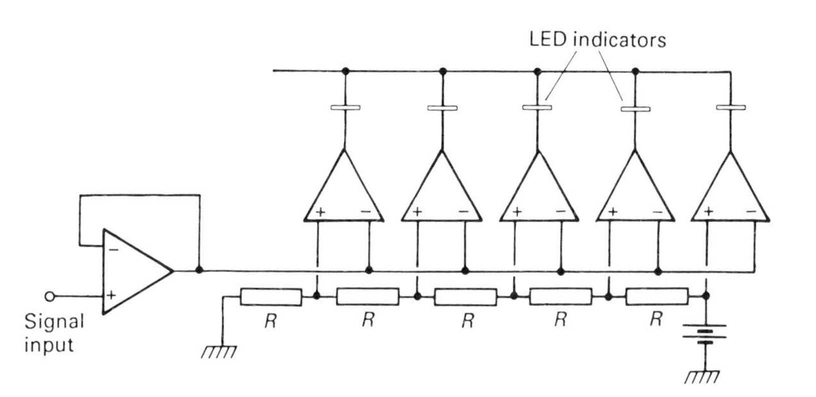

Figure 9.12

The principle of the bargraph display is illustrated in *Figure 9.12*. Here we assume there are five LED indicators in the display but there may well be more. Subminiature LEDs are often used for this purpose, but rectangular forms are available which can be stacked either horizontally or vertically into arrays. Complete assemblies can be obtained which contain a number of LEDs already stacked, together with the associated driving circuitry. In *Figure 9.12*, the input signals is fed through a high impedance buffer amplifier of unity gain, and the output of this is applied to the array of LEDs by way of five comparators, one comparator to each LED. The comparators are each biased to a different voltage level by a resistance divider chain. As the input level increases, the LEDs light up progressively from left to right, so giving an indication of the input amplitude, second by second. For linear relationship between display and input, the divider chain is made up from equal valued resistors; a logarithmic relationship can be obtained by suitably grading the resistor values.

This kind of display can be seen on most modern hi-fi amplifier equipment where it serves as a level indicator for the separate channels or as an output power indicator.

Problem 11 Describe an optoisolator module and give a few examples of its applications.

The combination of an LED and a phototransistor leads to an optically coupled system known as an **optoisolator**. If an LED and a phototransistor are separated a few millimetres and disposed so that they are facing each other, a transfer of signals can be made from one circuit system to another with complete electrical isolation.

Optoisolators are small packages, usually made up in 6-, 8- or 16-pin d.i.l. format, containing one or more infra-red-emitting LEDs as the light source(s) in close proximity to one or more phototransistor(s) as the light sensors. The spectral response of each source and sensor pairing is closely matched, and they are separated by a medium transparent to the radiated frequency band. This medium, which may be air or some other substance such as glass or polymer, is an electrical insulator, hence the isolation existing between the input and output

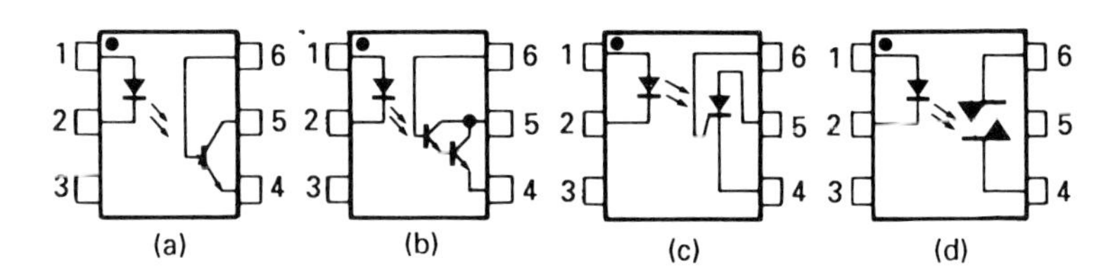

Figure 9.13

terminals of the package. Optoisolators can also be used as non-connecting interfaces in TTL and other logic systems. Voltage isolation up to 5 kV is provided by these units.

Figure 9.13 shows the general theoretical circuit symbols of four commonly used opto-isolators, each having an infra-red LED source but with different sensors. At (a) the sensor is a phototransistor; at (b) the phototransistor is part of a **Darlington** configuration and hence exhibits a considerably enhanced gain over the single transistor. At (c) the source is coupled to a thyristor which is triggered on reception of the light pulse from the source. And at (d) the isolator contains a triac which can be similarly fired.

C FURTHER PROBLEMS ON OPTOELECTRONIC DEVICES

(a) SHORT ANSWER PROBLEMS (answers on page 152)

1 Radiation is emitted from a LED when it is reverse biased. True or false?
2 A LED operates at a current of 15 mA from a 9 V d.c. supply. What series resistor would you use? Why is this resistor necessary?
3 Why are liquid crystal displays preferred to LEDs in battery operated equipment?

4 If a LED is to be operated from an a.c. supply, an ordinary diode should be connected in parallel with the LED with opposed polarity. Why do you think this is necessary?
5 A photoconductive light sensor is one in which the resistance changes when light falls on it. True or false?
6 Why is a photoresistor called a passive device and a photodiode an active device?
7 The viewing angle of a LED depends more on the packaging than on the *p-n* junction producing the light. True or false?
8 The dark resistance of a LDR is always greater than the lit resistance. True or false?
9 Why are 7-segment display modules made in common-anode and common-cathode forms? Which would you use with positive-going control pulses?
10 A phototransistor is essentially a photodiode with an integral . . . .
11 In a PIN diode, the depletion layer is artificially widened by a layer of intrinsic material. True or false?
12 LED displays can be damaged unless they are operated with low-frequency square wave supplies. True or false?

(b) MULTI-CHOICE PROBLEMS (answers on page 152)

1 The response time to rapidly changing inputs is best on (a) a LED, (b) a LCD, (c) a hot wire display.
2 The best form of display for a mains powered clock is (a) LED, (b) LCD, (c) gas discharge, (d) none of these.
3 To change from light to dark, a LCD character can take about (a) 1 μs, (b) 1 ms, (c) 100 ms, (d) 1 s.
4 The viewing angle of a LED can be changed by (a) increasing the diode current, (b) adjusting the shape of the package, (c) having a diffuser behind the lens, (d) having a highly polished lens.
5 A LED emits visible light because (a) electrons are emitted from the junction, (b) hole–electron recombination takes place at the junction, (c) the junction gets hot and glows red.
6 If a photodiode is reverse-biased, the effect of light striking the junction is to (a) forward bias the diode, (b) increase the leakage current, (c) reduce the leakage current, (d) switch the diode off completely.
7 The form of indication given by a bargraph is (a) analogue, (b) digital, (c) neither of these.
8 A hexadecimal display can show (a) numerals only, (b) alphabet letters only, (c) numerals and alphabet letters, (d) numerals, letters and a variety of other signs.

(c) CONVENTIONAL PROBLEMS

1 Make a list of light sources and light sensors. Are any of those you mention not much used in electronic applications?
2 List the advantages and disadvantages of LED readout displays. Does a LCD display overcome all, or any, of the disadvantages you have listed?
3 Why are LCDs not usually multiplexed?
4 Describe the construction and principle of operation of a light-emitting diode and give some practical applications of this device. (See page 112)
5 Sketch the form of the numerals 0–9 you would observe if the *a* and *b* segment connections on a 7-segment module were reversed. Which numerals, if any, would remain unaffected? (See *Figure 9.14*)

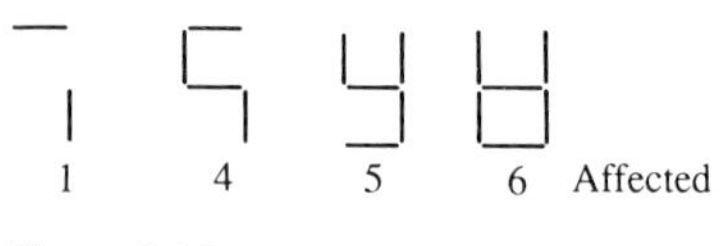

Figure 9.14

Figure 9.15

6 A wiring error has been made to a 7-segment LED display and the digits 0–9 appear in the forms shown in *Figure 9.15*. Deduce what is wrong with the wiring. (e, f reversed)

7 Describe the construction and explain the principle of operation of (a) a photoconductive cell, (b) a photodiode. (See pages 117–19)

8 Each segment of a 7-segment LED display draws a current of 10 mA. What currents will be drawn for each of the numerals 0–9? Assume that the numerals 6 and 9 have 'tails'.

[60, 20, 50, 50, 40, 50, 60, 30, 70, 60 mA respectively]

9 A light source and a photoconductive cell are arranged so that objects passing between them may be counted. What special precautions might be needed to avoid the possibility of false counting? The counter itself is capable of counting at any speed.

10 A 7-segment display is to indicate E for Empty and F for Full, and is to be controlled by two signals; see *Figure 9.16*. When the ON/OFF signal is at logic 0, the whole display is off; when this signal is at logic 1, the display is to indicate E if the FULL/EMPTY signal is logic 1 and F (FULL) if this signal is logic 0. What logic is needed between the signal sources and the display? (This problem is to check your logic as well as your optoelectronics!)

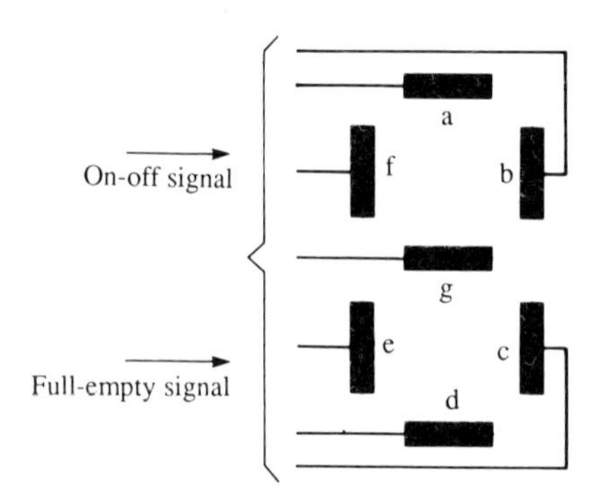

Figure 9.16

11 A motorised trolley must follow a painted white track on a smooth horizontal surface. The trolley carries a sensor system made up from a lamp and two photodiode detectors which respond to light reflected back from the white track. Suggest an arrangement whereby the trolley might be made to follow the track. Assume that only a simple steering control system needs to be activated; no precise details need be given.

12 Describe an opto-isolator module and give a few examples of its applications. (See page 122)

10 Fibre optics

A MAIN POINTS CONCERNED WITH FIBRE OPTICS

1 Electrical signals can be transmitted either as radio signals in space or as electrical variations along metallic conductors. By using a light signal, which is, of course, an extremely high frequency electromagnetic wave, a further means of communication is available, though in the past, the old methods such as a heliograph, were not associated in any way with electrical techniques. Light transmission has a number of advantages, however; it cannot be affected by either electrostatic or magnetic fields, two or more beams can cross without mutual interference or crosstalk, and outside the limits of the light beam, the signal cannot be 'overheard' or tapped.

2 If instead of sending a modulated light signal directly through space where it will of necessity travel in a straight path, it is directed along a suitable 'conducting' transmission line such as a flexible transparent **fibre** of glass or polymer, the signal remains secure and interference proof with the added advantage that it can be made to travel along a curved path to its destination.

3 The introduction of fibre optics offers solutions to a number of the problems mentioned above which are associated with traditional wire and cable links. It also has an advantage with regard to the transmission bandwidth. In coaxial cable the bandwidth available is proportional to $(\text{length})^2$, while in fibre transmission it is directly proportional to length. Because of this, a greater number of different signals (such as telephone conversations) can be multiplexed on to a single fibre link for a given distance than is possible with coaxial lines.

4 The basic principle behind optical communication is the refraction that light experiences whenever it passes from one transmission medium to another of different optical density. *Figure 10.1* illustrates this effect: at (a) it is assumed that medium B is optically denser than medium A; for instance, medium A could be air and medium B glass. The ray of light is passing from the less dense to the more dense medium and angle θ_1 (the **angle of incidence**) is greater than θ_2 (the **angle of refraction**). Diagram (b) shows the light ray passing from a dense to a less dense medium; here θ_1 is again the angle of incidence and θ_2 the angle of refraction, but this time θ_1 is less than θ_2. Notice that the angles are measured relative to a line (the normal) drawn at right-angles to the surface of the connected mediums. Notice also, in both diagrams, that a small part of the incident is *reflected* at the interface. Refraction occurs because light travels more slowly in any medium than it does in space.

The relationship between the angle of incidence θ_1 and the angle of refraction θ_2 is expressed in Snell's law:

$$\mu_1 \sin \theta_1 = \mu_2 \sin \theta_2$$

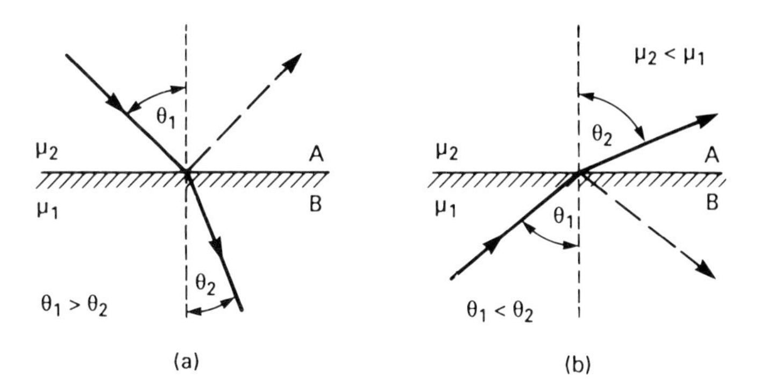

Figure 10.1

where μ_1 and μ_2 are constants (the **refractive indices**) of the particular mediums concerned. For glass and a number of plastics, the refractive index lies between 1.3 and 1.5.

5 In fibre optics we are mainly interested in the situation shown in *Figure 10.1(b)*. Light is bent *away* from the normal as it emerges from the dense into the less dense medium. Suppose the angle of incidence θ_1 is progressively increased; θ_2 will also increase, and since θ_2 is always greater than θ_1, a value of θ_1 will be reached where θ_2 becomes 90°. This is illustrated in *Figure 10.2*. Since the light energy in the incident ray is divided between the refracted ray and the *internally* reflected ray, and since at a greater angle of incidence no refraction can take place, the whole of the incident light energy will pass into the reflected ray. So for angles of incidence greater than the **critical angle θ_c**, the light is totally internally reflected and none of it escapes into the less dense medium.

Since $\mu_1 \sin \theta_1 = \mu_2 \sin \theta_2$, we can replace θ_1 with θ_c and θ_2 with 90°. Then $\mu_1 \sin \theta_c = \mu_2 \sin 90°$; but $\sin 90° = 1$ and so $\sin \mu_c = \mu_2/\mu_1$. If now the less dense medium is air for which $\mu_2 = 1$, then $\sin \theta_c = 1/\mu_1$ or $\mu_1 = \operatorname{cosec} \theta_c$.

6 If a parallel beam of light is directed into one end face of a long cylindrical fibre made, say, of glass, as shown in *Figure 10.3(a)*, the light will travel along the

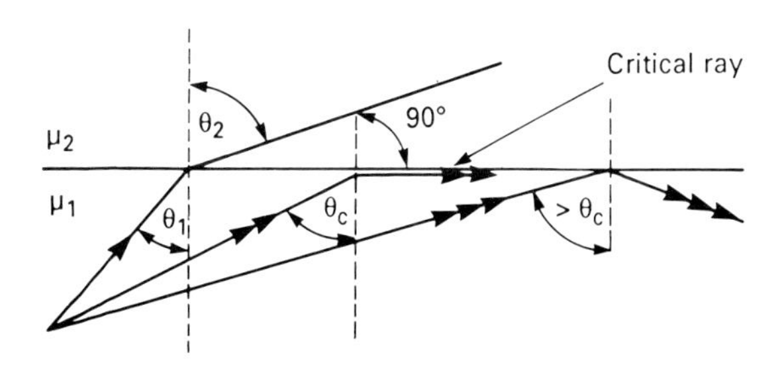

Figure 10.2

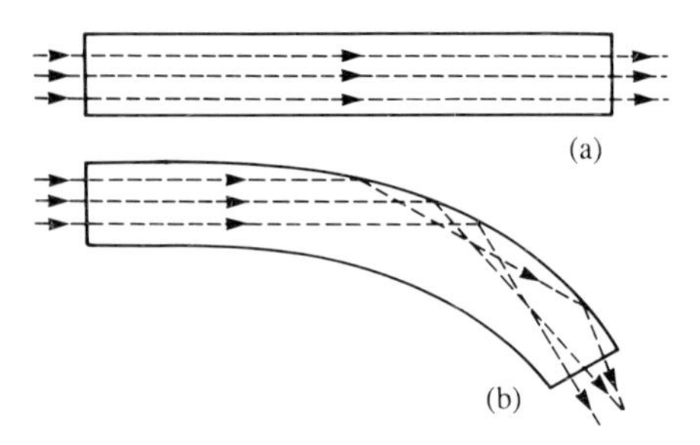

Figure 10.3

length of the fibre without being reflected from the side wall. If the fibre is formed into a curved path, it is still possible for the light to follow this curved track by the fact that internal reflection will take place at the glass–air interface provided the angle of incidence is always greater than the critical angle. In this way, as *Figure 10.3(b)* shows, optical signal transmission is possible along transparent fibres whether the path of the fibre follows a straight or a curved path.

7 Two types of fibre lines are in general use; a **step-index** fibre which is a plastic coated cylindrical core of silica or an acrylic polymer of constant refractive index throughout; and a **graded-index** fibre in which the density of the core and hence the refractive index changes throughout its cross-sectional area, from a high density centre to a low density perimeter. A single strand of optical fibre is made up of the inner core of flexible material, covered by a cladding layer (of smaller refractive index) this in turn being protected by an opaque polyurethane jacket. The total diameter of this assembly may well be no more than 0.1 mm. In a practical form of cable, a number of these strands are combined together and wound around a central metal support to provide strength with an overall polyurethane outer covering.

B WORKED PROBLEMS ON FIBRE OPTICS

Problem 1 What is the critical angle for (a) a glass–air, (b) a glass–water interface, given that μ-glass is 3/2 and μ-water is 4/3?

(a) For the glass–air interface where the refractive index of air is 1, we have:

$$\mu\text{-glass} = \frac{3}{2} = \operatorname{cosec}\theta_c$$

$$\therefore \qquad \theta_c = \sin^{-1}\frac{2}{3} = \mathbf{41.8^\circ}$$

(b) For the glass–water interface, we have

$$\sin\theta_c = \frac{\mu\text{–water}}{\mu\text{–glass}} = \frac{4/3}{3/2} = \frac{8}{9}$$

$$\theta_c = \sin^{-1}\frac{8}{9} = \mathbf{62.7°}$$

Problem 2 What are the units of wavelength measurement in optical work?

In optical work, measurements of wavelength are made either in microns (μ), the milli or nanometer ($m\mu$ or nm) or the Angstrom unit. These are defined from

1 μm = 10^{-6} metre
1 nm = 10^{-9} metre
1 A = 10^{-10} metre

The wavelengths of the visible spectrum run roughly from 430 nm at the blue end to 670 nm at the red end.

Problem 3 Why is the light from an ordinary electric lamp not suitable for optical transmission systems?

Refractive index for any material is not just a function of the material itself, it changes with the wavelength of the light. Red light is refracted less than blue light, so if a signal is made up of light having a wide range of wavelengths (as an ordinary bulb will produce), some of the frequency components will be lost in transmission by not being internally reflected. Monochromatic light, usually at the red end of the spectrum and beyond, that is, in the infra-red region, is used in fibre optical systems. Laser sources are commonly used because of their extremely narrow emission spectra. For experimental work, a red LED makes a suitable transmitting element.

Problem 4 Define the terms (a) numerical aperture, (b) acceptance angle, as related to optical fibre work.

In *Figure 10.3(a)* we assumed that light entered the fibre as a parallel bundle of rays, but in practice this is unlikely to be the case. Suppose we have a core of transparent material encased in some kind of cladding which has a smaller refractive index than the core, as shown in *Figure 10.4*. Consider now what happens if light rays enter the core at a variety of angles, two of which are indicated in the diagram. On entering the core, an initial refraction of all

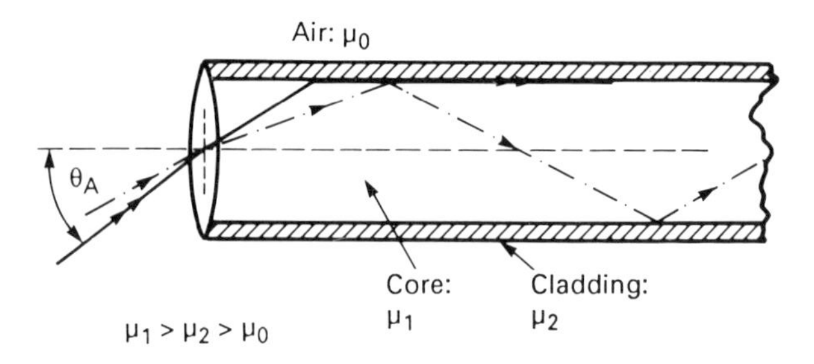

Figure 10.4

non-axial rays takes place but once inside, internal reflection from the core-cladding interface will only occur provided the incident angle of entry, θ_A, is less than some given value. At the critical point, and beyond, the ray will not be internally reflected. θ_A is called the **acceptance angle**, and the sine of this angle is the **numerical aperture**. Hence

$$\text{numerical aperture} = \sin\theta_A$$

$$\text{Then}\quad \mu_o \sin\theta_A = \mu_1 \sin(90 - \theta_c)$$

$$= \mu_1 \cos\theta_c$$

where θ_c is the critical angle at the interface. But from trigonometry $\cos^2\theta = 1 - \sin^2\theta$.

$$\therefore \qquad \mu_o \sin\theta_A = \mu_1 \surd(1 - \sin^2\theta)$$

But $\mu_o = 1$, so

$$\sin\mu_A = \mu_1 \sqrt{\left(1 - \left[\frac{\mu_2}{\mu_1}\right]^2\right)}$$

$$\surd(\mu_1{}^2 - \mu_2{}^2)$$

Thus, suppose we have a fibre in which the refractive index of the core is 1.5 and of the cladding 1.42. The numerical aperture will be $\sin\theta_A = \surd(1.5^2 - 1.42^2) = \sqrt{0.234} = 0.48$, and the acceptance angle $\theta_A = \sin^{-1}0.48 = 28.7°$. Any rays entering the free end face of the core at an angle greater than this (relative to the axial line of the core) will not be transmitted.

Problem 5 Distinguish between monomodal, meridional and skew propagation modes in optical transmission and discuss the characteristics of each.

There are three classifications of transmission modes (or the path followed by light) along optical fibres: these are monomodal, meridional and skew modes.

Monomodal transmission takes place along a fibre that has such a small diameter, typically 5–10 μm, that only one path or mode of light transmission is possible. **Meridional** and **skew** transmissions take place along a **multimode** optical fibre, typical diameters being 50–100 μm, in which many light paths are possible.

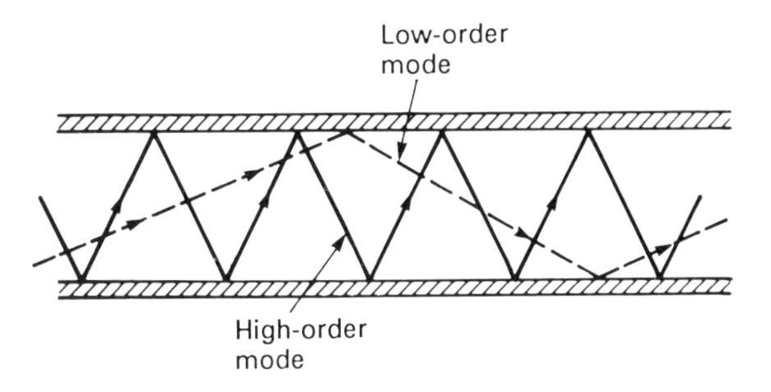

Figure 10.5

Meridional rays are those which pass through the axis of the fibre after each internal reflection, while skew rays never intersect the axis. Step-index fibre propagation is concerned with meridional rays. In *Figure 10.5* the effect of input rays launched over a range of acceptance angles is shown. High order modes are those rays which are reflected a great number of times (many tens of thousands) in their passage along the fibre, while low order modes are those which make far fewer transitions. This variation in path length and hence in the time taken for different rays to propagate along the fibre, although all rays start off simultaneously from the transmitting element, leads to a form of distortion known as **modal dispersion**.

This problem can be alleviated by the use of graded-index fibres whose refractive index changes gradually from a high value at the fibre centre to a lower value at the perimeter. As the speed of light through a medium falls as the refractive index increases, higher order modes which spend less of their time in the central parts of the fibre, travel faster than lower order modes; hence the time difference between high and low order modes is much less for a given length of graded-index fibre than it is for step-index fibre.

Modal dispersion is not the only form of dispersion; the velocity of light is a function of its wavelength when it travels through a homogeneous medium, consequently, unless the transmission is strictly monochromatic, the various frequency components of the signal may not arrive at the receiver simultaneously. This form of dispersion is known as **material dispersion**. The use of laser transmitters reduces it to negligible proportions.

Problem 6 Outline, and comment on, the forms of transmission loss experienced in optical fibre work.

As in any other communication medium, fibre optics has transmission losses. These come under three main headings: material absorption and scattering, curvature losses and coupling losses.

Material absorption results from the inevitable presence of molecular impurities within the fibre core which absorbs certain wavelengths. Inherent impurities also cause the scattering of rays (**Rayleigh scattering**) so that energy is dissipated by the unwanted break-up of the normal forward-moving light rays. Scattering also results from irregularities in the core-cladding interface so that perfect reflection from this interface is not achieved. *Figure 10.6* shows these effects. Fibres constituted from very pure glass are least likely to have serious losses from absorption and scattering and are used over long distance transmission paths in preference to polymer fibres.

Curvature can affect attenuation because if the bending radius at any point along the link is too small, part of the energy will be lost due to rays striking the core-cladding interface at angles less than critical. The makers of the fibres recommend a minimum bend radius. In general, polymer fibres will work with tighter bends than will glass fibres, though some of this has to do with mechanical problems.

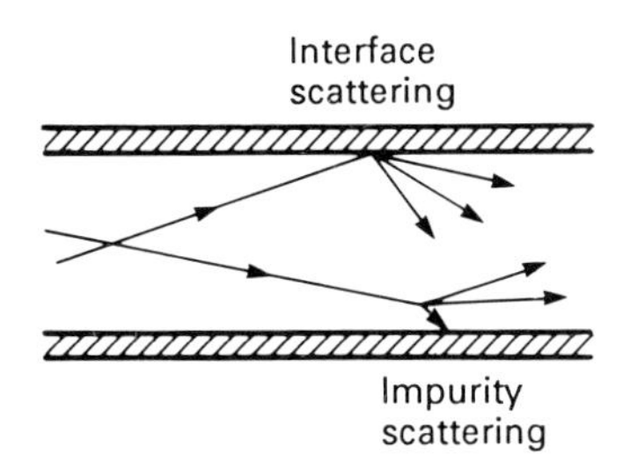

Figure 10.6

One of the few disadvantages of fibre optics is the difficulty of joining lengths of core together and of making connections to the transmitter and receiver head units. Special connectors are available but they are expensive and they do introduce the bulk of the transmission loss in a particular system. The losses arise from misalignment of the joined faces, reflection at the faces, and the separation of the faces. A small gap is necessary to avoid scratching the faces when the coupling is made; such marks would make matters much worse.

Problem 7 Make a list of the advantages that fibre optics has over the coaxial cable for communications purposes.

(a) Low interference. As we noted earlier, optical fibres are almost completely immune to all forms of electrostatic and magnetic interference. There is also no cross-talk between adjacent fibres.
(b) The input and output circuits of optical systems are isolated from one another electrically.
(c) Illegal tapping into optical systems is virtually impossible.
(d) Because the loss experienced with modern monomodal systems is very low, especially when used with laser sources, the necessity of repeaters (amplifiers) along the line is reduced to a much lower number than is usual with coaxial cables.
(e) Fibres offer a very large bandwidth which means that many different signals can be multiplexed on to a single fibre.
(f) Optical fibre cables are much smaller in size and lower in weight than equivalent coaxial cables.

Problem 8 Describe a simple optical link that might be used as a school or college experiment.

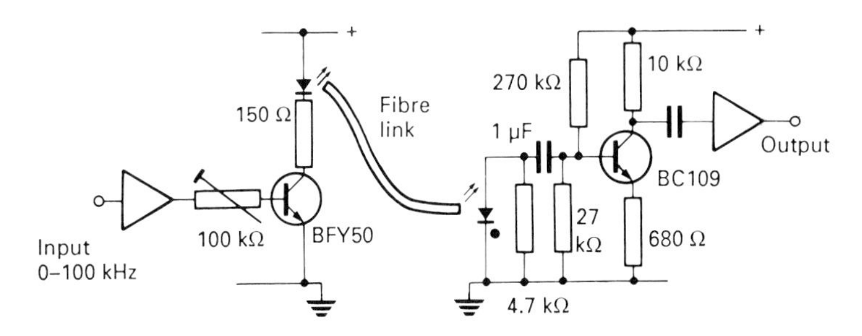

Figure 10.7

A simple laboratory set-up is shown in *Figure 10.7*. This uses visible light and a polymer optical fibre. Suitable transmitter LEDs and receiving photodiodes are available from a number of suppliers, together with a length of matching fibre. A 1 m length is suitable for this kind of experiment. Only the relevant parts of the transmitter and receiver are shown as the remaining amplification can be quite conventional. Some experimentation may be needed on the component values indicated, dependent upon the actual LED and photodiode used.

By feeding the input from a low-frequency generator, the effect of frequency variation can be studied, as can the effect on attenuation of bends in the fibre being made greater than the recommended minimum.

C FURTHER PROBLEMS ON FIBRE OPTICS

(a) SHORT ANSWER PROBLEMS (answers on page 152)

1 The angle of incidence equals the angle of refraction. True or false?
2 Define the terms: (a) refractive index, (b) critical angle, (c) total internal reflection.
3 What are the three main categories of optical fibre design?
4 In a multimode fibre, light rays can follow many hundreds of different paths. True or false?
5 The wavelength of light is measured in nm. What is a nm? What are the extreme wavelengths at each end of the visible spectrum, approximately?
6 What are (a) meridional rays, (b) skew rays?
7 Explain the difference between a step-index fibre and a graded-index fibre.
8 The refractive index of space is unity. True or false?
9 State Snell's law of refraction.
10 What is meant by Rayleigh scattering?
11 Why can excessive curvature of a fibre link increase the attenuation of the system?

12 What is/are the advantage(s) of a laser source transmitter over a gallium-arsenide LED transmitter? Why are lasers used exclusively on monomode fibre links?

(b) MULTI-CHOICE PROBLEMS (answers on page 152)

1 Refraction of light occurs in passing from one medium to another because (a) its velocity changes, (b) it loses energy, (c) it is scattered inside the material, (d) none of these.
2 Light can be internally reflected when it encounters (a) a lower density medium, (b) a higher density medium, (c) an opaque medium.
3 When internal reflection takes place, the angle of incidence (a) is equal to the angle of reflection, (b) is less than the angle of reflection, (c) has no connection with the angle of reflection, (d) is equal to the critical angle.
4 Material dispersion can be reduced by using (a) a broad spectrum of light, (b) a narrow spectrum of light, (c) infra-red light, (d) ultraviolet light.
5 Monomodal optical fibres allow (a) the transmission of a great number of light modes, (b) the transmission of only one light mode, (c) the transmission only of skew modes, (d) none of these.
6 Losses occur in optical fibres because of (a) high ambient lighting, (b) temperature increases, (c) molecular impurities, (d) scratched faces at connecting points.
7 Fibre optic strands are cladded so that (a) the signal cannot escape from the core, (b) the core is insulated, (c) mechanical strength is provided, (d) no ambient light can interfere with the signal.
8 White light is unsuitable for transmission over an optical link because (a) it is too bright, (b) it has too many frequency components, (c) it travels more slowly than infra-red, (d) it is difficult to modulate.
9 Rays entering a fibre at an angle less than the acceptance angle (a) will not be transmitted, (b) will be transmitted, (c) will travel only along the axis of the fibre, (d) none of these.
10 Optical fibres are made of two types of glass or polymer parts because (a) the outer part insulates the inner part, (b) the outer part has a lower refractive index than the inner part so that internal reflection occurs, (c) the outer part reflects any unwanted light away from the inner part.

(c) CONVENTIONAL PROBLEMS

1 Explain the terms: critical angle, acceptance angle, numerical aperture, meridional and skew transmission modes, as relevant to optical transmission work.

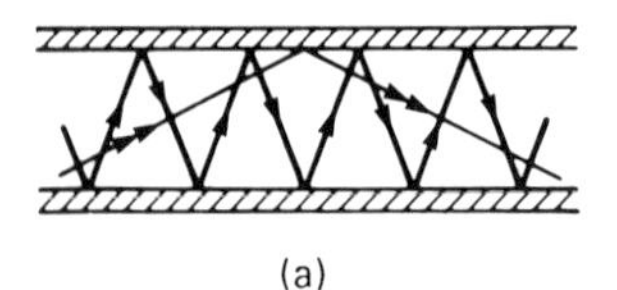

(a)

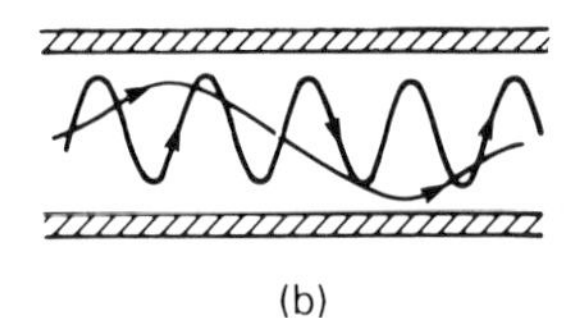

(b)

Figure 10.8

2 The core and cladding of an optical fibre have refractive indices of 1.47 and 1.43 respectively. What is the critical angle for this cable? What is the acceptance angle and the numerical aperture?

[76.6°, 0.34, 19.9°]

3 A certain fibre has a core of refractive index 1.46 and an acceptance angle of 30°. What is the refractive index of the cladding?

[1.37]

4 A polymer fibre has an attenuation of 200 dB/km at a frequency of 660 nm. A 10 m length of this fibre is used to connect together a transmitter unit which introduces a loss of 3 dB and a receiver unit which introduces a loss of 2.5 dB. If the maximum power injected into the transmitter at 660 nm is 24 μW, what power is delivered to the photodiode at the receiver?

[4.4 μW]

5 If the spectral sensitivity of the photodiode in the previous problem is 0.5 A/W at 660 nm wavelength, what current will flow in this diode under the operating conditions given in the problem?

[2.2 μA]

6 Explain the difference between step-index and graded-index optical fibres. *Figure 10.8* shows the path of light rays in these two types of fibre at (a) and (b) respectively. Explain the form that these rays take.

7 Transmission losses in fibres can be caused by (a) material absorption, (b) Rayleigh scattering, (c) curvature radiation, (d) coupling alignment. Briefly explain the mechanism of each of these causes.

8 Make a sketch showing how the curvature of a fibre can affect the critical angle condition for rays passing along the fibre.

9 An experimenter makes up a simple optical link using a red LED as his transmitting element and a germanium phototransistor as his receiver. Although he has no faulty components or incorrect assembly in his system, it fails to work. Explain why.

11 Power amplifiers

A MAIN POINTS CONCERNED WITH POWER AMPLIFIERS

1 A **large-signal** or **power amplifier** is a converter of d.c. power drawn from a power supply source into a.c. or signal power delivered to a load. As such amplifiers are designed to deliver power to their respective loads, the main consideration must be the elimination of power wastage which implies a high conversion efficiency. All wasted power has to be dissipated by the output converter and this sets a limit to the useful power output that a particular device may supply.

So that the power delivered to the load should be as great as possible, it is necessary to ensure that the load is properly matched to the output impedance of the active device, whether bipolar transistor or FET. Maximum power is transferred from a generator to a load when the proper relationship is established between their impedances. The particular value of load impedance which suits the device being used is known as the **optimum load**.

2 Power amplifiers are operated normally in either **Class-A** or **Class-B** (strictly Class-AB) bias conditions. For specialised purposes there is a further class known as **Class-C**. There are also intermediate classes such as the Class-AB mentioned.

In Class-A working the input signal causes collector or drain current to flow during the whole 360° of the input cycle. This is determined by the d.c. operating point P being set approximately at the centre of the dynamic load line or transfer characteristic as shown in *Figure 11.1(a)*. Small signal amplifiers are nearly always operated Class-A as are low power output stages in inexpensive radio receivers where powers of the order of a watt or less are involved. The maximum theoretical efficiency of a Class-A amplifier is 50%; this low figure results from the fact that there is power dissipation at the collector or drain electrode equal to $V_Q I_Q$, even when there is no signal input.

In Class-B working, the operating point P is moved to the lower extreme of the dynamic characteristic as shown in *Figure 11.1(b)*. Collector or drain current then flows for about one-half-cycle of the input or approximately 180°. This means that the quiescent power dissipated when there is no signal input is zero. However, two transistors or FETs must be used in a Class-B amplifier, each conducting in turn so that the full input cycle is restored at the output. The maximum theoretical efficiency of a Class-B amplifier is 78.4%.

A Class-C amplifier has its operating point positioned beyond collector current cut-off so that current flows over only part of the input half-cycle, the angle in general being about 120°. Class-C is used only for r.f. amplification with tuned circuit loading, thereby producing sinusoidal outputs at the resonant frequency of the load.

3 The loads used with power amplifiers may be loudspeakers, servomotors or aerial systems, to name only three. The impedances of such loads vary

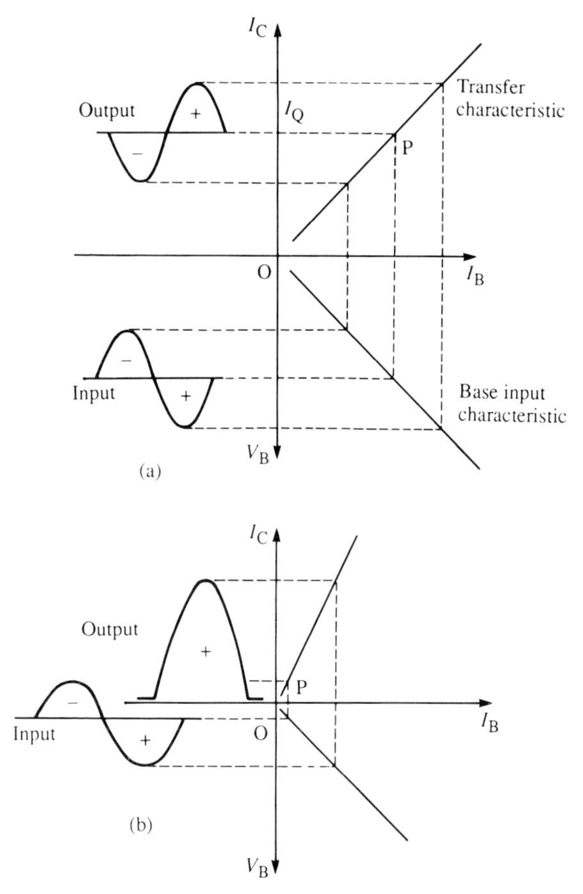

Figure 11.1

considerably and unless they are matched to the output requirements of the amplifier, the useful power developed will be seriously reduced. A method of matching used in low power situations is that of transformer coupling, the turns-ratio being suited to the particular load requirements. This method is not much used nowadays, certainly not in large power circuitry, but it is instructive to mention the method before turning to the alternatives.

Figure 11.2 shows a basic power amplifier stage with transformer coupling to a load impedance Z_L. The effective impedance seen at the primary terminals is then $Z'_L = n^2 Z_L$ where n is the turns ratio of the transformer in the direction indicated. Hence Z_L can be matched to the required collector loading Z'_L by a suitable choice of turns ratio. A stage using a single transistor in this way is known as a **single-ended** output.

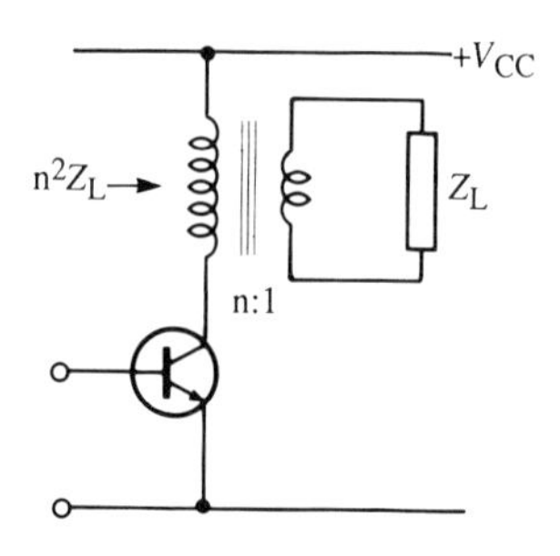

Figure 11.2

4 If the d.c. resistance of the transformer primary winding is neglected and there is no input signal, the power supplied to the amplifier at any instant is $P_{dc} = V_{cc}I_Q$ where I_Q is the quiescent collector current flowing at the selected Class-A bias point, see *Figure 11.3*. But the power dissipated at the collector at any instant is the product of the collector voltage and current, hence $P_c = V_QI_Q = V_{cc}I_Q$, since the voltage drop in the transformer winding is negligible. Hence the *whole* of the power supplied is dissipated as heat at the collector. The maximum collector dissipation as stated by the transistor manufacturers must never be exceeded in a practical design.

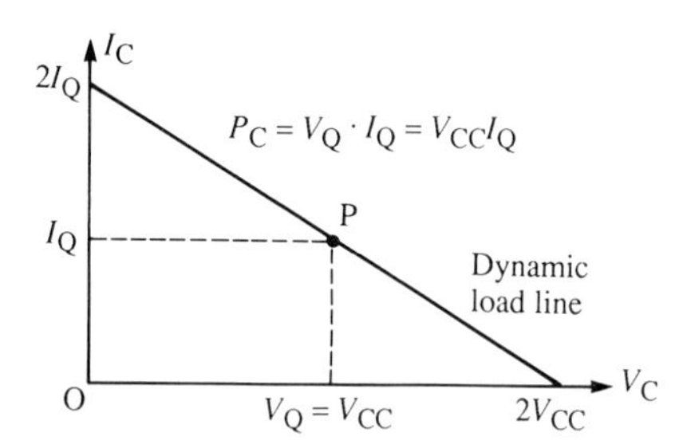

Figure 11.3

When the signal is superimposed on the quiescent current the power supplied will be unaffected; the mean current will still be I_Q. So the signal power developed in the load must be drawn from the collector dissipation. Then $P_{dc} = P_c + P_L$ or $P_c = P_{dc} - P_L$. Hence, in Class-A working, the collector dissipation is greatest when the input signal is absent.

5 If the transistor in Class-A working could be driven to the limits of its output power capability by utilising the whole of the characteristic, the collector current swing would be $2I_Q$ with a corresponding collector voltage swing of $2V_{cc}$. The output signal power would then be expressed by

$$P_L = \frac{I_Q}{\sqrt{2}} \times \frac{V_{cc}}{\sqrt{2}} = \frac{V_{cc}I_Q}{2}$$

But the d.c. power supplied $P_{dc} = V_{cc} I_Q$, hence efficiency

$$\eta = P_L / P_{dc} = \tfrac{1}{2} \text{ or } 50\%$$

In practice, efficiencies of 40–45% are possible. This theoretical maximum of 50% applies only to cases where the effective d.c. resistance of the collector load is small and the inductive effect of a transformer primary allows the collector voltage to swing to approximately $2V_{cc}$. If the load is a resistance, the collector cannot swing beyond V_{cc} and the efficiency has a theoretical maximum of 25%. A conversion efficiency of this order (less in a practical circuit) is not desirable where powers over a watt or so are concerned. It is here that the low output impedance of the emitter-follower comes in useful in combination with Class-B working.

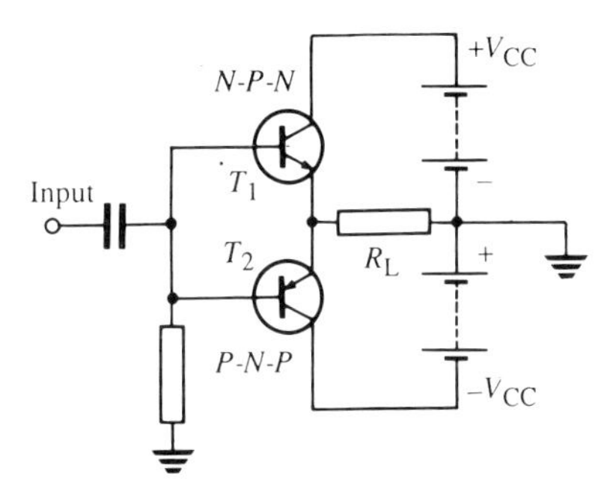

Figure 11.4

6 The emitter-follower principle is used to construct a Class-B power amplifier without the use of transformers, and a basic circuit is shown in *Figure 11.4*. Two transistors are used, one a *p-n-p* and the other an *n-p-n* type. An input signal is applied to the common base connection; the *p-n-p* transistor then conducts only for the negative half-cycle of input and the *n-p-n* conducts only for the positive half-cycle. The resulting output current flowing in the load is then a reconstruction of the complete input waveform of base voltage. This is a Class-B **push-pull amplifier**, the transistors taking it in turn to push and pull the current through the load. Circuits using mixed transistors like this are known as **complementary** circuits.

The efficiency of this circuit is much superior to that of Class-A. For a start, if there is no signal input both transistors are very close to cut-off and there is no wasteful dissipation at either collector. In an ideal situation we can assume the whole of the transfer characteristic to be used, and the current and voltage pulses for *either* transistor will be as shown in *Figure 11.5*. For sinusoidal variations the power output into the load will be

$$P_L = \frac{1}{2} \frac{\hat{V}_c \hat{I}_c}{\sqrt{2}.\sqrt{2}} = \frac{\hat{V}_c \hat{I}_c}{4}$$

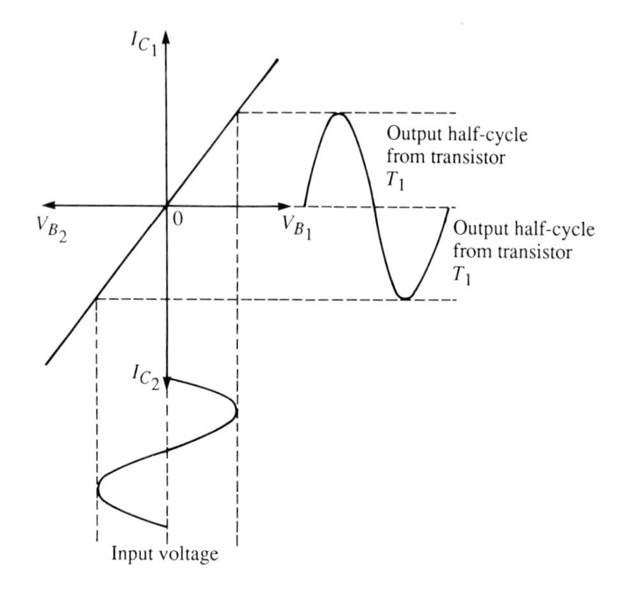

Figure 11.5

The factor of $\frac{1}{2}$ comes in because we are dealing with a half sinewave, and $V_c = I_c R_L$. The *average* value of the d.c. power delivered to either transistor will then be $V_{cc}\hat{I}_c/\pi$; recall that the average value of a unit half sinewave is $1/\pi$. The efficiency is therefore

$$\eta = P_{L/P_{dc}} = \frac{\pi V_c \hat{I}_c}{4V_{cc}\hat{I}_c}$$

But for full drive in the ideal case we are considering, $V_c = V_{cc}$, hence $\eta = \pi/4$ or 0.784 (78.4%). In a practical case, a voltage swing of V_{cc} is not possible, but efficiencies of 70% can be easily obtained.

7 Power amplifiers are now commonly available in integrated form, very often incorporating their own heat sinks and having output powers ranging from a few watts to 100 W or more. These packages require only the addition of a few external components plus the loudspeaker, to make a compact, high quality audio power amplifier. The output stages are, of course, of complementary design and the load is fed directly from the package, so making the clumsy transformer coupling a thing of the past.

An example of a low power integrated amplifier is the LM386N; the circuit diagram of this amplifier is shown in *Figure 11.6* with an application diagram showing the required external components in *Figure 11.7*. There is a differential input arrangement using Darlington pairs and the output circuit takes the complementary form. An internal feed-back resistor connected from pin 5 (output) to pin 1 can be bypassed by external components to modify the gain and

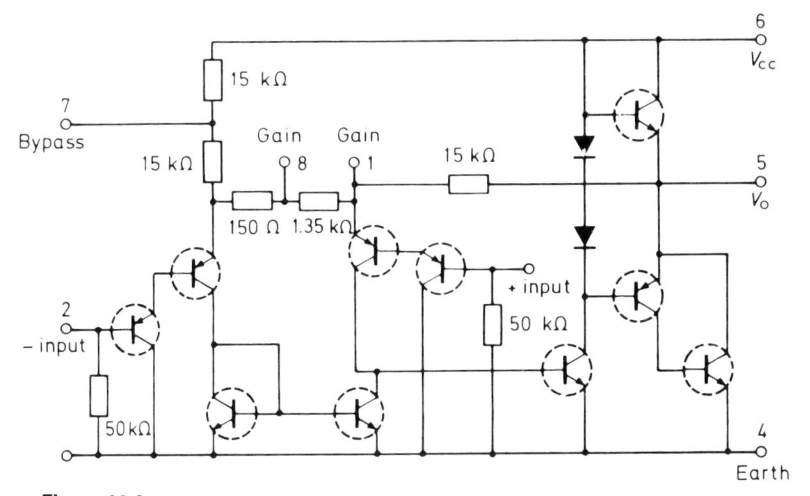

Figure 11.6

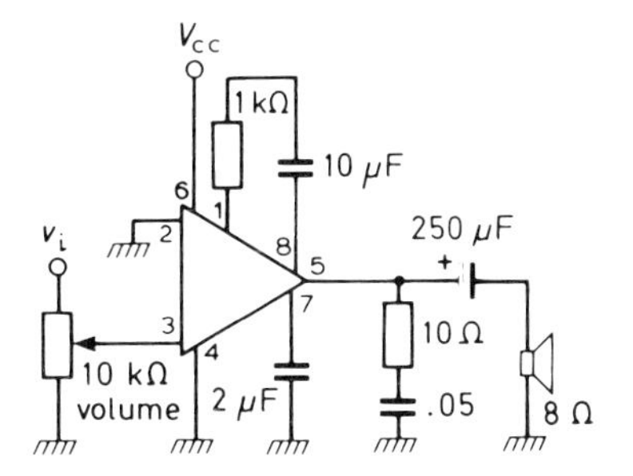

Figure 11.7

frequency response. Pins 1 and 8 are made available as a gain adjustment device; with the pins open the 1.35 kΩ internal resistor sets the gain at 20, but by bypassing this resistor with an external capacitor, the gain can be increased to 200. In the application circuit, the gain is set by the 1 kΩ resistor and the 10 μF capacitor to be between 20 and 200.

8 Some integrated circuits are fitted with their own heat sinks (for example, the HY60) but for circuits using such types as the 2030 an external heat sink is necessary (as it is with discrete power transistors) to remove the surplus heat and thus increase the power dissipation of the device. It is found experimentally that the steady state temperature rise at the collector junction is proportional to the power dissipated at the junction, that is $T_j - T_a = \theta P_c$ where T_j is the junction temperature, T_a is the ambient temperature and θ is a constant of

proportionality. This constant is known as the **thermal resistance**, and its value depends upon a number of factors: the type of transistor, the conduction of heat from the junction to the case of the transistor and the conduction of heat from the transistor to its surroundings. Hence θ can be looked on as the resistance to the removal of heat from the collector junction. The actual resistance associated with the transistor casing is known as the **intrinsic** thermal resistance, θ_i. By mounting the transistor on a heat sink (a piece of aluminium, say, of large surface area) the thermal resistance between junction and surroundings is greatly reduced and the junction runs cooler and much more safely.

Thermal resistance can be treated as analogous to electrical resistance, hence an 'equivalent' thermal circuit can be established from which all basic calculations may be made. Electrical resistance has the units volts-per-ampere, the voltage dropped across a unit resistance being equal in magnitude to the current flowing. In the same way, since thermal resistance $\theta = (T_j - T_a)/P_c$, θ is defined as the temperature rise for unit power dissipation, and is clearly seen to have the units of **degrees C per watt** (°C/W).

B WORKED PROBLEMS ON POWER AMPLIFIERS

Problem 1 A Class-A amplifier draws a quiescent current of 1.2 A from a 20 V supply and delivers a signal power of 8 W to the collector load. Calculate (a) the efficiency, (b) the collector dissipation for this amplifier.

Power supplied $P_{dc} = V_{cc} I_Q = 20 \times 1.2$

$= 24$ W

(a) Efficiency $\eta = P_{L/P_{dc}} = 8/24$

$= \mathbf{0.334}$ (33.4%)

(b) Collector dissipation $P_c = P_{dc} - P_L$

$24 - 8 = \mathbf{16\ W}$

Problem 2 A Class-B push-pull amplifier uses two transistors each rated at a maximum collector dissipation of 6 W. For a sinusoidal input signal, calculate the maximum power obtainable for an efficiency of 70%.

$P_{dc} = P_L + P_c = P_L + 6$

For $\eta = 0.7$, $\dfrac{P_L}{P_{dc}} = \dfrac{P_L}{P_L + 6} = 0.7$

$\therefore 0.7P_L + 4.2 = P_L$

$P_L = \mathbf{14\ W}$

Problem 3 In the diagram of *Figure 11.8*, calculate the required turns ratio of the transformer if the effective collector load is to be 20 Ω and the load resistance R_L is 4 Ω. If under these conditions, the input resistance of the transistor is 10 Ω and an input signal of 150 mV rms produces 3 V across R_L, calculate the power gain in dB.

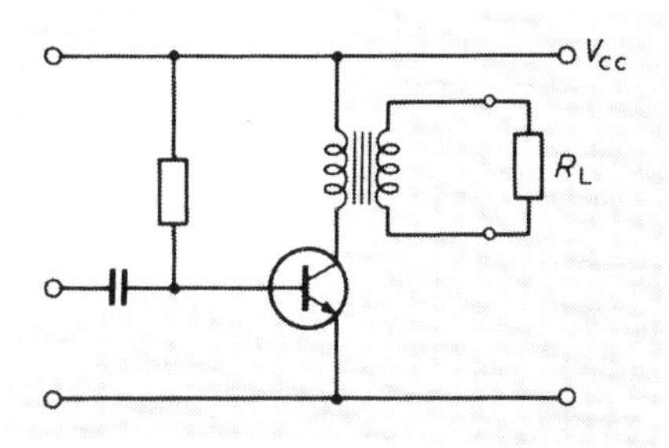

Figure 11.8

The effective collector load resistance $= n^2 R_L$

$$\therefore\ n^2 = \frac{20}{4} = 5$$

and $n = \sqrt{5} = 2.24$

The transformer will be 2.24:1 step-down.

The input power is $\dfrac{(150 \times 10^{-3})^2}{10} = 0.00225$ W

$= 2.25$ mW

The output power will be $\dfrac{3^2}{4} = 2.25$ W

Power gain $= 10 \log \left[\dfrac{2.25 \times 10^3}{2.25} \right] =$ **30 dB**

Problem 4 Explain a disadvantage of strict Class-B working. How might this be overcome?

When Class-B was discussed earlier, it was assumed that the transfer characteristic was a straight line. In real transistors, the characteristic is curved, particularly at the lower end. As a result of this, what is known as **crossover distortion** occurs as the input signal passes from one base to the other of the two transistors involved. *Figure 11.9* shows the effect of this. The transfer

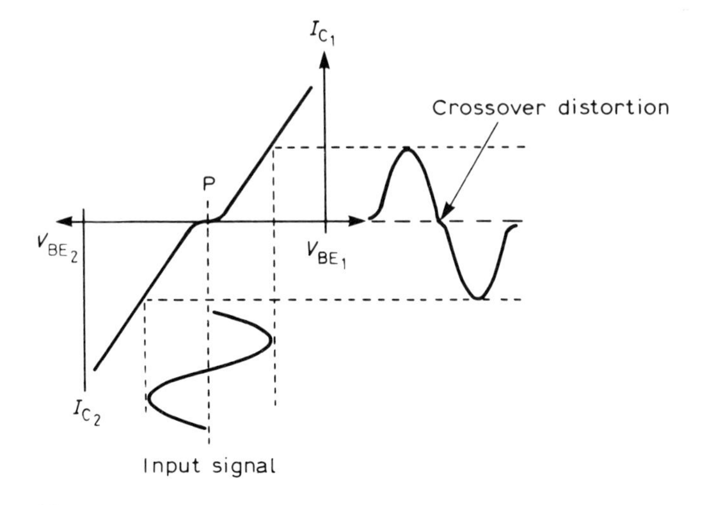

Figure 11.9

characteristics are drawn back to back and the base bias point is at P, exactly at collector current cutoff. As the input signal passes from one base to the other, little base current flows until the input exceeds a value which carries it beyond the lower curved regions of the characteristic. Past this, I_c is closely proportional to base current and there is no further problem. At the crossover point, however, any one of the transistors is not fully turned on by the time the other is turned off. The output waveform is consequently distorted in the manner shown. The effective extent of the trouble increases as the signal amplitude decreases.

To overcome the difficulty it is necessary to ensure that the region very close to the origin point P is not effectually used. This can be done by applying a small forward bias to the transistors so that a small quiescent collector current flows all the time. As one transistor is then driven off through the non-linear region of its

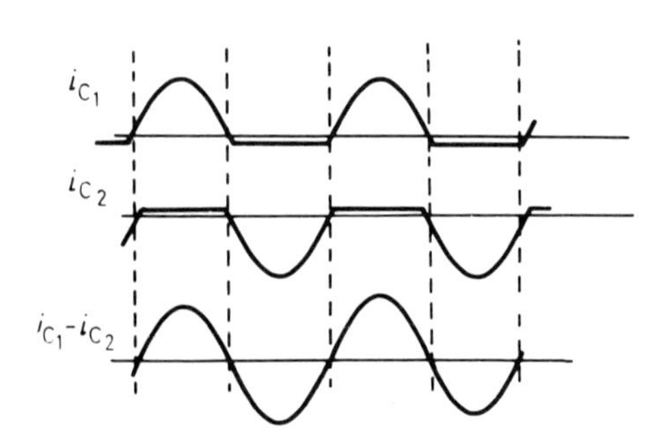

Figure 11.10

characteristic, the other transistor is already well into conduction and its collector current swamps out the distortion introduced by the first. The corresponding waveforms and their combination in the load resistance is then as shown in *Figure 11.10*. Slight forward biasing in this way comes under the classification of **Class-AB** working.

Problem 5 Give an example of how the required forward biasing for Class-AB working might be achieved in a practical circuit.

It is found that a forward bias of about 0.7 V is sufficient to eliminate crossover distortion, this voltage being roughly the same as that required to get the base-emitter diode of the transistors to conduct. The use of the voltage drop across a conducting diode therefore seems to be a way of providing a compensating bias.

A common biasing circuit is shown in *Figure 11.11*. Resistors R_1 and R_2 are selected to ensure that the silicon diodes D_1 and D_2 are always forward biased, so that there is approximately 0.7 V across each of them. The transistors are therefore provided with just enough bias to make them conduct under quiescent conditions. The resistors R_3 and R_4 which are included in the emitter leads provide some current feedback and help to improve the d.c. stability of the circuit.

The quiescent conditions are closely dependent on the forward bias on the transistor bases, and changes in temperature can affect this; this in turn leads to a

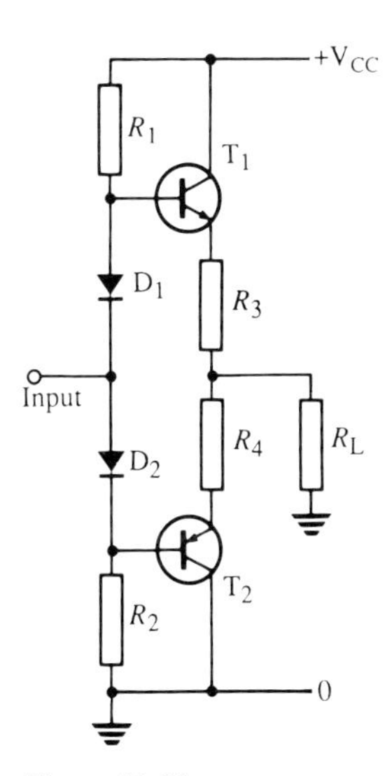

Figure 11.11

large change in collector current. The advantage of using diodes for the bias supply is that as the temperature changes, causing changes in V_{BE}, the voltage across the diodes also changes and in the same direction, so tending to keep the quiescent collector current constant. A number of integrated amplifiers such as the popular 741 use this kind of output stage.

Problem 6 How can distortion be produced in a transistor power amplifier when large signal excursions are applied?

Distortion in the form of various harmonics being added to the required signal output arises in power amplifiers because the characteristics relating current to voltage in such devices as diodes and transistors possess a marked degree of non-linearity, and any signal traversing those parts of the characteristics which depart from linearity suffers varying degrees of distortion. The problem does not arise in small signal amplifiers because the restricted portions of the characteristics over which the signal currents act are substantially linear. Signals traversing the extremes of characteristics where curvature is most evident are the cause of the problem.

In *Figure 11.1* earlier, the base input characteristic (relating base current to base voltage) and the transfer characterisic (relating base current to collector current) we considered to be linear. In a real transistor both are non-linear; the base input characteristic follows a parabolic or **square-law** relationship, typical of diodes, while the transfer characteristic follows a **cubic law** resulting in an S-shaped curve.

The typical square-law characteristic is shown in *Figure 11.12(a)*. For a sinusoidal input signal of sufficient amplitude, the output exhibits distortion in the form of either flattened or sharpened peaks. This kind of output can be shown to follow from the addition of even-order higher frequency components in which the **second harmonic** is dominant. A typical cubic characteristic is shown in *Figure 11.12(b)*. For a sinusoidal input here, the output exhibits a flattening of both peaks, although not necessarily in a symmetrical form. The resultant output is of a shape which comes from the addition of odd-order higher frequency components, of which the **third harmonic** is dominant.

Problem 7 Describe a circuit using an integrated power amplifier module which will provide an output of up to 20 W in a 4–8 Ω load.

There are a number of integrated power amplifiers which will give outputs of this sort; we will look at the TDA2030 which is a monolithic audio amplifier having a very low harmonic and crossover distortion, and which features short-circuit protection and thermal shutdown in the event of serious overheating. It is supplied in a 5-pin T0220 package which can be bolted on to a suitable heat sink and when used with a single supply rail does not need an insulating washer. The internal circuitry consists of a differential voltage amplifier stage having inverting and non-inverting input terminals, followed by further amplifying stages which

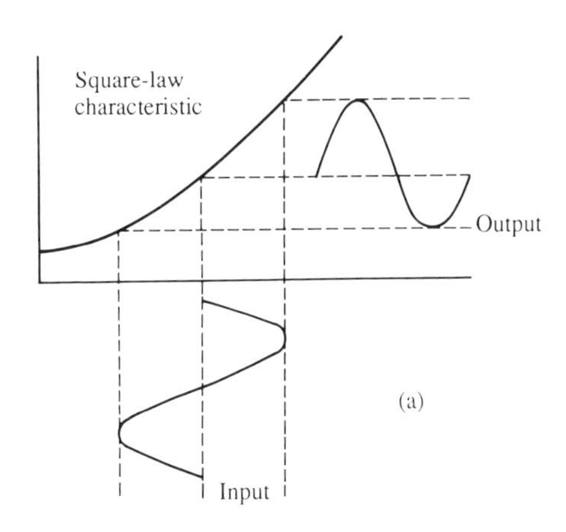

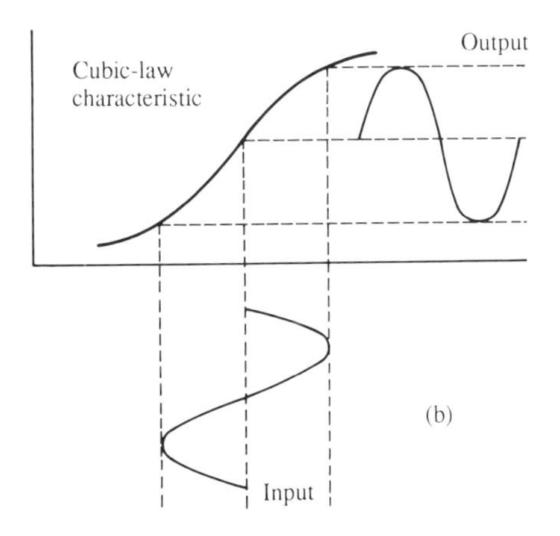

Figure 11.12

feed finally to a Class-AB output stage. *Figure 11.13* shows the form of the package and its connections.

A suitable circuit using a single power supply rail is given in *Figure 11.14*. All the external resistors may be of $\frac{1}{2}$ W rating. The input goes to the non-inverting input terminal, and feedback is made by way of R_5, R_4 and C_2 from the output to the inverting input. The input is bootstrapped, so making the input resistance of the order of several megohms. A Zobel network consisting of R_6 and C_3 is wired across the output load. The power supply should be extremely well smoothed and, if possible, regulated though this is not essential if its output resistance is small. The decoupling capacitors C_4 and C_5 are best mounted as close to the

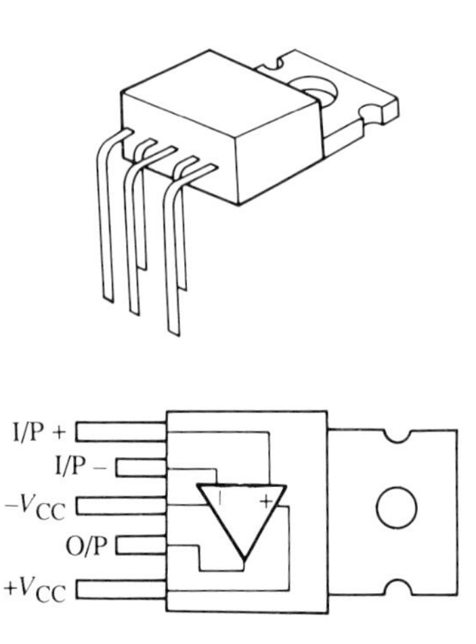

Figure 11.13

appropriate pin of the integrated circuit. For power outputs up to 10 W and an 8 Ω load, the third harmonic distortion is approximately 0.1%.

The circuit incorporates a number of protective devices which ensures that no damage results if the junction temperature becomes excessive or surges occur on the power supply or if the output load is accidentally shorted out.

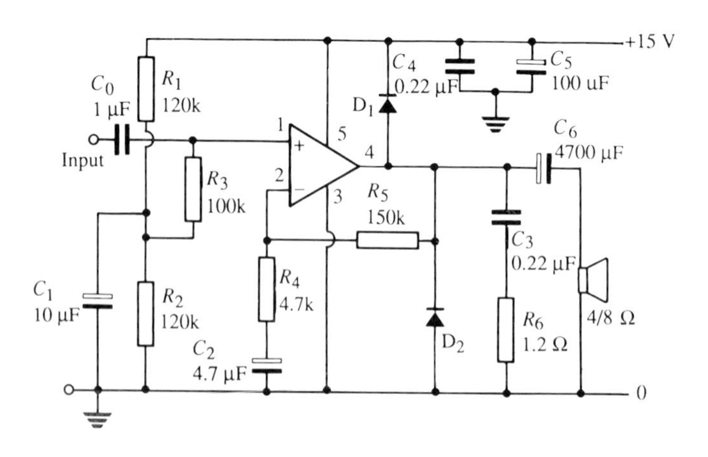

Figure 11.14

Problem 8 A certain silicon transistor having a maximum permissible junction temperature $T_{j\,max} = 180°C$ will dissipate 40 W when its case is maintained at 100°C. What is its intrinsic thermal resistance?

$$\begin{aligned}\theta_i &= (T_{jmax} - T_a)/P_c \\ &= (180 - 100)/40 = 2°C/W\end{aligned}$$

Notice that we have taken T_a as 100°C. This is clearly not the actual ambient temperature of the air surrounding the transistor, but 100°C is the *case* temperature and we are interested here in the temperature *gradient* across the transistor.

Problem 9 A piece of blackened aluminium plate has a thermal resistance $\theta_s = 1.5°C/W$. The transistor of the previous problem is bolted to this plate. What will be the maximum collector dissipation that can be tolerated if $T_a = 25°C$?

The total thermal resistance θ_T from junction to ambient is the sum of the individual thermal resistances making up the conducting path, that is, $\theta_T = \theta_i + \theta_s = 3.5°C/W$. The overall temperature drop is $180 - 25 = 155°C$ so that $P_{c\,max} = 155/3.5 = 44.3$ W. Since the dissipation has gone up from 40 W to 44.3 W, the effect of the heat sink has been to keep the case temperature below 100°C, so making available a greater useful output power.

Make a point of noting that although the thermal resistances are *added*, the heat sink effective 'parallels' the case-to-air resistance and hence reduces its value. Putting a mica washer in as insulation leads to a reduction in the permissible dissipation.

C FURTHER PROBLEMS ON POWER AMPLIFIERS

(a) SHORT ANSWER PROBLEMS (answers on page 152)

1. A power amplifier is working with an efficiency of 60%. What class is it probably working in?
2. The optimum load resistance of a certain transistor is 200 Ω. What turns ratio of transformer is needed to match the transistor to a 16 Ω loudspeaker load?
3. A power amplifier has an efficiency of 28% and delivers a power of 3 W to the output load. What d.c. power is drawn from the supply?
4. An amplifier operating at 30% efficiency is required to provide an output of 5 W. What is (a) the collector dissipation, (b) the supply power?
5. If the amplifier of the previous problem (with the same dissipation) is used in a circuit where the efficiency is 70%, what will be the new output power?
6. What are the advantages of operating in Class-AB?
7. Explain the tems: (a) optimum load, (b) collector dissipation, (c) single-ended, (d) complementary output, in connection with power amplifiers.
8. Explain why a Class-A power transistor runs hotter as the input signal is reduced in amplitude.

9 Power amplifiers supply both current and voltage to a load. True or false?
10 In Class-B operation, both P_{dc} and P_c are negligible with no signal input. True or false?
11 Class-C is not used in audio amplifiers because it is no more efficient than Class-B. True or false?
12 Why are output transformers no longer used except possibly in very low power applications?

(b) MULTI-CHOICE PROBLEMS (answers on page 152)

1 A circuit with $\eta = 40\%$ is providing a 12 W useful output power. The internal dissipation is (a) 6 W, (b) 12 W, (c) 18 W, (d) 30 W.
2 In Class-B operation the transistors are biased (a) at collector current cut-off, (b) just above current cut-off, (c) well beyond collector current cut-off, (d) none of these.
3 A Class-A amplifier is one in which collector current (a) flows all the time, (b) flows for one half-cycle of input, (c) flows for longer than one half-cycle of input.
4 A 100 W amplifier working in Class-AB will waste as heat at least (a) 10 W, (b) 25 W, (c) 50 W, (d) 78 W of the input source power.
5 The Class-AB operating point is chosen (a) to equalise the collector currents, (b) to minimise crossover distortion, (c) to reduce the collector dissipation, (d) to prevent overloading.
6 The efficiency of each of the transistors used in a Class-B amplifier is 50%. The overall efficiency of the two transistors is (a) 25%, (b) 50%, (c) 78%, (d) 100%.
7 In Class-A operation, the collector dissipation is greatest when (a) the input signal is continuous, (b) the input signal is intermittent, (c) the input signal is zero.
8 One of the transistors in a Class-AB amplifier fails. As a result (a) the output power falls to one-half with little distortion, (b) the output power falls to one-half with gross distortion, (c) all output vanishes, (d) the remaining transistor overheats and rapidly fails also.
9 Due to misadjustment, the base bias on the transistors used in a Class-B amplifier is set to bias the transistors slightly beyond collector cut-off. Because of this (a) the crossover distortion will be increased, (b) the crossover distortion will be reduced, (c) the crossover distortion will be unaffected, (d) no output will be obtained at all.
10 Unbalance between the output swings from a Class-AB amplifier indicates (a) failure of one of the transistors, (b) second harmonic distortion, (c) odd order distortion, (d) incorrect load impedance.

(c) CONVENTIONAL PROBLEMS

1 Draw a circuit diagram of a Class-A transformer coupled power amplifier using a single transistor. Find the maximum theoretical conversion efficiency of such an amplifier and explain why this value cannot be obtained in practice.
2 What are the advantages of operating an amplifier in push-pull? Draw a circuit diagram for a simple Class-AB amplifier and explain its operation.
3 Derive an expression for the maximum theoretical efficiency of a Class-B amplifier stage. Is your argument also true for a Class-AB stage?
4 A Class-AB amplifier has two transistors each rated at 10 W dissipation. If the efficiency of the amplifier is 70%, what is the maximum power output? [23.3 W]

5 A power transistor is supplied from a 15 V supply and feeds a 10 Ω loudspeaker load by way of a 2:1 step-down transformer. When a sinusoidal input is applied, the collector current varies between 0.65 A and 0.05 A without distortion. Find (a) the output power, (b) the d.c. power supplied, (c) the efficiency, (d) the collector dissipation. [(a) 1.8 W, (b) 5.25 W, (c) 34%, (d) 3.45 W]

6 What is the purpose of diodes D_1 and D_2 in the circuit diagram of *Figure 11.14*? Explain also why the input resistance of this circuit is greater than a megaohm. Why is a polyester 0.22 μF capacitor connected in parallel with the 100 μF electrolytic in the input power line?

7 A silicon transistor having T_{jmax} rated at 150°C will dissipate 20 W when its case is maintained at 100°C. What is its intrinsic thermal resistance? [2.5°C/W]

8 A heat sink made of aluminium has a thermal resistance of 2°C/W and the transistor of the previous example is bolted to it. Find the maximum collector dissipation that can be accepted if $T_a = 25°C$.

Suppose now that a mica washer having a thermal resistance $\theta_w = 1°C/W$ is interposed between the transistor and the heat sink to act as insulator. What now will the maximum collector dissipation be? Comment on your answers.

[27.8°C/W; 22.7°C/W]

Answers to problems

In general, only the answers to numerical problems are given. Descriptive answers will be found within the text.

Chapter 1 (page 6) Short answer problems: 7 40 dB; 8 10 W, 9 0.5 V 10 120 dB.
Multi-choice 1(b), 2(c), 3(a), 4(c), 5(d), 6(a).

Chapter 2 (page 16) Short answer problems: 2 (a)T, (b)T, (c)F, (d)F, (e)T, (f)F, (g) F.
Multi-choice: 1(b), 2(b), 3(c), 4(b), 5(c), 6(b), 7(a), 8(d), 9(b), 10(c).

Chapter 3 (page 27) Short answer problems: 8 (a)T, (b)F, (c)T, (d)F, (e)F, (f)T, (g)T, (h)T, (j)F.
Multi-choice: 1(a), 2(c), 3(b), 4(c), 5(c), 6(b), 7(d), 8(b), 9(a), 10(b), 11(c), 12(d), 13(b).

Chapter 4 (page 40) Short answer problems: 12 (a)F, (b)F, (c)F, (d)T.
Multi-choice: 1(a), 2(b), 3(b), or (d), 4(c), 5(b), 6(b), 7(b), 8(c), 9(b), 10(c).

Chapter 5 (page 58) Short answer problems: 1 T, 2 F, 3 T, 9–5 V, 10–33.
Multi-choice: 1(a), 2(a), 3(a), 4(c), 5(a), 6(c), 7(b), 8(c), 9(c), 10(c).

Chapter 6 (page 76) Short answer problems: 2 2.5 KΩ 4 63, 7 1.56% 9 16; 15.
Multi-choice: 1(b), 2(c), 3(b), 4(b), 5(c), 6(c), 7(d), 8(c), 9(d), 10(a), 11(b).

Chapter 7 (page 93) Short answer problems: 1. (i) power, (ii) breakover, (iii) gate.
Multi-choice: 1(c), 2(d), 3(b), 4(a), 5(c), 6(a), 7(b), 8(c), 9(a), 10(a).

Chapter 8 (page 108) Short answer problems: 1 F, 5 T, 6 T, 7 F, 8 T, 9 F.
Multi-choice: 1(c), 2(a), 3(b), 4(e), 5(a), 6(c), 7(b), 8(a), 9(b).

Chapter 9 (page 122) Short answer problems: 1 F, 5 T, 7 T, 8 T, 11 T, 12 F.
Multi-choice: 1(a), 2(a), 3(b), 4(b), 5(b), 6(b), 7(a), 8(d).

Chapter 10 (page 133) Short answer problems: 1 F, 5 T, 8 T.
Multi-choice: 1(a), 2(a), 3(a), 4(b), 5(b), 6(c), (d), 7(a), 8(b), 9(b), 19(b).

Chapter 11 (page 149) Short answer problems: Class-B or Class-AB, 2 3.54:1, 3 10, 7 W, 4 11 7 W, 16.7 W.
Multi-choice: 1(c), 2(a), 3(a), 4(b), 5(b), 6(b), 7(c), 8(b), 9(a), 10(b).

Index

THE ROAD TO EDEN

Vance Royal Olson

ISBN: 978-0-9931299-0-2

Published by:
The Blacksmith Arms Publishing House
London, England

Print design and cover layout:
Simona Meloni

email:
blacksmithpublishing@gmail.com

In loving memory of David and Florence Olson.

On their wedding Day: July 24, 1953

TABLE OF CONTENTS

Introduction

At present we are on the outside of the world, the wrong side of the door. We discern the freshness and purity of morning, but they do not make us fresh and pure. We cannot mingle with the splendours we see. But all the leaves of the New Testament are rustling with the rumour that it will not always be so. Someday, God willing, we shall get in.
(C. S. Lewis "The Weight of Glory")

Disintegration

Whatever we believe about the Genesis story, no one can deny that mankind longs for paradise. From Avalon to Asgard, Heaven to the Happy Hunting Ground, Shangri-la to Svarga Loka, Valhalla, Nirvana or Jannah; every culture and tribe throughout world history seems to have some variation of desire to exchange the ordinariness, the hurtfulness, the alienation and the sheer hard work of everyday life for the elusive paradise of our hopes and dreams.

Somewhere in the human soul, it would seem, there lurks a distant memory of paradise that once was, or should have been; but is now lost. Why else would a desire, so unfulfilled, be so universal? The Bible affirms that mankind did indeed begin life in paradise; a perfect and beautiful garden: the Garden of Eden. The tragic spectre of Adam and Eve's expulsion from that garden, heads hung in shame and despair, resonates with the sense of loss and alienation already present in our own hearts. To well do we know that road out; to poorly the road back in.

As for the Garden of Eden itself: there are two commonly held views. One: that there really was a literal, physical, historic garden – perhaps on the borders of modern day Turkey and Iran, between the Caspian Sea and Lake Urumiya, near the head waters of the Tigris and Euphrates rivers – and that Adam and Eve lived there, circa 5400 BC. (Or by Ussher's chronology: 4004BC) And two: that the whole Genesis story including Adam and Eve eating forbidden fruit and being expelled from the Garden of Eden is poetry and metaphor with profound meaning; a 'spiritual' story. Some thinkers have polarized strongly one way or the other; but the mainstream of Judeo-Christian thought has always held that both are true.

The separation of the two views is itself caused by being out of and away from Eden. Eating from the tree of the knowledge of good and evil, the last event in Eden, was an act and a choice 'out of' the God who is life and truth and 'into' the kingdom of a liar and murderer. The 'out of Eden' predicament is of living under the curse of lies, where words, meaning and reality can be separated; and the curse of death, where the spiritual departs the physical. The separation of words, knowledge and reality – the road to death and the road out of Eden – is the essence of the forbidden tree.

Understanding or even agreeing with this, however, can only be the very beginning of a solution or way back in. Good art, literature, cinema, music, poetry, culture and the beauty in nature may seemingly weld a temporary union of meaning and reality, taking us back to Eden for a moment; but in the end the vision fades, the music cadences almost out of memory and the grasp of 'life' or 'spirituality' flows out as surely as the tide.

The Two False Loves

Additionally two false solutions present themselves. One: covetousness (equivalent to idolatry) which seeks a direct connection – through ownership or worship – with the physical world we find ourselves in, but feel disconnected from or 'outside' of. (Poetically this is an attempt at 'over the hedge' access to the garden.) And two: lust, which seeks direct connection – through worship of another person – with the God, Creator and Father we no longer know, but sorely miss. (Poetically this is 'eggs in moonshine' or 'the land behind the looking-glass' because people were after all created in the image of God.) These two false solutions – I call them: 'the two false loves' – find ubiquitous repetition in the pagan fertility cults, and in everyday life and culture, from ancient times to the present day. Far from being slight corruptions of love, these false loves are in fact diametrically opposite to love.

The Four Loveless Powers

Nor do these false loves ever turn into love, but rather always flow in the direction of power; friendship with the 'world' instead of friendship with God; choosing again and again with Adam and Eve, to eat the forbidden fruit of knowledge without reality, words without truth, theology without anointing; and power without love. These worldly, loveless and serpentine aspirations throughout history have taken the form of: military power, religious power, political power and financial power. They are sometimes called 'the four winds of the earth'. (Ref: Rev 7:1) I call them: 'the four loveless powers'. John lays out the choice starkly: "*If anyone loves the world, love for the Father is not in him.*" (I John 2:15) Others have succinctly framed our choice: 'the love of power; or the power of love'.

The Cross & Resurrection of Jesus

For those who seek love and reality, the only true solution to our 'out of Eden' dilemma is found in another great literal and historic event which also has profound poetic and spiritual meaning: the death of God's own son Jesus Christ on a Roman 'tree' and his subsequent resurrection in a garden near Jerusalem, circa 33 AD. The original and perennial lie is about the character, motives and love of God; the serpent's lies in Eden and subsequently, have cast God as a self seeking despot who puts himself first. (The four loveless powers are founded in this deception.) The death of Jesus on the cross; God in Christ, the greatest of all loves laying down his own life, demonstrated the exact opposite and established the truth literally and historically as well as spiritually, metaphorically and poetically throughout earth and heaven for all time. In the death of Christ, we find sight at last for our blindness to God's goodness; relief from the lie. And there too we find death for our own covetous and lustful selfishness; relief from slavery to loveless desire, the false loves.

The resurrection of Jesus proves in every time and realm that the love of God the Father is greater than the power of sin and death. In the resurrection of Jesus, the emptiness, fear and heartbreak of death is finally vanquished; and the grace of all-conquering, unfailing love is given. The gates of death that Jesus broke through, on the way in as he died and the way out as he rose again, are the same gates that we can now also walk through into life and love; the vast world of the heavens, and in the fullness of time, the unity of heaven and earth. (Ref: Ephesians 1:10) This is where meaning and words and action and reality are always in perfect union and harmony. And that, even beyond Eden as it once was, is Eden as it was always destined to be.

The Holy Spirit

Opening our eyes to see God as he truly is and the transformation of desire, from selfishness to true love, is the road to Eden and the subject of this book; but the journey is not just a mental, theological exercise. It is life in the Holy Spirit, which, by its very nature, defies being tidily framed or communicated in plain statements alone. Knowledge without reality and theology without Holy Spirit anointing is the lie; it is dead religion; it is hypocrisy; it is the old 'out of Eden' problem – even, ironically, if that knowledge is about the death and resurrection of Jesus. (The definition of a door is a very different thing than actually opening a door and walking through.) The early disciples who literally saw Jesus physically and historically live, die and rise again, were not changed by the facts alone; they were still 'out of Eden' until the coming of the Holy Spirit with the fiery anointing at Pentecost.

The Holy Spirit, also called the Spirit of 'truth' or 'reality', came to Earth in power fifty days (i.e. Pentecost) after the Cross and brought the reality of Jesus' death and resurrection to the first disciples. The same Holy Spirit can bring the reality of Jesus death and resurrection to us now. He resists the separation of ideas and reality (lies) and the separation of the physical and spiritual (death and meaninglessness) with joyful diligence and precision. And He awakens us with His own beautiful desires so that we can live life from the heart as we were meant to; rather than trying to keep external laws and principles out of fear, determination or pride. Through the Holy Spirit the tangible love or 'desire' of God is poured into our hearts. The road to Eden is to walk in this Holy Spirit love; the anointed reality of Jesus death and resurrection.

There are three ways life in the Holy Spirit is the Road to Eden. One: in Him we have legitimate direct access to God our Father. (Ref: Ephesians 2:18) Now we can walk in the twilight of this world in friendship with him as Adam and Eve did at the beginning. Two: in the Holy Spirit we can have lust free and un-controlling connection to, and true love for, all the beautiful people around us that we previously desired wrongly – i.e. as slaves to our covetousness and lust. Now they can be our brothers and sisters in Christ; true family, bonded to us in the Holy Spirit. And three: in the Holy Spirit we can join with the longing, the elusive memory, beauty and fading glory in the physical world around us, and gain at last legitimate connection with the creation; not 'over the hedge' by covetousness and idolatry, but through friendship with the creator, designer and restored head over it all; Jesus. For in Him "... *are hid all the treasures of wisdom and knowledge.*" (Colossians 2:3) And those treasures are more than just facts and information; they are the realities that all knowledge and wisdom pertains too – that is: everything! In Jesus; heart and treasure, heaven and earth, can finally truly unite for us. (Ref: Matthew 6:21, Ephesians 1:10)

Time & the Father's Heart

In the union of heaven and earth and the treasures in Christ, there is a mystery to do with time and space that I will be teasing out as best I can. That mystery – as I experimentally understand it today – exists within the synergy, fellowship and love we share with our brothers and sisters and God; Father, Son and Holy Spirit. Ultimately God is not in time or space – still less, outside them – but all places and all times are in Him... and more even than that: all places at all times are in Him – though He doesn't hold them in a single conscious thought, as is sometimes suggested, but in the vast expanse and depths of His heart.

The crux of the mystery plays out like this: because true love, by definition, does not control responses but gives freedom, the future is not absolutely established in every detail; there are still some 'future contingencies' dependant on human (and perhaps angelic) choices. This means the future is not 'set in stone' and cannot be known in every detail; God 'holds it open' we might say – though of course many things have already been settled, established and already "...fixed by his own authority". (Acts 1:7)

The idea of future contingencies can be challenging enough but let me take it a step further. Those friends who walk the twilight of Eden with Father, Son and Holy Spirit also begin to know that the past holds contingencies as well. This means that even some things in the past (far fewer than in the future I think) are held open by

God and are therefore unknown, intrinsically unknowable; and contingent upon choices and decisions unmade, and prayers not yet prayed.

The death and resurrection of Jesus – the Cross – remains the 'evergreen' centre in the heart of God upon which all these contingencies, future and past, are being resolved. Much of history future, and some history past, remains a living dynamic field that God our Father opens up to his sons and daughters; those walking in the death and resurrection of Jesus in the Holy Spirit. For this reason we should preach the Cross in the present, preach the Cross into the future and – peradventure we are given access in the Holy Spirit – preach the Cross into the past as well. The desire and mystery of this seems to me like the deep pool, the over flowing fountain, and the tree of life itself at the very heart of Eden.

INTEGRATION

Here, logic ever yields to love, and words can do no more than bring us to the beach beside the great sea; the edge of those eternal waters upon which all that we are, and know, and ever could know has been founded. Beyond that we must each weigh our own anchor and set our own sails; for the only way to truly know God's love is to live God's love in the Holy Spirit. Indeed the last word of the 'Word' himself (i.e. Jesus) brought Peter to that beach and launched him onto a journey that offered the eventual attainment of the deep true Godly love (Greek: 'agape') he desired. (Ref: John 1:1, 21:1-19)

In what follows, I seek to merge poetic meaning with literal history: my own story, journey and testimony woven with understanding and experience of God. Thankfully, marriage – the only pre 'out of Eden' institution we still have – and the God ordered family of dad and mum raising children, remains as a tangible and visible starting point in our journey. Opening eyes; seeing God our Father as He truly is – His very face – began for me with the sight of my own earthy father's face. And transforming desire – true love – in the Holy Spirit began for me in the atmosphere of my mother's worship of God.

CHAPTER 1
OPENING EYES

Do you think I am trying to weave a spell? Perhaps I am; but remember your fairy tales. Spells are used for breaking enchantments as well as for inducing them. And you and I have need of the strongest spell that can be found to wake us from the evil enchantment of worldliness which has been laid upon us...
(C. S. Lewis "The Weight of Glory")

FIRST OPENINGS

Five times I have seen a human child open its eyes for the first time. The third child looked at his mother and me with a knowing little smile that seemed to say: "So that's what you look like!" The fifth child, with wide eyes, surveyed each of her four brothers in turn as if at last to connect a face to the known voice – it had become a game to say to the maternal bump: "Hello little baby in there!" Now each of the brothers gave their welcome and baby answered each of them, one after the other, in her own way, with big acknowledging baby eyes.

From the moment of first light each of us humans

grows into the consciousness that we have entered a scene, not at the beginning, nor at the end, but at some indeterminate point in the middle. We are given no explanation and are hardly likely to understand one if we were. If we are fortunate, our early hours, days and years will contain answers for our hearts that our minds will not soon, if ever, grasp. Namely: the distinctive beauties of masculine and feminine love, shining in the faces and flowing through the hands of the man and woman who hold us. If our minds do grasp answers, it is logical to expect those answers to fit with what our hearts already know. The creator behind it all must be something like mum and dad – the creator's love must be both masculine and feminine – or why else would life be ordered as it is?

But of course not every individual life was ordered in that way. Some lives began without love. For some babies there was neither face nor hands of love – masculine or feminine. The deficit of that love compounds our 'out of Eden' predicament by reinforcing the lie that God doesn't love us. To begin to unravel that lie and to find the road back into Eden, we will go back to the day the human race got itself outside in the first place.

The Temptation and Fall of Adam and Eve

When Eve stood by the forbidden tree, listening to the serpent one day in Eden, she was shown something very similar to what Jesus was shown many years later in his wilderness temptation. This follows from the truth that Jesus was 'the second Adam' and that he faced and overcame everything the first Adam failed in. What Eve and Jesus saw was a vision of "*...the kingdoms of this world and the glory of them...*" (Matt4:8) And Luke adds: "*...in a moment of time.*" (Luke 4:5)

Perhaps the fantasy and spell the serpent wove went something like this. Eve saw a future scene: the glory of the world's kingdoms in a moment of time; a glimpse into a possible future. She saw all her children and children's children before her with their great accomplishments in culture, the arts, exploration, learning, and the conquest and development of hitherto untamed nature. Great cities filled the earth and in each: palaces and households of great beauty. Tables were laden with plentiful feasts. The charming, the beautiful, the dignified, and the wise mingled, chatted and danced. Then she saw herself honoured above all: Eve the mother of all the living; eternally young and beautiful; always the first and most honoured... or was it... worshipped? Oh yes, and the pleasures in her own body, the sexual delights of endless variation were beyond her wildest dreams. But what was it just to the side of her view? Who was building all these cities and palaces? Who was bringing all that delicious food and drink to the great feasts? Who was clearing up after it all? Who was taking care of the babies and children? And what were those empty blank stares that she sometimes almost saw?

But she couldn't quite bring these thoughts into focus. Already the intoxication of being worshipped was dulling her thoughts and melting away the concerns. Always just out of view in her vision, too, was that serpent and his underlying desire: "I *will make myself like the most high.*" (Isaiah 14:14) And so she bought into it. She chose it. She ate the fruit. Then: "Hey Adam, try this – you'll love it."

Seeing her ecstasy, he takes the fruit in his hand, enters her fantasy, and adds some of his own; the great conqueror – throwing himself down from a high temple, vindicated by supernatural intervention... Is this a man or a god? He was the great priest, the guru of wisdom sought by many; they delighted in every word he spoke. What devoted worshippers these are – temples in every town and city – and yet they almost look... well drugged or controlled... but no, surely not. Ah... more food! Where is all this delicious food coming from? Why the very stones are bending to his will to become not just bread but every delicious dish, spiced to perfection; the glorious foods of the nations. The people bow to his will as well; he is the great father; the great king over all. Thousands serve him and obey his every word. And he is one with the great kings of the nations; not just one wife but hundreds of beautiful women are indulging his every appetite and whim. Palaces filled with wives and concubines... And so he takes the bite: "I choose this!"

Eyes Opened to Knowledge

Then their eyes, which had been half opened to see the glory and beauty; the knowledge of good (so called); were now fully opened to the knowledge of evil as well. At this point they saw to the left and right of the fantasy. They saw the broken bleeding backs of the slaves building the cities and palaces. They saw the hungry children of those who served at their feasts. They saw weariness and strife and heartache. They saw young men suffering and dying on the battlefields of the wars required to maintain the wealth and beauty that seemed only a moment earlier to flow so freely. For every extra lover, a broken sweet heart wept somewhere else and beyond that cruelty, domination and obsession – the blank stares of loveless lust and the worship of genitalia. The little selfish whims that had seemed so cute a moment ago were garishly ugly in this new light.

Even the stones weren't turning into bread by magic. In truth they were being shaped by craftsmen into millstones. By the labour of another they were grinding grain that yet another was growing and harvesting in dust and sweat. Someone was baking that flour into bread and someone was carrying it to their tables. Who were all these servants, slaves and 'lovers' on the sidelines? These were their own children and children's children of course. And in that moment they knew it; their own precious children – still just a promise – sacrificed to lusts and vanities and covetousness.

The foul stench of serpent breath swept away every pleasant fragrance; they tasted the poison in the worm eaten fruit; their open eyes were starring into a horrid serpent face; fangs still dripping; eyes dark, despising and condemning. They were ashamed of their beauty, their glory and especially their own bodies. They ran to cover themselves and hide. Such is the kingdom of darkness; that kingdom headed up by the ancient serpent, the devil. And so also has become the world... apart from the goodness, God manages still, to get into it.

Judgement of Mercy

When God came into the Garden of Eden that evening, Adam and Eve were lurking in the twilight among the trees; ashamed to be seen by anyone – especially him. Contrary to one popular teaching, it is not God who cannot look upon sinners; but sinners who dare not look upon God, and who are mortally ashamed in the light of his loving gaze.

God on the other hand insists on coming into their presence – yes with judgement, but with such judgement, so full of mercy, that it could scarcely be called judgement in the usual sense. In their fantasy Adam and Eve had perceived themselves as the great king and queen that others – their own children of course – bowed down to, served and indulged. This is the essential heart of the serpent's kingdom: get to the top, stay at the top, and have others serve you. This is what the serpent wanted and this was the scenario that Adam and Eve chose for themselves. At the heart of God's kingdom of love, on the other hand, the greatest king is the greatest servant; the greater always serves the lesser.

Adam's judgement; having to work and provide food by sweat and toil was the precise antidote to the deluded fantasy of having his children serve him. They would not serve at his table; he would serve them. Eve's judgement is parallel; the children were not the servants of her vanity and pleasure, but she would put herself through pain to bring them forth. Such is the eternal order of God's kingdom and the true foundation of masculine and feminine love at its most basic level.

Jesus embodies both these loves as the suffering servant; ultimately in his suffering on the cross and his resurrection, where all judgement ends and rebirth is given. For Adam and Eve, though they incurred judgement, the overflowing mercy is that redemption began the very same day with the promise of a 'seed' or 'descendant'; Jesus who would crush the serpent's head. (Ref: Gen 3:15)

The physical expulsion from the garden was a precise and true reflection of where Adam and Eve had arrived spiritually. Having chosen to believe the serpent's lie and disobey God they had crossed into the realm of lies and death – and the first step to getting out of that realm is to know that you are in it. Put another way: the worst thing about deception is being unaware of it and the first step into truth, or the first truth you need to know, is that you are in fact deceived. Coming out of Eden into the world where lies and death literally happen was for the human race, the first step on a path of truth or reality that would one day lead back in.

So Adam and Eve were sent into a dusty, resistant,

boring, meaningless and dying world; but with the silver lining of a promised child who would somehow get them back into life as it had been designed to be. And even in the 'meaningless world' little glimpses, memories and hints of what once was and would be again were left as a witness and encouragement by God. These lingering hints of Eden are all around us and none are more poignant than the birth of children; ever the reminder to Adam and Eve and their descendants of the promised child to come. Even in our times – now 2000 years since that child came, conquered lies and death, and reopened the ancient gates to Eden – each birth speaks of God's certain faithfulness to the human race.

Visible Love: the Enduring Relevance of Family

We have come back now to our starting point; to the babes in the arms of a man and woman. It is to this beauty that God desires every human eye to first open – before the fall to sin, yes, but even more so now in the fallen world. His judgement on the human race, seen here in Adam and Eve was for the purpose of preserving that witness; of keeping that channel open. And it was not just so the children opened their eyes to love but also for the sake of the parents attaining their destiny of being in God's own image. In later chapters we will consider in greater depth how this love in the face of human parents is intrinsic to us seeing the face of God the Father himself.

For when all the philosophical, aesthetic and theological paths have been followed up, we return to the starting point; we discover that behind it all is that face of purest love and beneath it all are those arms of unfailing love; that person 'God' of whom man and woman together are a visible image.

"*God is love.*" the apostle John states (I John 4:8), putting into simple words, what I would call the ultimate point; the distant star and bedrock of all reality. That which we knew in our hearts as a babe in arms becomes the zenith of all human intellectual and artistic attainment. And those disciplines, if honestly pursued, become a quest to refract the white light of love into the beauty and glory of its seven rainbow colours; from fiery reds to deep brooding indigos, and to every shade and hue between. Joining that quest, these words are my offering of worship to the One – God: Father, Son and Holy Spirit – who, in my most clear sighted moments, I perceive to be more worthy of honour, love and worship than we humans will ever fully know.

CHAPTER 2

IMMORTAL LONGINGS ACROSS THE GENERATIONS

Apparently, then, our lifelong nostalgia, our longing to be reunited with something in the universe from which we now feel cut off, to be on the inside of some door which we have always seen from the outside, is no mere neurotic fancy, but the truest index of our real situation. And to be at last summoned inside would be both glory and honour beyond all our merits and also the healing of that old ache.
(C. S. Lewis "The Weight of Glory")

SKUDESNES LUTHERAN CHURCH

Lars Knutson paused from his preaching and looked up from his notes. A few moments passed and other faces also looked up; a faint glimmer of hope in their eyes. But, instead of seeing the preacher close his books, they saw him extend a shaky hand forward to where a glass of water stood on the pulpit. A thoughtful usher, who had filled it only half full with foresight of the perilous shaking it would soon endure, now looked on with quiet satisfaction of his well warranted precautions. All other breathing stopped. The water sloshed up the sides of the glass – a drop or two escaped before the elderly reverend managed to steady it against his lips. He then took a drink so tiny that one could wonder if it had all been worth it. Indeed the glass appeared no less full than it had been, even with the drops that had escaped as he had picked it up and the several more that had flown as he set it down. The preacher's throat was still dry and raspy and he made a half hearted attempt to clear it – to no effect. The congregation, not having glasses of water to hand, summoned what saliva they had, and swallowed hard.

I was desperate for a drink. The sermon went on. A little girl in front of me, standing on the pew facing back, offered me a salty cracker; but at a sharp glance from mum, I withdrew my hand. In any case my eye was on her bottle of apple juice; but it was never offered. I looked back up at the preacher and longed for that water that was just sitting there – or even for him to take a good big drink. This event, when I was about 4 years old, is my earliest memory of 'desire' in church. I would now interpret this as a 'desire of the Spirit'; God's desire that his ministers should drink much more deeply of both heavenly and earthly life. The kind of sparse, moderate, unenthused approach to life that passes for holiness in some religious contexts is as great a weariness to God as it is to the poor people sitting within its dreary precincts.

One day, Pastor Knutson came out to our home on his annual visit. "The preacher's here!" set the whole family in a stir, and as he was driving slowly up the driveway and into the farmyard, mum sent dad out to stall him while she tidied the house. Around the meal table, as we kids were wolfing down our food, he launched into a monologue about how unhealthy it was not to chew your food properly and how especially bad it was to wash it down with water. Apparently doctors concurred with this. And, a distant relative of his, on his mother's side, had died young of stomach ailments with suspicion pointing to his habit of drinking water while eating. I remember washing a bit of food down as Pastor Knutson was saying this, and wondering how bad it could really be. I could also see by the slightly pained but polite smile on my father's face that he wasn't entirely accepting Pastor Knutson's teaching on this point. How else should a boy judge truth, but in the light of his father's face?

The Deep Well in the Pump House

After church we often went to grandpa and grandma's farm for dinner (as we called the midday meal). First stop for me and my three older brothers was the pump house where ice cold, fresh water, straight from a deep well, awaited us. The small wooden pump house was built around a galvanized iron tower that had, in earlier days, supported a wind mill wheel of the classic western style. A rickety old door opened inward to reveal a rustic interior that could have been a scene from the old west. Ancient tools, an old wheel with wooden spokes, coils of hay wire and various machine parts lined the walls and hung from the ceiling. A scythe with its great curved blade and wooden handle hung conspicuously, along with a couple of sickles, above a cluttered work bench. An old barrel, its top covered with old bolts and rusty nails, stood on the floor. Fencing tools stood in one corner; crowbars, spades, hand augers and pincer shovels. Scrap iron and pieces of wood leaned between the open wall studs. Binder canvasses (conveyor belts made of canvas with wooden slats) filled high shelves on one side – probably never used since a harvest at least 30 years earlier – but saved never-the-less against some unpredictable future day. My grandfather, a survivor of the great depression (which was very harsh in that part of the world), did not lightly throw things away. The cast iron pump, originally designed for a handle, had been modified; first for wind power and now for an electric motor and pump jack – homemade from an old stationary engine. We switched it on.

On a nail above the pump hung a small hemispherical dipper. Grandpa had hammered it out on his anvil using aluminium from a small piece of a WWII training plane wing that had found its way into his workshop scrap pile. My oldest brother got it first and filled it directly from the pump spout. It was very thin and dented and had a lot of holes that made it drip everywhere as you drank but that didn't matter on the dirt floor. Eventually I got my turn. The water was delicious, cold and satisfying, and the holes in the dipper – looking from the inside as you drank – appeared as little sparkles of light, like stars shining through.

I remember a moment of confusion around that time when standing under a starry sky and my older brothers were talking about the constellations – particularly the big and little dippers. We had always called the dipper in the pump house 'the little dipper' and the larger one beside the drinking water bucket in the kitchen 'the big dipper'. For a moment I thought my brothers were talking about the stars that shone through the bottom of the little dipper... but the big dipper was a solid thing with no leaks... eventually I figured out that they were talking about star shapes in the sky. Even so, there were deep wells of a heavenly origin in the heart of the family – in the hearts of the fathers – that a little boy could see.

THE MEAL

Back in the house grandma was getting on with the cooking, mum was helping, dad was napping in a chair, and grandpa was bustling over his various papers and church magazines. The conversation began between grandpa (dad's father) and my mum. Theology came first. Perhaps we would have to leave the Lutherans and join the Alliance Church if the German influence toward drinking and pastors wearing robes prevailed over our Norwegian stance. Of course we were right and everyone knows that the wine Jesus drank (and made) was a lovely fruit cordial without alcohol – so much more refreshing! And wasn't it the Pharisees who wore robes?

Then there was the charismatic thing – surely it was just another passing fad like rock and roll music and long hair – even the Pentecostals didn't like it – perhaps they weren't so far off in their doctrine after all. Uncle Maynard and Auntie Laura had finally been forced to leave their charismatic experiment behind after a recent meeting had dissolved into a shouting match and acrimonious name calling. Auntie Laura had actually been called a... what? In regards to a conversation of this type, it is a well know fact that a child's sensitive ears ignore the louder words but zero in on anything whispered. I was no exception. "What's a bitch?" I asked innocently enough... obviously not a child friendly concept, I concluded, by the glare I received.

The conversation moved over to the 'jolly old cat'lics', as my grandfather called the Catholics in his Norwegian accent and a twinkle of humour. Humour or not they were roasted along with the World Council of Churches; the wretched purveyors of 'Churchianity' – one of my grandfather's words which he said with the same look of disdain and disgust on his face as you might expect of a man holding a dead rotting fish, at arm's length, by the thinnest part of its tail.

On this occasion the World Council of Churches had made a statement that sounded a bit soft on communism, which, to my grandfather's ears, could have been anything short of consigning them to a slow and painful death followed by eternal flames. To be fair, though, he did say: "Love the sinner and hate the sin" – quoting Richard Wurmbrand, the Romanian pastor who had suffered years of torture in a communist prison because of his Christian faith. And, had any actual communists showed up at my grandfather's door cold and hungry, I have no doubt they would have been treated most graciously.

Perhaps the KGB had infiltrated the Vatican or the World Council of Churches... probably both. At this my father roused himself and mum diverted her attention to the dinner preparations. Would the Americans defeat the Russians in an all out war if it came to that? What about the missiles? I suppose they would try to shoot them down over us here in Saskatchewan – if the Russians sent them over the North Pole...

Now, grandma entered the conversation for the first time: "I suppose the Lord would look after us as he always has," she said in a scolding tone; not being one to cower under speculative fears or elaborate conspiracy theories. In truth I suspect she doubted any communist was quite as clever as her husband's numerous theories would seem to imply.

"Yah, yah, mama," said grandpa looking over his glasses with twinkling eyes. It was time to eat. A final flurry of dishes and chairs and we were all seated waiting for grandpa to pray.

He began his prayer with some scriptures and thanks and a general survey of the gospel. Then he took up some of the issues of the preceding conversation; the suffering Christians under communist oppression; the church; various relatives, family and friends – in some cases he was unwinding himself from positions he had taken earlier but which now, in the place of prayer, seemed harsh and grating. And, as in all of his longer prayers, he would have said in his normal (in prayer) mix of King James English with a Norwegian accent: "And Father we do pray for the children and the children's children even unto the third and the fourth generation." His father had charged him to pray this prayer for the children as he in turn had been charged by his father. For how many generations this charge had been carried, was not known.

Many years later, near the end of Grandpa's life, when his speech was slurred from a stroke, grandma asked him, as in old times, to pray before we ate. The prayer was short, the voice unfamiliar; but then he prayed that prayer for the children and for a brief moment – just for that one line – his normal voice returned: "And Father we do pray for the children and the children's children even unto the third and the fourth generation." That was the last time I saw him.

Greenwich Time Breach

A few years after that, when I was church planting in London, England (Greenwich area), I had what I consider my first experience of what I'll call a 'time breach.' (In modern sci-fi this would be called a 'wormhole' – a direct link between two separate points in space-time.) It seemed at first like a vision. I was seeing an old man, humbly dressed, stepping out of a small rustic wood shed on a stony hillside. I knew it was Norway and I thought it looked like the early 1800's. The shed seemed like his place for prayer. He had joy on his face – a big toothless smile – as if he was seeing, in the Spirit, the answer to his prayer. Then for a moment our eyes met and I knew he was seeing me, just as I was seeing him, and we both knew that we were seeing each other. Then the opening closed as smoothly as it had opened.

I've always thought that this old man was one of my fathers and the source of the fatherly charge to pray forward down the generations. Perhaps, someday, something will turn up in the history books to verify my thoughts. My brother has researched the family tree back many centuries and I've taken a guess at who I think this man was – it's a good guess though because not many lived to his age which appeared to be around 80 years. Also the time period is interesting because it was a time of spiritual revival in Norway, when such phenomena are most likely to happen.

The revival had begun in 1797 when a 24 year old farmer named Hans Nielsen Hauge had an encounter with God while ploughing in his father's field. He spent the next 8 years travelling up and down the country teaching and preaching, usually in homes, until in 1804 he was imprisoned in Oslo and held without charge for ten years on the vague accusation of preaching without a licence. Jealous bishops in the Lutheran Church were behind this of course and the laws of the day were easily abused by those with political influence. Never-the-less the revivals that he initiated, the leaders and preachers he trained, and the business enterprises that he started and encouraged are seen by many as a turning point for Norway in that era. I had heard of this man because he was still a legend in the lore of our family and my grandfather often referred to him as an example of genuine Christianity.

My ancestor – the one I think most likely from the vision – was a man named Kristoffer Larsen Syre, who lived from 1758 to 1842 on the island of Karmoy in Norway. This man was 84 years old when he died which fits with what I saw and he would have lived through the whole period of the Haugian revivals. Being four generations back from my grandfather, his own prayers would have included my grandfather. ("Father we do pray for the children and the children's children even unto the third and the fourth generation.")

The heart behind this stands in stark contrast to fathers like King Hezekiah, for example, who when Isaiah told him that there would be peace in his own time but defeat by an enemy later and even that his own sons would be slaves to the king of Babylon, comforted himself with the thought: "*Why not, if there will be peace and security in my days.*" (2Kings 20:19) This selfish choice and attitude of Hezekiah is in keeping with that old choice made by Adam and Eve as they chose the serpent's kingdom that terrible day in Eden. Little wonder then that Hezekiah's son Manasseh was one of the worst kings Israel ever had. I thank God for fathers who walked against that grain, generation after generation by the grace and help of Jesus and the Holy Spirit. A selfish man or women doesn't 'care very far' we might say but true hearts hold a burden of love and care to eternity; not walking in the way of the serpent but in the kingdom of our eternal heavenly Father. (In the epilogue of this book I indulge in a little speculation and fiction about Kristoffer's final days... and beyond.)

THE MEAL CONTINUED...

Back at the old farm house, the food was cooled to a child friendly temperature by the time grandpa had finished his prayer. Everything was home grown; the roast chicken had probably seen the chopping block the previous day, boiled potatoes, creamed vegetables, and homemade pickles – even the dill was grown in the garden. The bread was home baked and the butter had been churned in the hand churn that I remember cranking until my arms ached. The cream and milk were from the cows on the farm. I won't name all the delicious berries we had with fresh cream for dessert – some were local wild berries – because then I would have to describe them, and how can you describe a totally new taste?

THE BLACKSMITH SHOP

We didn't work on Sundays, other than the necessary care of animals and cooking, so the adults retired to the front room after dinner for sleepy half conversations, dozy reading or straight naps. The boys – me and my brothers – went out to play in the farm yard. Besides the buildings already mentioned there was the barn, the pig house, the oil-shed, about 8 granaries and 'the shop' – a blacksmith shop with a few modern additions like an arc welder and some power tools. The original blacksmith equipment; still intact, though no longer used, consisted of a forge, anvil, grinding wheel, post drill and trip-hammer – a heavy mechanical hammer that my grandfather had made after his right arm had reached its limit of hammering ploughshares. A line shaft, powered by a gasoline stationary engine, ran across the back of the shop and from this shaft, flat drive belts drove the forge blower, the trip hammer, a grinder, and the post drill. The forge still had coal in it and everything was ready for action though in fact most of it had hardly been used since electricity had become available just after the war (WWII). Hand tools: tongs, hammers, chisels, punches, wrenches, files and numerous other tools were piled on the benches and secreted throughout the shop. Scrap iron, old machine parts, jars of bolts, small boxes and various mechanical devices filled every crevice and cranny.

Within the lore of knights and quests, and many spiritual or fairy tales generally, the blacksmith shop has a certain mystical credence – as if the roar of bellowed fire and the hammering of glowing iron, followed its own crescendo and cadence to the very edge of music. Even as the shop lay in the silence of Sunday afternoon, the memory and echo of iron and fire and hammer lingered in shadowy corners and hung expectantly in the air. I felt it as a child at play before I had ever encountered any of the stories – not powerful and dramatic; just an elusive sense of significance that faded in direct proportion to the attention it was given. In the years since I have come to recognize this kind of thing as the playful mystique of God; a desire of the Holy Spirit. But, giving it a 'name' is a long ways from understanding it. In many unlikely places, the heart of a child will touch the presence of God in this way, but not be able to articulate it. It is a type of 'knowledge' or 'wisdom' at home only in the heart and mind bowed in unconscious worship of God the Father. The moment we lose the mystery of it, or try to make it a settled static truth, it slips away to find rest in a child's heart... until men and women have grown young enough once more, to receive it.

THE HARVESTERS

A large section of the farmyard was given over to several rows of harvesting machines; no longer used, but kept for some future possible day. There was the old threshing machine and two binders, for cutting grain and binding it into sheaves. The canvases for these binders were stored in the pump house as I mentioned earlier. These would have dated from the turn of the century (1900) and had been used into the thirties during the depression years. A 1928

Massey Harris model 9B combine which had replaced the old threshing crew system of harvest and made the threshing machine and binders redundant, stood in the centre of the collection. There was also several other old combines, an old Ford grain truck with a Chevrolet cab on it and an even older grain wagon box on the back, a few tillage machines and a 1926 Buick car in an advance state of dilapidation. From the scythe and sickles in the pump house, to the latest John Deere Model 55 combine still in use, the whole farmyard could be viewed as a history of grain harvesting. My grandfather even kept all the accessories and seemed to have the idea that these machines could still be used one day. I remember a day when my father wanted to take the canvas for the 9B combine to adapt it for use on a newer machine. This canvas was stored in the oil shed – a more weather proof and secure building than the pump house – because it was deemed more valuable than the binder canvasses. Grandpa gave in reluctantly to this request because he still wished to keep the machine runable, though it hadn't been used in 30 years. In fact, that particular machine was one of my favourites as well. It had iron platforms and iron stairs to access the various areas – a bit like a Victorian industrial plant. The engine could still be turned over with the crank and I longed to start it and run it. I begged both dad and grandpa to get it started but they always declined; saying the fuel tank would need cleaning and the carburettors would be gummed up and all manner of other reasons. Of course if they had started it; I would have wanted to see the whole machine run –

and then working in an actual harvest. It never happened.

My longing to see this machine run and my grandfather's reluctance to let the possibility completely slip away, and indeed the whole collection of vintage harvesting machinery, seem to me now to represent a longing for ancient harvests, the great harvest at the close of the age and even harvest universal. Grandpa, with his long heritage in revival and renewal, longed for the old days, not, I think, from nostalgia or sentimentality, but rather he desired that cutting edge of spiritual reality – the combine sickle slicing into a thick stand of ripened grain. And he longed for it in the future, not just the past. The actual yearly physical harvests and the spiritual harvests he had experienced, though highlights, were still only tokens of that for which he longed.

Abraham's Time Breach

Like Abraham, Grandpa was a pilgrim in search of a city whose builder and maker is God and like Abraham he only saw it and greeted it from afar as a stranger and exile on the earth. His farm was a record of his heart – perhaps a little cluttered but at the centre, pure faith. This kind of desire, however it may be manifest, is desire born of the Spirit, and will bear fruit, even if those prophets who walk in it never see that fruit in their lifetime. Indeed they may not even think it is anything other than the nostalgia of which they are often accused.

Did Abraham know, as he walked that long and dusty road, that he shared in the travailing desire of God for the Son of Promise to be born? That same son or 'seed', who would bruise the serpent's head, that God had promised to Adam and Eve all those years earlier in Eden? (Ref: Gen3:15) I believe Abraham did know that he shared and participated in that 'desiring' of God (desire of the Spirit) but not until his experience at Mt Moriah when God stopped him from offering his son Isaac as a burnt offering. The concluding remarks about this experience are found in Gen 22:14. "*So Abraham called the name of the place The Lord will provide (or 'see'); as it is said to this day, 'On the mount of the Lord it shall be provided' (or he will be seen)*" I've put in brackets the alternate reading which appears in the margin or footnotes of most translations of the bible. It seems that there is a word play in the original language connecting the ideas of provision and seeing. In keeping with normal prophetic understanding I would interpret this by saying there were two levels of seeing or provision going on in Abraham's experience. At one level there was a literal ram caught in a thicket which Abraham could offer instead of Isaac; the Lord's provision... by 'chance' so to speak. At a deeper level Abraham saw the provision of a saviour to come – the long awaited son of promise – the big picture salvation which Abraham and Sarah had participated in prophetically and by faith as they brought forth their own son Isaac.

Was Abraham's 'seeing' just a mental realization – a sudden grasping with the mind of the meaning of what had been going on for the last few decades? Well, yes, there was that but I think it goes deeper. Jesus said in John 8:56 "*Your father Abraham rejoiced that he was to see my day; he saw it and was glad.*" It is instructive to read on: "*The Jews then said to him, 'You are not yet fifty years old, and have you seen Abraham? (Alternate reading: "has Abraham seen you?") Jesus said to them, 'Truly, truly, I say to you, before Abraham was, I am'.*" (John 8:57-58)

Did Abraham experience a 'time breach' that day on Mt Moriah? Had Abraham seen into the future all the way to the promised son and had that son, Jesus, seen him as well? Had Jesus and Abraham seen each other, face to face – and both known it? It would explain why Abraham called the place: "He will be seen." I personally believe that is what happened – though I wouldn't necessarily dogmatically defend it. Jesus must have already understood this from the scripture or experienced his side of it, in the first century AD, by the time he was having this conversation with the Jews. The knowledge he would have had of it as God from heaven had been laid aside at his incarnation when he emptied himself to become a human child. (Ref: Phil2:6-8) One thing I'm certainly not saying that my experience of seeing my great, great... grandfather was as significant as Jesus' and Abraham's experience but I am at least saying that such experiences have this biblical parallel and precedence.

Incidentally, this story of Abraham and Isaac on Mt Moriah is the one and only story I can remember from Pastor Lars Knutson's preaching – he too, must have had some kind of miraculous 'breach' from his normal dryness that day to capture my attention. And I remember questioning dad about it on the way home from church, but he only reiterated that it was a very interesting story and then fell into deep and distant thought about it himself.

Bossy Cows and Pussy Cats

Late in the afternoon, when we saw grandpa walking to the barn with the milk pails we would run to join him from wherever we happened to be playing; it was time to milk the cows – by hand. Three or more cats lived in the barn with the cows and got the first share of the milk in some old sardine tins that served as bowls. Several of the cats became quite skilled at catching a jet of milk straight from the cow when grandpa decided to amuse us children. As the jet was directed higher and higher the cat would stand on its back feet and stretch as far as it could to keep its mouth in line – eventually falling over with its face covered in milk.

The cows, which had descended from my great grandfather's herd, had names like: Fairy, Blue Bell, Brindle and Half-ears (ears half frozen off because her mother had managed to slip away and hide in a grove of trees on the winter night she was born) to name a few – there were about 20 in all. They came to my grandfather when he called them by name, individually, but when he wanted them all to come, he used his own generic cow name,

'Boss'. The call of, "come boss," would bring them all in for milking from a quarter of a mile away. And at the words, "take your place bossy" they would each go to their own stall in the barn next to the pump house. Disobedient and defiant calves were labelled, "sinner calf" and were liable to be the first to market. Several years after grandpa retired and the herd was at our farm the cows would still come running over to him when he came to visit and when we took them to his farm for summer pasture the older ones still remembered their place in the old barn.

Fairy, Blue Bell's mother, had given birth to her after sneaking away one snowy night but we all searched and found them in time to save the little ears from frost bite. They were both rare blue roan throw-backs and became favourites – almost pets. (Great grandfather's herd had had some blue roans several generations earlier.) My grandfather let Fairy grow old and near death before taking her to market because he couldn't bear to sell her earlier; and my father, who inherited Blue Bell, actually let her die of old age.

"Do cows go to heaven grandpa", I asked one day. He mumbled something I didn't understand, but, to the prying stare of a little boy, his face seemed to say yes and his eyes were soft and distant. For long years, he and his fathers, in the Pietistic tradition, had fought for the necessity of conversion as a tangible experience and I suspect it was hard for his mind to accommodate the idea of even a cow walking into heaven on any other basis. Apparently, no theology the mind can frame will ever exactly fit the heart. I never asked Grandma the same question – but thinking of it now – if I had she would have probably said: "yes"... or if she thought to long: "I don't know." The tension of those two responses and other little tensions I've touched between grandpa and grandma are like snapshots of masculine and feminine beauty; tiny glimpses of God and – dare I say it – the tensions of heart and mind within God Himself.

Of Craftsmen and Artisans and the Rise of Mammon

The fresh milk was brought in and taken down into the basement for separating into skim milk and cream with the hand cranked separator. The basement was as full as every other place on the farm. In the back corner there was a potato bin large enough to hold potatoes to last through the year to the next harvest. Along the wall on the right were shelves lined with jars of preserved fruit, vegetables, pickles and even chicken. On the opposite wall stood a wood lathe and a harness maker or shoemaker's sewing machine. The wood lathe was my grandfather's second; an older lathe, homemade and foot powered – by means of a recycled sewing machine treadle – was stored in my dad's workshop. Grandpa had made this first wood lathe and then made a spinning wheel on it with only a small picture from a post card as a pattern.

The crafts of the farmer, the blacksmith, the shoemaker, the harness maker, the woodworker and the lay preacher had all been plied by my grandfather though when I knew him he was only farming and preaching a couple of times a year. He had witnessed the decline of the others crafts in the first decades of the 20th century with the rapid technological advances of that period and the wars that had accelerated them. More than just technological, that process had brought about a change of mindset; an axial shift toward utilitarianism which meant that people measured the cost of a product in reference to the time it took to make. The eye was always on the clock; not much room for the heart. Who could spend several hundred hours making a spinning wheel? Who would spin and knit for many hours to make socks or mittens or a sweater? In his own way, my grandfather protested the change but it was inexorable and he knew it. Never-the-less the old blacksmith shop, the shoemaker's sewing machine, the wood lathe and even the spinning wheel, though not used, retained their place of honour on his farm; silent sentinels of a passing age.

The desire for the honour and joy of the old crafts continued to haunt my grandfather as it does many others on earth – even to this day. There's a mystery here. On one hand it was the descendants of Cain (Adam's first born) who were the first craftsmen and musicians in their East of Eden land of Nod; but then, God's own first born, Jesus, grew up the son of a craftsman and worked as one himself for at least 15 years. Perhaps the crafts are the best training for an integrated 'back to Eden' life – or maybe they are part of Eden in a more intrinsic way. After all, the first man in the Bible said to be filled with the Holy Spirit was Bezalel; an artisan charged with the task of replicating Heavenly designs on earth for the wilderness tabernacle of Moses. (Ref: Exodus 31:1-5)

In a big picture prophetic view it seems to me that a loss in the arts and crafts has occurred over the last century or so, as money, that arch enemy of true hearts, has become an increasingly treacherous slave driver. From its beginning money has increased in power: pushing aside lesser powers – such as those wholesome 'graces' given to artisans and craftsmen and musicians – until it reaches a day of fullness where there is room for little else. John Galsworthy, in a 1911 essay, *Quality*, eloquently grapples with this same dynamic in telling of the demise of the shoemaking Gessler brothers. These brothers literally starved to death making their fine quality individually fitted boots while bucking the market trend of cheap mass produced shoes.

My grandfather, though more moderate than the Gessler brothers, remained, whatever the difficulties, pretty indifferent to money all his life. He would straighten and reuse a bent rusty nail to avoid waste but then lend thousands to a distressed relative who had never been frugal – knowing he would never get it back – and not caring except that he worried in case he was 'spoiling' someone. He lived frugally, gave generously, and when he retired passed his farm over to his children.

He was at all times, that I saw, a humble servant. To keep the record real, I should point out that my father and other older members of the family remember a time when he was a severe and maybe even tyrannical disciplinarian – but that was before my time. I remember, as a little boy, asking my mother if there was anyone except Jesus who never sinned. The question put her into deep thought so I suggested grandpa; to which she replied – her eyes still far away – that perhaps he didn't sin very much.

"Would he go all day without sinning?" I pressed.

"Yes" she thought.

"A whole week" I continued.

"Well, maybe sometimes".

The question still seemed open so I asked grandpa next time I saw him: "Oh yes, I'm an old sinner," he answered.

"Why grandpa, what do you do?" I asked.

He was working on his John Deere 55 combine harvester at the time and without looking up he said in a half teasing, half serious tone: "Oh, I can't tell you that". His face gave away nothing either as he was intently focused on getting a pin in behind a pulley in an awkward place. There's a little prophetic 'anagram' hidden in the name and model of the machine: 'Five' is a number usually associated with grace and 'John Deere' reminds us of John, the beloved apostle who lay nearest to the breast of Jesus. Double five (55) is double grace: sin cleansed away and its memory hidden – even from the prying eyes of a little grandson.

The Evening

After a light supper, mainly of leftovers with a few extras, we retired to the living room. In the corner stood the spinning wheel, against the wall an antique pump organ and on another wall – one of my favourite things in the house – a cuckoo clock. It looked like a small woody cabin covered with carvings of leaves and birds. A little cuckoo popped out once every half hour and at the appropriate number of times to mark the hours throughout the day. Two heavy metal pine cones descended on chains beneath it to power the clock and cuckoo mechanisms. (Only grandpa was allowed to reset them!) I would stare at it and long for that world of which it was a part; that place where musical mechanism, woodland, and fairy-tale converge in some ancient Germanic valley – a dwelling-place of the smiths and craftsmen of old.

Eventually we would pressurize grandpa into getting out his guitar or autoharp and sing to us. I remember staring deeply into these instruments as he played; searching again for that mysterious merging of mechanism and music and heart. He sang old Scandinavian worship songs, that originated, as perhaps all else that his heart prized, in the revivals and visitations of the Spirit from earlier days. They carried you to deep cold woodlands, ever green, and to skies blue and severe; but also to warm firesides, to hand-crafted pinewood furnishings and to the love of all that is both humble and noble. And they carried the name of Jesus into Norway and its pioneering offshoots –

past and future – redeeming the glory of that nation for its heavenly destiny, when on that day, the kings of the earth bring the true glory of all the nations into the eternal city; free at last from the shadow of the serpent's tree, they will walk in the light of God himself. (Ref: Rev 21:24)

A common feature of my grandfather's worship with Scandinavian and other folk music is that it was filled with immortal longings and desires which are echoes from the heart of God and a joy to experience. (Of course I'm speaking about something deeper here than just taste or preference for a particular kind of music and I certainly don't deny that there may be sentimental counterfeits.) My view, and my recurring chorus, is that many of the desires that flow throughout the earth, in millions of places within both nature and culture, are from the Holy Spirit and that if we would be awakened, as Isaiah was, we would see that indeed: "... the whole earth is full of his glory." And if we are truly awakened we will become contributors, along with that great company of the faithful from all the ages, to the ultimate reality of the whole earth being "*...filled with the knowledge of the glory of the Lord, as the waters cover the sea.*" (Habakkuk 2:14) Of course that is the day that the whole earth becomes 'Eden' – glory and beauty and meaning and life and reality; heaven and earth in perfect unity.

Chapter 3
Mum and Dad

See that you do not despise one of these little ones; for I tell you that in heaven their angels always behold the face of my Father who is in heaven.
(Jesus Christ, Matthew 18:10)

Dad

"How long do people live?" I asked my dad one day shortly after my 4th birthday. I was standing beside his seat on the farm tractor, sometimes watching his face, and sometimes watching the disc we were pulling as it turned over the spring growth of weeds to reveal the rich black soil beneath. Our little black Cocker Spaniel, Pepper, was running along behind sniffing at anything of interest in the fresh earth, his tongue hanging out as he trotted along; cheerful and unfailing as ever. Like me, he was happy to spend the whole day with dad – round and round the field we went hour after hour; just the three of us. We didn't talk a lot – because of the noisy tractor engine – but there were plenty of smiles, a few songs and a lot of deep thinking as we gazed out over the wide prairie around us.

Distant clumps of trees, which sometimes indicated a farmstead, were all that broke the flat sweep of fields in the miles and miles to where the horizon blended with the shimmering mirage like haze created by the summer sun. To the east and west the land sloped gently upwards forming a wide shallow valley whose slight concave meant that the horizons were even farther away than they would have been had the land been exactly flat. It seemed to my eye then and to my memory now to be of vast extent and uncertain limit. Such was the world in which my consciousness of life's limits was emerging that day when I asked my father the simple question: "How long do people live?"

I can clearly remember why I asked the question. I had been thinking about how much I liked to be with him and I had just made the decision that I didn't want to live any longer than he did. If an adult thought that way about someone it is highly likely to contain elements of self pity and pathos but in the heart of a four year old it is no more or less than love. I remember the thought exactly – there was no self consciousness or thought of how sad I would be without him – at its depth it was: "I want to be with him."

"Well, you never really know" dad said, "but in the bible it says three score and ten or four score if you're strong – that's 70 or 80 years."

"How old are you dad?"

"Thirty four" he said, looking at me with a slight question in his face. I, however, was deep in calculations – calculations out of my depth and beyond my years. I was visualizing two rows of numbers; one of his years and one of mine.

"Could someone die when they're 50?" I asked at last. (I had decided he could live to 80 and me 50 – another 46 years)

"Yes, some people do."

"Ok that's good." I said – having settled the matter in my own mind that I would die at the same time he did. I never spared much consideration for the thoughts that might lay behind the quizzical look on my father's face.

I've been reluctant to write this chapter not because I have anything to hide about him but because he was such an outstanding and exceptional father that I fear a backlash of criticism that I'm a terrible idealist or ancestor worshipper or something like that. I have no memory of him ever behaving selfishly for example... can that be real? He chose relationship above money every time... is that really possible? He wouldn't have qualified as very wise by many of the parenting books I've read – he'd give you his last dollar to buy something he didn't want you to have simply because you asked for it. He never tried to control with moods or self pity but actively compensated against any such control by being cheerful even at times when he would have been perfectly within his right to be sad – who does that? I feel sorry for people who never had a father like him and I have no wish to rub salt in the wounds of those who never had a father at all; or who had abusive fathers; absent busy fathers; or who have experienced breakdown, dysfunction or divorce in family life.

My father remained faithful until he died, to God and mum and to us children.

Many of the books and teachings surrounding fathers are testimonies of how God has healed and helped those who have suffered in this area. I am thankful for those testimonies but my purpose here is to bring another side to the story and the time has come to set aside my reluctance because there is a Father in heaven who has never faltered or failed and there have been fathers on this earth who have revealed more of his heart than has yet been taught in any of the books and conferences I've heard of.

The first prayer I ever remember praying happened on one of those days riding on the tractor with dad. I was studying his face and noticing the deep lines he had in his forehead and I remember thinking how I liked them – his face just seemed so beautiful to me – and so I asked God if I could have lines like that when I got older. (And I got them, though I had to wait thirty years or so.) I don't really know why I loved him so much but I assume it was because he loved me from day one, and that nothing but love ever came through his face to me. I can remember seeing his delight with my little brother and sisters when they were babies – sometimes he would pick them up off a blanket and just hold them up to have a good look. There was something distinctly masculine about it and mum might scold a little if the baby cried perhaps from a little chill or the insecurity of not being held tight. There was no cruelty in it though – just a delight in seeing them, even if it meant a little nudge out of the totally safe and cosy. The Holy Spirit has at times reminded me of my father doing this kind of thing when I've felt pushed out of an unnecessary comfort zone. In general though, I don't think true masculine love is very easy to define (nor of course is feminine love) and in many ways, on the face of it, my father was full of mystery and contradiction.

Let me give you a different kind of example. My father had grown up on a farm and learned to hunt and shoot from an early age. As a boy, using just a homemade slingshot, he could bring down a wild prairie chicken, with a shot to the back of its head, as it flew up from the grass in front of him. Later he became an excellent marksman with a rifle, and had a collection of guns which he taught us how to use when we were old enough.

The guns were tools for controlling pests such as the large population of ground squirrels (commonly called gophers) which plagued the pastures. Shooting these pests was one of the summer chores that we boys enjoyed, after school and in the holidays. Sometimes we hunted rabbits or took pot shots at birds but dad never approved of shooting anything except for food or if they were harmful pests like the gophers. Reasonable farmer attitude so far... right?

But then... he hated to see hawks kill baby rabbits – it offended him the way they picked them up and just took a bite out of their head – he agreed with me (3 – 4 years old) that the mother rabbits would be pretty sad about that. In fact he was sometimes a veritable defender of baby rabbits. During harvest a baby rabbit would occasionally get picked up by the combine harvester and dad would make dramatic efforts to shut the machine down before it got drawn into the machinery – and he was genuinely grieved if he didn't succeed in time. On seemingly random days when we boys had got the guns out to shoot some gophers or whatever, he would come out – especially if it was a very nice sunny day – and say: "Don't go shooting today boys; nothing wants to die today." On those days he carried such solemnity and depth that we put the guns away without a word. I can't speak for all farmers of course but I don't think that fits the stereotype. And as far as good management is concerned... or parenting books... well, what can be said?

There is, however, a window here into a mystery of our heavenly Father's heart. Our Father is bearing with a situation that is far from ideal because not everything on earth is according to his will as it is in heaven: which is why Jesus taught us to pray as he did in the Lord's Prayer: "Your kingdom come, your will be done on earth as it is in Heaven." The large painful gap between what God desires and what actually happens on earth is something that He 'puts up with' or 'forbears.' In his pragmatism of managing such a situation we should not be surprised if what is a good thing to do one day is a bad thing on another day. Obviously I'm not talking about fundamental moral commandments here but the more variable realms of wisdom, emotion and relationship. As a farmer trying to provide for his family my father knew we had to deal out death to pests at times but in his heart he still had a place for all the cute little furry creatures and some days he just had no stomach for anything but life.

Similarly, we get glimpses of our heavenly Father's 'forbearance' (as Paul calls it in Romans 3:25) in the Old Testament when God is agonizing over judging those he loves such as the tribe of Ephraim in the book of Hosea. "*How can I give you up, oh Ephraim? ...My heart recoils within me; my compassion grows warm and tender.*" (Hosea 11:8) We see here the tension in judgment and mercy. We should not expect from God that kind of monolithic perfection of psychology currently exalted as 'professional' or 'appropriate' – broken hearted lovers are not likely to be quite so emotionally balanced. Our heavenly Father bore the sin of the whole world in his heart from the fall in Eden until the day Jesus died – and even now Father and Son in heaven are engaged together with the Holy Spirit in the sometimes agonizing travail of intercession.

In Eden, God the Father took the decision and bore the responsibility for not visiting upon Adam and Eve summary death; but it was a sword into his own heart to bear such disintegration. We could say he was dying for the sin of the world from the beginning. That sin was like a cancer tearing away at His tender heart and it was into this context that Jesus – doing, as ever, what he sees the Father doing – chooses to bear that sin into death on the cross. The cross reveals to us in stark physical reality what sin had been doing to the Father's heart from the beginning. The only motivation of the cross was love; a mysterious, masculine kind of love blended with a beautiful feminine kind of love prophetically paralleled in Mary as she also felt the sword through her own soul in fulfilment of the prophet's word. (Simeon in Luke2:35.) Perhaps this pain of Mary corresponds to the pain borne by the Holy Spirit who was also there at the cross?

My dad, at the age of about twenty, had lost a fifteen year old brother to leukaemia which had no doubt, along with other sorrows and hardships of life, wounded his heart. He told me once about how he had given direct blood transfusions (circa 1950) to this brother – which would perk him up for a few days followed by a gradual fading away again. In the way that I've been speaking I suppose it's fair to say that even that heartbreak enabled him to reveal something of the broken hearted Father in heaven.

I'm not saying my earthly father was perfect – he had weaknesses and failings too, but there are some failings that are small enough to be comfortably covered by the overflowing love that is liberally rained from heaven upon the just and unjust. Or perhaps I should say that the righteousness of a father – or a Christian in general for that matter – is not so much a question of faults as of overall 'condition of the heart' on the long term. My father's love did not fail in all the years I knew him.

Once when I was about 16 or 17, I had upset my mother to the point of tears by some typical teenage attitude and words – the details of which I no longer remember. This had brought subsequent pressure on dad to take some kind of disciplinary action – maybe even revenge. I'm not saying this was wrong of my mother – quite the reverse – there is a necessary tension between facets of love and righteousness with which the masculine and feminine grapple on earth; a reflection of the same tension in the heart of God in heaven. Dad's face was grieved, as he agonized in that pincer movement so many fathers (including God) have been in before and since; but without a trace of self pity, control or desire to dominate. He spoke a few words which I don't remember as he handed me his keys – he never used vehicles or money as a control mechanism – and I took his pickup truck and drove off defiantly. His face, however, continued to loom in my consciousness, and these words formed in my mind from it: "I would die for you 3 times."

That was the thought that lay behind the face and it melted my heart. I was seeing beyond his face into the innermost heart of God the Father in heaven. I was seeing into the painful paradox of unfailing love; that heart which loves the other more than it loves itself. Of this love Jesus spoke when he said, "Greater love has no man than that he lay down his life for a friend" and in the case of God, who is three persons, the love is triply great because the dying, in terms of the painful cost, was three fold. This 3 fold dying happened when Jesus died on the cross – the Father and Holy Spirit, in their perfect love, bore equal pain in the experience of death... how could it be otherwise? The vague line of thought that suggests the Father and the Spirit could be unmoved, as Jesus suffered, denies their love for him by the very definition that Jesus himself gave for love. Besides, the Apostle Paul states that, "*God was in Christ reconciling the world to himself*" (2Cor 5:19). Of course I am not saying my father loved as much as God loves; but it cannot be denied that he loved enough that his face could convey – or be a window into – the deepest and greatest love there is.

I'm sorry if you never had a father like that, because it was truly precious to experience his love, and even my memories of him are precious to me because they remain as an open window into heaven. I am convinced that it is the heavenly Father's wish that every child born on this earth should have a father to provide that window into heaven. Never-the-less hear this: it is the Father in heaven who is the source of that love and, through the cross and the Holy Spirit who lives now on earth, this love is tangibly available to every one of us, even if we never had the kind of fathers that God wanted for us. It is my prayer as I write these words, that the window my own father opened for me might also open for you, by the Holy Spirit, through hearing my story.

Additionally, those of us who are fathers should ask God to help us be the kinds of fathers whose faces are windows to God's face. Consider Jesus words: "*...in heaven the angels of these little ones always behold the face of my Father who is in heaven.*" (Matt 18:10) These angels fully cooperate with every earthly father who wishes to participate in the noble task they have of bringing the Heavenly Father's face into focus for their children. We could say it this way: there is an anointing from heaven to help us be those kinds of fathers that bring the best of heaven to earth; for nothing is greater than His face and indeed nothing ever could be greater than his face; for from His face flows the radiant beauty and power of a love that cannot fail; the very person, God our Father himself.

Later on we will look at other times when, through the memory of my father's face, I saw into the heavenly Father's face – not just as insight or wisdom but as an immediate dynamic connection. Those mystical and miraculous moments are of course only the tip of an iceberg whose main bulk consists of the day in day out serving of a heart which prefers the other to itself; and even in that, never 'playing the martyr' or making a show of it.

Dad died unexpectedly in his sleep one night at the age of 61. A few hours earlier he and mum had woken and, unable to sleep, had sat up in bed reflecting on where all their children had scattered to and how each one was doing. In the morning he was gone. I was working in London at the time but was able to fly home for the funeral. I had always felt a little bit sorry for my brothers because they didn't seem to be as close to my dad while we were growing up – and I was a bit worried that I was dad's favourite. I happened to mention this to my brothers, as we stood around talking after the funeral service, and was amazed to hear that each of them (all four – three older and one younger) had felt the same. Each one of us felt that he had in some way been treated more specially than the others and we each had specific reasons for thinking that way – and had always been a little bit worried about it in case the others were upset or jealous.

One of the things I had seen on my father's face and had also heard him say many times was how terrible a thing it was for brothers not to be on good friendly terms all their lives. My father, for his part, succeeded in laying a foundation for that, but not by some fastidious effort at fairness – in fact he treated us all quite differently. For example, I spent far more time with him during several periods than my brothers did – because I wanted too – but he gave more to them at other times and in other ways that never interested me. He bought one of my brothers a car, but for me, he helped repair one I had bought myself. Of course we were all of different ages with different needs and interests and desires so it should hardly be surprising if no text book on parenting can help a father with the inner secrets of the trade as it were.

This chapter is not intended as practical advice on being a father; it is only some glimpses of the heart that I'm trying to sketch. For example you might think my father was a kind of soft 'push-over' but in fact there were times that he was a strict disciplinarian and he wasn't one to be trifled with if he had set a punishment – but even in this he wasn't 'consistent' in the way some parenting books demand. My own view is that 'perfect' parenting is a legalistic illusion. God himself is wrestling out his 'parenting' with a heart that is often breaking; with threats of judgement that he retreats from at the first sign of repentance; with accusations – as every parent hears in the back of their mind – of indulgent leniency from the legalists and of cruel harshness from the liberals; and with consequences that he ultimately bore in his own heart and soul and body on the cross at Golgotha. The fruit of this labour will be the vindication in the long term, "*...he shall see the fruit of the travail of his soul and be satisfied.*" (Isa 53:11)

In remembering my father, and being a father now myself, I can't think of any way that an earthly father could be more 'vindicated' and 'satisfied' than that little moment at dad's funeral. He wouldn't have claimed to be a good father and certainly wouldn't have aspired to teach anyone about parenting. He had no significant academic accomplishments, no great financial success, and no famous ministry; but his five sons discovered, in the end, that each had felt like the favourite son. As far as God is concerned we can each live like the favourite. Perhaps the most sinful of us, like the prodigal, will discover this more easily but hear also the Father's words to the elder son: "*Son, you are always with me, and all that is mine is yours.*" (Luke 15:31) If your heart feels cold and you are weary of the journey, let these words sink in: "Son, you are always with me, and all that is mine is yours." Of course if you join the Father in rescuing the lost sons and daughters you can enjoy the party as well!

In case I have painted an overly rosy picture of my father let me touch on one other point. I can remember a day – probably several – when we were all in the house and had been fighting and bickering as children do. Mum was in a bad temper with it all and dad too had reached the limit of his patience. He then broke all the good parenting instructions by lamenting out loud that he wished he had never had kids. I personally doubt if a parent has ever lived who has not felt that way at certain times so we might as well be honest about that. The pain of those impossible situations, where there really is no right way to respond, drive us all to those kinds of feelings like nothing else can.

But as a child I was not at all bothered by this. I can remember exactly the situation and my feelings. I looked at his face even as he was saying: "I wish I had never had kids!" In his face I could see a depth and complexity of love that was actually comforting in the very moment of him saying the words. It was that terrible paradox of love itself that drove him, as it has many parents, to such frustration. I can imagine the inverse situation where a 'well trained professional parent' is speaking outwardly sweet and measured words, but with a face radiating cold hatred. I know which I would rather have. And let's not forget that God himself has also spoken words of similar frustration when his children were behaving badly such as in the days of Moses. Therefore, even in this apparent fault my father opened a little window into heaven.

This has been my attempt to sketch a tiny glimpse of the heart and life and thought of a man who opened, through his own life and journey in God's grace and despite all the complexities and perplexities, access to the very face of the Father in Heaven. One effect of having such a father has been to create a peculiar prejudice and idealism – a certain high expectation of what the universe must be at its root, before ever considering other facts.

I'm a prejudiced idealist because I first opened my eyes to beauty, truth and goodness and now I can never possibly believe that anything less could – rather someone greater must – be at the foundation of all reality. My whole life – in its best moments – has been a journey and search to apprehend, to understand and to love that 'someone.' At its core, this is the road to Eden.

Mum

I confess from the outset that what I'm about to say about mum is not to be compared with what I've written above about my father. I do not claim to understand feminine love even to the faltering level that I've attempted to define masculine love. Feminine love, by definition, is indefinable; paradoxical, elusive, derived, and hidden – in plain view. There she was in front of Adam after all words and logic and 'names' had been established – even the much vaunted 'feminine mystique' is only an out of time male attempt to name the unnameable. A wise man would probably say even less than all of that, but a child can see what he can see and say what he wants so I'll indulge a brief contrast between mum and dad – which doesn't fit any of the stereotypes anyway.

Whereas my father felt deeply, mum thought deeply; where he was an idealist, she was practical; to his grieving she brought resolution; to his procrastination she offered decisiveness; to his deep love she added faith and action. If he was a poet then she was a philosopher; if he a patriarch, she a theologian; or if he a pastor (shepherd), she a teacher. And of course she was a teacher, in a one room school house on the prairies; and he one of those old fashioned farmers who, besides growing grain, kept animals.

Outwardly, mum was as sociable and effervescent as my father was self effacing and retiring. She had a keen sense of humour – especially for the paradoxical and ridiculous – and all who knew her will remember how she laughed and laughed until her eyes streamed with tears; and how she would dry her eyes with a tea towel if she happened to be in the kitchen.

Inwardly, she was a deep thinker and a selfless lover; always loving and serving and somehow finding ways to out-love all who loved her. After I had moved away to a city about 300 mile away, I remember bringing friends home for the weekend, arriving at 3A.M. and her getting up to cook and serve us all food and drinks. At the time I thought that was a perfectly ordinary thing for a mother to do, but now I can see that it was one of many times that I took her for granted and imposed on her love. She never complained or looked for a payback though; and until her last day, her love for me remained greater than I ever reciprocated.

But she was never a 'push over'; her mind was sharp, her imagination shrewd and resourceful. She could not easily be fooled – certainly never by children or teenage boys. However, she sometimes took the posture of naiveté if it served a higher purpose. No doubt such wily ways can also be used for 'lower purposes'; but in mum's case, the virtue of any feigned naiveté flowed from her loving and humble heart. Indeed in her case, the general truthfulness of her character always shone through every wile and, as a result, even we 'dumb boys' learned to discern 'wily ways' in safe home waters; an inoculation, as it were, for the more treacherous waters in the world – and the Church for that matter.

Perhaps that part of motherhood which involves the driving of folly from the heart of a child is the legitimate province of those skills, which, in other contexts, might be denigrated as the devices of fallen woman. One way that good women redeem and purify those devices ('desire' Genesis 3:16) is through the painful and seemingly impossible task of birthing and raising children. And in case that sounds ridiculous to 'liberated' ears, I'm not saying there aren't other ways to develop genuine humility and a serving Godly heart – any true ministry for God will take a man or woman in that direction. I do think, however, that the pendulum of current thought and values has swung too far away from recognizing and honouring the self sacrificing mother-heart, with its wide panoply of creative resources.

When I was about 4 years old, mum purchased an old piano from the closing down sale of the local one-room school house (not the same one she had taught in – that was about 10 miles away). She refinished it herself, had it serviced and tuned, and then taught herself to play – which didn't seem to take her very long. She played and sang out her heart in worship, loud enough to drown out the sound of squabbling children. (There were 5 of us under 7 when she got the piano.) She sang beautiful love songs to Jesus, songs of the journey, and heart searching songs of calling and commitment. Some of these songs, as I remember her playing and singing them, are still the underlying architecture of my deepest longings and spiritual perceptions.

She was a good teacher of the bible, the gospel and good morals – in fact she loved all learning and past that on to us. Here's an example of her imparting a love for truth that is also relevant to our general themes of desire and seeing the Father's face. When I was 4 or 5 years old, mum came one evening in our big bible story book to the story of God's glory, and a view of his back, passing by Moses at Horeb. "What does God's back looked like?" I asked. (The picture in the book was only of a vague bright light.) This question put mum into deep thought and she reiterated – while trying to think of a good answer I suppose – that Moses was not allowed to see God's face but only his back. Then she hesitantly and thoughtfully conceded that she didn't know what his back might look like – but with such a curious and intelligent depth in her face that it gave me an even greater thirst to know. And so the question whittled away in a distant corner of my mind for about fifty years before I began to get an answer.

Here's the text: "*Moses said, 'I pray thee, show me thy glory.' And he said, 'I will make my goodness pass before you, and will proclaim before you my name 'The Lord' [Hebrew: YHWH or Yahweh]... and I will cover you with my hand until I have passed by; then I will take away my hand, and you shall see my back; but my face shall not be seen.'*" (Exodus 33:18-23)

Perhaps we could paraphrase: "You can see my glory in your mind's eye – have mental knowledge of it; my back. But you will not be able to experience the complete imposing presence of that glory – full experiential knowledge of it; my face." Of course seeing God's back was no doubt an intense experience for Moses too; but it was something less than what he had experienced earlier. It still allowed him to maintain some duplicity and imperfection without facing immediate judgement; it was a pragmatic adjustment made by God with a view to something better in the future. This is God's patient grace in operation. Within the big picture of Moses' life and ministry and the witness that he is to God, he becomes from this time on a symbolic type of the law; a definition of goodness (God's back) but not the imposing actuality of goodness (God's face).

And here's the cruncher: we still have a lot of 'God's back' theology in the church to this very day. The 'God's back' theology that we have in the church did not come directly from the stories of Moses – though I think it originates there – but from the influence of Greek philosophy. (Perhaps Solomon, who received and taught delegations from many nations, was the link. Ref: I Kings 4:34). 'God's back' is what Moses was left with after Israel's fall to idolatry in Exodus 33 and it is basically what the world was left with from that time on. We might say that it was the best that Israel was able to 'dish up' to the surrounding nations to whom it was called to be an example and a light.

That light, such as it was, was perhaps most clearly grasped by the Greeks. To the Greek philosophers like Socrates and Plato, for example, God was the very 'nth' or 'infinite' degree of goodness – and even Jesus affirmed that this was true (Ref: Luke 18:19) – but of course it is 'God's back' and, to tell the truth, insight that preceded these Greeks by a thousand years in Exodus 34. In the Church age we should know better but early on we also settled for 'God's back' theology with such concepts as: omnipresence, omniscience, omnipotence and so on. These are all philosophical extrapolations and projections of the mind – true in a way – but far from the truth of his very person; his face. These and all the other great pillars of theology, even with their very best definitions and explanations, can never be more than a glimpse of God's back. They are the extrapolations by intelligent minds; not the seeing by pure hearts.

Unfortunately mum died a few years before I understood this; but if she hadn't she would have enjoyed an in depth discussion of it and would almost certainly have remembered the day I asked her the question in the first place. And she never stopped learning herself. After Dad died she took a job as the hostess of a Mission house providing hospitality for missionary families on furlough. She also managed the finance and salaries for the Canadian office and overseas personnel; learning computers and computer bookkeeping for the first time on an old MS DOS system when she was over sixty.

She was intelligent and yet humble; wise and yet fun; determined and yet flexible; uncompromising and yet gracious; comforting and yet uncomfortable; a hospitable, loving and serving hostess... and yet sometimes an awkward guest. She had known her share of sorrows too; having lost her father in her teens and her husband in her late fifties. But she had a certain toughness of heart to carry on anyway – I'm not sure if a professional counsellor would give her a clean bill of health or not; but I know she poured herself out in love, and that it is easier for me to picture God – perhaps more the Holy Spirit – as feminine, because of her.

We know of course that God is both masculine and feminine because we are told that he made both male and female in his own image. (Ref: Genesis 1:27) The tensions of masculinity and femininity are necessary to ultimate, perfect love being revealed and expressed. We see this in God's design of family, but also at the cross and resurrection. The weeping Marys – whether anointing Jesus' feet at the foreboding table, enduring the grisly crucifixion at Golgotha, or braving the rarefied morning airs at the garden tomb – are no mere ornament to the great narrative; rather they are the pierced heart of a love that is forever mysterious and forever feminine.

CHAPTER 4

CYCLES, STARS AND C.S. LEWIS

Here, then, is the desire, still wandering and uncertain of its object and still largely unable to see that object in the direction where it really lies.
(C. S. Lewis "The Weight of Glory")

PURITY AND SEEING

According to a story of uncertain origin, when the Soviet cosmonauts returned from space the first time, they declared to the delight of their atheist superiors, that they saw no sign of God in the heavens. A little Russian Christian school girl, on hearing this from her teacher, responded with the question: "But Miss, did they have pure hearts? Because the bible says only the pure in heart will see God!"

The reference of course is Jesus' teaching from the beatitude in the Sermon on the Mount: "*Blessed are the pure in heart for they shall see God.*" (Matthew 5:8) Jesus' own heart was absolutely pure so that he not only saw God himself but provided a perfectly clear window for others to see. A window so clear that he could say: "*He who has seen me has seen the Father.*" (Jn 14:9) However, many people who saw him did not really 'see' him or the Father because they were blinded by the impurity of their own hearts. In this context 'purity' means a whole or single hearted love for God. Idolatry, which is to love alternative gods – perhaps as well as God – is impurity. Jesus had a pure and perfect love for the Father. Having a pure heart, he lived his life on earth with the Father in constant view. Because he knew what the Father thought and desired in every situation, and because he acted accordingly, he perfectly revealed the Father's heart and thought. Further on in the Sermon on the Mount, Jesus expanded on the beatitudes that he had spoken at the beginning.

The connection between purity of heart and seeing is developed like this: "*Do not lay up for yourselves treasures on earth, where moth and rust consume and where thieves break in and steal, but lay up for yourselves treasures in heaven, where neither moth nor rust consumes and where thieves do not break in and steal. For where your treasure is, there will your heart be also. The eye is the lamp of the body. So, if your eye is sound, your whole body will be full of light; but if your eye is not sound, your whole body will be full of darkness. If then the light in you is darkness, how great is the darkness! No one can serve two masters; for either he will hate the one and love the other, or he will be devoted to one and despise the other. You cannot serve God and mammon.*" (Matt 6:19-24)

One reason these verses are misunderstood is because they are often broken down into separate paragraphs giving the impression that the teaching is not interconnected. If we recognize them as one interconnected expansion of the beatitude: "*Blessed are the pure in heart for they shall see God*" (Matthew 5:8); their meaning comes quite naturally into focus. The impure heart is divided: treasure here and heart there; trying to embrace opposites; straddling a contradiction; worshipping one but saying and thinking you are worshipping the other; trying to love both. Adam and Eve did this when they made their choice in Eden; believing they were choosing goodness, but actually choosing evil.

'Worldly treasure' or 'mammon' is more than just money; it encompasses all the loveless powers and false loves of this world which lie at the heart of the serpent's kingdom. In many ways, though, money sums them all up; a power in itself but also the easiest doorway to all the other powers: military, religious and political. The god behind worldly treasure is the serpent and the person who sees it – chooses it – as his treasure (the eye not sound) will see not the face of God but the face of the serpent; the selfish dominator (masquerading as God of course).

All of the treasures and powers that are 'of this world' as opposed to 'of heaven' are discernible by the simple test of whether they are headed up by a generous servant, God; or a selfish dominator, the serpent – who is also sometimes called a dragon. (e.g. Isaiah 27:1, Revelation 20:2) And the thing with dragons is this: if our eyes are on the dragon's treasure then a dragon we are sure to see; and if we fill our hearts with a dragon's treasure then a dragon is what others will see in us. (Everyone knows of course that a dragon's treasure and a dragon's heart and the dragon himself are in the same place – that's basic Dragonology 101!) Against this inevitable consequence Jesus gave his sober warning: "*If the light in you is darkness, how great is the darkness.*" Summing up his point, he makes the categorical statement: "*You cannot serve God and mammon.*"

God and mammon are opposites; the heads over opposing kingdoms. One: the great servant; the other: the great demander of service. Only an impure and divided heart tries to love and serve both. In the words of Elijah on Mt Carmel: "*How long will you go limping with two different opinions?*" (I Kings 18:21) And from James: "*Purify your hearts you men of double mind.*" (James 4:8) The eye that is sound can clearly see that earthly treasure is not worthy of trust; not worth putting our heart into... why, even at the most basic natural level, it is subject to moth, rust and thief!

Beyond that natural level, Jesus touches the spiritual level indirectly when he warns against the greatness of the darkness, but Paul directly states that covetousness is idolatry. (Ref: Colossians 3:5) To love money or what it can buy – that is, worldly treasure – is the same as bowing in worship to a false god. Indeed, it is bowing to a false god, the serpent. And of course that was Adam and Eve's sin; but also the same sin that Jesus overcame in his wilderness temptation.

The second commandment which forbids idolatry carries with it the warning that the consequences of idolatry effect children to the third and fourth generation. (Exodus 20:5) Children know intuitively – in their hearts – whether their fathers or mothers love them more or less than money; what a joy to a child to be loved most; what wounding and pain to be second place – even if it's clothed in all sorts of words like 'wisdom', 'common sense', 'being practical', and 'good stewardship'. A mother and father are windows to a child – either to light or to darkness. A child can see through the attitudes, words, actions, and the subtle nuances of facial expression into the heart; discerning true heavenly love or the lack of it – light or darkness; God or serpent. The father and mother who love God in the first instance and with Him their child, more than money or any other worldly treasure have a pure heart – unmixed – whereas those to whom worldly treasure sometimes takes precedence have divided – mixed – impure hearts. The consequences, for good or ill, run to four generations.

All of us whether fathers, mothers, sons or daughters, have, since the fall of Adam and Eve in Eden; been plagued by impurity of heart and double mindedness. Grace to overcome this impurity is given to us in the gospel through Jesus death and resurrection and by the coming of the Holy Spirit who works into our innermost being – right into our hearts to the root of the problem. At the root of our hearts are the same conflicting desires that Adam and Eve had, and of course the same old serpent still pulls and tugs on those heart strings with all his cunning deceit; the same old fruit; the same old tree; the same old treacherous enticements; and the same old heartbreaking consequence if we choose wrongly. "*If the light in you is darkness: how great is the darkness!*"

This 'darkness' is the serpent's so called 'light'; the 'knowledge' he had and still has to offer. We might say that darkness is dark, but darker still is that darkness that presents itself as light; it is an active empowering darkness. More than just the absence of true light; it is a dynamic suppressor of truth; a living, breathing, in your face liar. That, I think, is the meaning of Jesus' words: "*...how great is the darkness!*"

Fortunately we have help against this darkness. We have the true and dynamic light of the world: Jesus who overcame every temptation and enticement to worldly power that the serpent threw at him. And the actual reality and power of Jesus' victory is brought tangibly to us by the living, breathing, in your face Spirit of truth. Paul talks about the conflict of darkness and light in our hearts in terms of desires; the 'desires of the flesh' being in opposition to the 'desires of the Spirit'. (Ref: Galatians 5:17) While getting our thinking right on these things is important, it is the dynamic truth of the Holy Spirit in our hearts; sifting and shaping our desires and especially bringing to our hearts his own beautiful desires that brings victory, real change, and purity of heart to us.

As a little boy I benefitted from the purity of my father's heart and the purity that had been forged by forefathers who loved across the generations; an inheritance far more precious than any amount of earthly wealth or power. The prayer that my grandfather prayed is the antithesis to the warning attached to the second commandment which forbids idolatry. Where the curse went to four generations;

now a blessing was prayed: "And Father we do pray for the children and the children's children even unto the third and the fourth generation." For me that has remained a treasure to this day of course but it did not and still does not preclude the necessity of my own hard choices and battles and mistakes and starting over again and again within all the usual vicissitudes of life. I will pick up my story at an era in my early twenties when I first got into motorcycles to illustrate God's grace within the conflicting desires of the heart. The Holy Spirit patiently and persistently works in our hearts to purify them so that we too can see the Father – and be a window for others to see him in us.

First Motorcycle

The motorcycle thing for me all started when my younger brother, Duane, bought a brand new 1979 Yamaha XS400, in deep navy blue. We shared an apartment and worked at the same machine shop: 'Tero Welding and Machine'. Some of the other guys at the shop began to get bikes about the same time and I also had a friend, Clayton Mellsen, who was a bike enthusiast to say the least, and who, but for his wise and gracious wife, Brenda, would have spent every penny on bikes and every waking moment riding them or talking about them. (Of course that's a bit of light hearted exaggeration about Clayton but as far as Brenda is concerned she was the wisest and closest to God in our circle of friends.)

I had always regarded motorcycles and convertible cars as unsafe and uncomfortable; a view that I had assumed from my father without much analysis. In fact, dad had had, in his youth, a brief flirtation with some of the great American classics like: Indian, Matchless and Harley-Davidson but times were hard and his father had discouraged it strongly. Thinking of it now with sons of my own I am much more acutely aware of the safety aspect and I think that may have been the real cause for my father's reticence. As far as grandpa is concerned, a closed in car was viewed as superior because the memory of all weather horseback riding and the discomfort of open sleighs and wagons was still quite vivid.

But, the desires of fathers – suppressed or otherwise – have ways of resurfacing in their sons, for better or worse, and at this point in my life both generational yearnings and the hot bed of desire in my environment conspired that my eye should fall upon a 1978 Yamaha XS400, like my brother's, but red in colour. Desires, both good and ill, act upon us from many sides and are highly contagious; but we are still responsible for our choices. James' word that each man is tempted when he is lured and enticed by his own desire precludes any possibility of our choices being someone else's fault. (Ref: James 1:14)

And so, I bought the bike. In any case God was fine with it; my brother was happy; my friend Clayton was delighted; his wise and gracious wife, Brenda, did no more than cast a light hearted glance heavenward; and even my father had a good close look at it – commenting that it was 'pretty nice' – when my brother and I took our first road trip out to the farm on the weekend. And of course we all had a good chat about motorcycles old and new!

In spite of all this good will, the bike was a disappointment. It was one of the last years of the old electric ignition and the points and condenser needed replacing about every 5000 miles – even after 4000 it was impossible to start without the most elaborate efforts. Mine was one of those personalities that would be inclined to make the efforts even more elaborate and extreme, as if to shame the designer of the machine. On one occasion I got one of my co-workers to tow it behind his truck up to about 60 miles per hour; before realising I hadn't switched the key on – a little push by hand probably would have done the trick. Surely Mr Yamaha was wishing he had done better that day!

It was in one of those moments of advanced psychological warfare that the very reasonable voice of God suggested that filing the ignition points with my nail file would only take a few minutes. I say 'reasonable voice of God' but it came, not vocally or in words, but as a fleeting pictorial memory of my grandfather in a parallel situation with one of his farm machines; patiently repairing rather than angrily trying to force. I knew in my heart that this was more than just a memory; it was a word from God. When I filed the points – and it worked of course – a great peace came over my soul and the Lord indicated that he wanted to talk to me about something else; my longer term career plans. First: a little background.

About Stars and the Father's heart

I've always loved the stars and there are few places on earth where they are more beckoning and beautiful than the wide open prairies of western Canada; my childhood home. On a clear night a boy could lie on his back, looking straight up, and feel part of – or at least have the pleasure of longing to be a part of – the vast and lovely world of the heavens. For this reason I had always intended to study astronomy and had even started on a general physics degree a few years earlier. Financial and other considerations had resulted in a detour; apprenticing as a machinist/engineer, but at the time of the first motorcycle I only had a year left and was planning to return to university. I did already know by then that the thing a person desires when looking up at the stars can never really be met by studying astronomy or even by interstellar space travel (should that ever become possible). I suspected then, and still do, that actually travelling to such places would be no better than a good science fiction film as far as the fulfilment of the desire, but astronomy still seemed like the closest thing available – what else was there? It was against this background that God wished to speak to me after he had helped me get my motorcycle started.

He showed me the scene of a large and serious traffic accident; cars and trucks piled up and smashed as far as the eye could see. It was not a situation that could be ignored. There were many injured. The Lord was asking me to get involved in the kind of tone that reminded me of the way my father might have asked me to come away from something I was enjoying, to help in an emergency on the farm. Never one to pull rank; my father, would meekly appeal to reason in those kinds of situations – without a trace of the manipulative self pity found in the falsely humble. But this was a different degree altogether. I have never seen anyone close – not my father or mother or any other man or women however blessed – to the meekness and humility of the Lord on that occasion. "Astronomy is a noble study," he seemed to say almost apologetically, "but we have a situation here that's very pressing, and there are so many wounded."

Should the maker of heaven and earth speak to a grumpy and impatient young man that way? Could it be that God the Father is the meekest and humblest person anywhere ever? I believe he is. I cannot doubt that the Father is, and has always been, the greatest in meekness, in humility and in serving – and that he will remain so, for all eternity. It was this greatness that Jesus was referring to in John 14:28 when he said: "...the Father is greater than I". For the love of this greatest Father he could even face the cross because he saw that the Father had been bearing the sins of the whole world in his own tender heart since the fall in Eden. I would also go so far as to say that the flashing lights and thrones and rainbows and flying creatures and clouds and smoke and all the normal portrayals of heavenly throne room glory are actually concealed glory, and that the humble hearted father, is actually revealed glory, rather than the other way around as it is often taught.

Standing by my XS 400 that day, I was in awe of God because he was so humble; He was so much more like my dad than any person that I had imagined from what I had been taught in church. And I could not refuse his request to set aside studying astronomy and join his rescue work. I had no desire to refuse his request; my only desire at that moment was to be with him.

Peckham 'Space Breach'

To complete one strand of this story I'll jump ahead to a day less than 3 years later. I was standing on a doorstep in Peckham, London, England, explaining the gospel to a young man who belonged to a coven of witches and warlocks that met in Greenwich. In spite of belonging to such a group, the young man, claimed to be quite a good person. His reason: he preferred white magic whereas some of the others in the group practised black magic and were – his term – 'malevolent'.

He then gave a more detailed account of their evil and I said something about knowing and loving Jesus. He countered with: "Well... you can't really 'love' Jesus".

So I said: "I Love Jesus" and then the Holy Spirit said (through me): "I Love Jesus" and then the Father said (also through me): "I Love Jesus".

At this third statement I was one with the Father for a moment and I felt the whole universe with all its beautiful stars stretched out across my heart – right inside me: vast and lovely and beautiful and good. If a man were to study astronomy all his life or travel the galaxies in a star ship he would never lay hold of the stars like I did in that moment – "*when the morning stars sang together, and the sons of God shouted for joy.*" (Job 38:7) To the man on the doorstep this threefold declaration of loving Jesus – with crescendoing authority – was so overwhelming that he staggered back and slammed his door.

An explanation here may be helpful and it also affords an opportunity to emphasize one of my underlying themes of how God integrates symbol and life and reality in a way that we could never contrive and often don't even understand – I didn't see what I'm about to say until I started writing this story many years after the event. There are two things.

First, a story that begins with me being 'adjusted' by God in the midst of angry 'psychological warfare'; ends with victory over a witchcraft spirit. God overcomes evil by humility, grace and love; not by being more dominating than the dominator. Human anger does not work the righteousness of God (Ref: James 1:20), more likely it is an expression of frustration at not being the one 'on top' and an attempt to get there. The expression of anger in this way: "...a path to the dark side surely is!" (As Yoda of Star Wars, might say.) Much that appears – even in the church – as ordinary behaviour, is really a subtle form of this same angry psychological warfare aimed at putting someone else 'in their place'. Religion, control and witchcraft are all united in this angry assertion of self-will; and by it the ministries of individuals or churches can cross over from light to darkness in a single choice of the heart.

Second, vehicles are often symbolic of a person's ministry or calling and the way this calling from God was associated and integrated with my motorcycle is an example of how God weaves symbol and reality. On this, I will say more from time to time as we proceed.

Enter: C.S. Lewis

For the winter of 1980/81 I took my red Yamaha out to the farm to store it until spring and my brother put his, newer, blue one beside the sofa and arm chair in the living room of our apartment. It made a nice ornament and was handy as an extra seat, but irritating too, when certain meddlesome and fiddly kinds of persons came to visit. Clayton Mellsen, of course, thought it was great; his wife, Brenda, thought it was silly (being already a little weary of motorcycles); and the landlord restrained himself (vacancy rates were high and we always paid on time) – though I caught his eye surveying the carpet just beneath it.

It was in this same room that winter that I first read C.S. Lewis: all seven of *The Chronicles of Narnia* and also *Surprised by Joy*. I had often been on the edge of discussions involving C.S. Lewis and had been introduced to his thought by my cousin Carolyn in discussions of all things literary and spiritual. In reading for myself, however, I found at last some kind of articulation for the complexity of desires, longings and joys that variously buffeted my heart and mind. Lewis describes in *Surprised by Joy* his own experience of desire that is at once joy and sorrow; a sheer joyful longing evoked by – in his examples – a model 'toy' garden (that his brother made) , the idea of autumn (from Beatrix Potter's *Squirrel Nutkin*) , and 'northernness' (obliquely from Longfellow's *Tegner's Drapa*).

I used the expression: 'Joy-Sorrow' to describe my feeling which I now understand to be a desire of the Spirit – a desire from the heart of God touching directly to our hearts by the Holy Spirit. The reason there is sorrow mingled is because these desires always have a direction and purpose; a longing at one level for what once was, and can be again; but more than that too. There is also a longing for what has always only been potential – that deep ancient desire in the heart of the Father for children wise and mature enough to share deep friendship with Him. This longing in the Father's heart came before Eden and before the fall of man in the garden. "Deep calls to deep" says the psalmist in Psalm 42 as he grapples with conflicting joy and sorrow; with hope and despair; wrestling to find that not yet defined deep place.

Of course in the heart of a young man many desires try to claim attention and it takes some experience to sift them because – though some are clearly good and some are clearly bad – some desires that may at first seem trivial or even wrong are, in truth, doorways into a very good and wider world. I think my desire for motorcycles and many other desires that don't have any religious tone at all are in that category. If we get too religious – too negative about desires – we miss most, if not all, that the Holy Spirit is trying to do with us.

One day I was discussing the concept of joy and desire from C.S. Lewis's 'Narnia' and *Surprised by Joy* with Les Klemensen; an old school friend of Clayton Mellsen who had moved into the area sometime during that winter after having spent a year working in London, England. Les also got a job at Tero Welding and Machine and found himself, on all sides, immersed in motorcycle talk – though he was more of an intellectual philosopher type. On this day, as if to draw our discussion of joy and desire into a wider synthesis, Les, as he stood facing my brother's motorcycle (still there by the sofa) , with an open handed gesture and an affected English intellectual accent, said: 'The Joy of Motorcycle'. This saying (always with an intellectual air and English accent) along with, 'The Call of the Open Road', became the light-hearted watchwords of our little motorcycle gang. And of course Les bought himself a bike at the first sign of spring.

FORBIDDEN FRUIT

After a long, cold and snowy winter, during which, in response to my heavenly Father's request, I had made plans to attend Bodenseehof Bible School in Germany in preparation for missionary work, my brother, loathing even the thought of ignition problems, took his bike from the living room straight down to the motorcycle dealer and traded it on a new Yamaha 750 Seca. Then my little red XS400 was the smallest in the gang, never reliable, and in truth, not able to stimulate the least bit of desire or joy in my heart – though from the safety of 30 years in the future, I remember it quite fondly.

Martin, the motorcycle dealer, was a podgy little balding man with glasses, and though not likely a rider himself, was one of those salesmen able to impart desire into their customers. Always busy and bustling with all the little grunty breathing noises that a certain type of salesmen has – just giving you enough attention to keep you hooked – and everything he had, seemed precious. This motorcycle shop was not a good place for a man who had committed himself to serving God and who was planning to attend bible school in a few months to be spending much time. But with my brother's new bike, and the accessories, and the parts, and the problems with my own bike, and all the talk: there I was. Even if I had been able to resist the general atmosphere of desire and Martin's salesmanship with the help of Brenda Mellsen's wise council – for by then some of us treated her as a sort of pastor and she had already helped me to see the wisdom of not buying a boat a few weeks earlier – I was no match (or didn't try to be) for a new Honda CX500 poster of a young man standing beside one in the twilight overlooking a city.

My heart was smitten and handily Martin had one in stock – a very low mileage used one – same beautiful blue as the poster – but it was likely to be gone at any moment!

I had to act quickly – $500 for my old bike and the rest on a credit card. "Money for bible school? – well I'm sure it will all work out somehow if it's meant to." (This kind of determinism is such a handy old doctrine for sinners! Indeed sinners make very creative theologians generally – but of course there's usually something available 'off the shelf'!)

The Lord said nothing. Clayton, Les and my brother Duane all rejoiced with me in the bike's qualities; shaft drive, water cooled and such a nice throaty roar to the transverse V-twin engine. Brenda did her best to hide her thoughts; but her weariness of motorcycles and of people who ignore Godly advice was plain enough in her eyes.

Perhaps I should note here, so as not to confuse anyone, that there is nothing wrong with having motorcycles; indeed God had used the first one to teach me some good lessons. However, there are moments on our journey with God that the choice we are making – to serve God or mammon – hangs on seemingly very small things. Sometimes it's only a matter of timing as far as it appears externally, but ultimately the real issue is about the inner choices of the heart. Similarly: there was nothing wrong, in general, with a hungry Esau having a bowl of soup but to choose it above his birthright on one particular day was hugely consequential. (Ref: Gen 25:27ff)

ENTER: THE MOTH

Within a few days a disconcerting knock developed in the engine of my new bike. Martin was impossibly busy just then but his parts man suggested I check with the Honda dealer to see if it was still under warranty. The Honda dealer looked up the serial number – unfortunately, no warranty and... again unfortunately, it appeared that the serial number indicated that the bike was one of the early 1979 models that were prone to a camshaft fault... a warranty repair would have been available to the original owner. Back to Martin: "...Well, I might be able to take it on trade on a new one once the problem is sorted out."

"Lord?"

I went back to the apartment and flipped open my bible. There on the page in front of me was Psalm 39:11 "*When thou dost chasten man with rebukes for sin, thou dost consume like a moth what is dear to him...*" There was no getting around it; I was caught in the cold glare of the Lord's searchlight. I found out later, that the word translated 'dear' is the same word translated 'covet' in the 10th commandment and the same word used by that wretched Achan who took forbidden treasure from Jericho, but even if I had known then, I would have been no more or less convicted. I was guilty. More remorseful than repentant, I went out to the Mellsen's house for a coffee.

Clayton, a mechanic and ever the loyal friend, suggested that we could make the necessary repairs and offered to help, but I knew I needed a deeper answer than that. Brenda was tired; the baby had been up in the night and was still unwell and crying a lot. There wasn't much she could say anyway because the ball was obviously in my court and, being a nurse by profession, she wasn't one to sit around discussing every nuance of pain (ask any nurse's spouse if you don't believe me). I went home and laid it all before the Lord – I won't say repented – I don't know – just laid it out. Somehow there was grace to actually repent in the end.

First thing next morning I went down to Martin's motorcycle shop and offered it back to him for just the amount I had put on the credit card – he wasn't interested. Then I went back to the Honda dealer to see what the repair would cost, and the Lord was with me, and he seemed pretty cheerful. (Perceived as a fleeting memory of my dad's cheerful face on a grey and chilly winter morning when everyone else seemed to be indulging grumpiness.) There was a different man working in the Honda dealership that day and when he looked up the serial number he said: "Well sir, your bike should be fine – it should have had the upgraded cam. Those engines have a few knocks and clunks but they're bullet proof – I wouldn't worry if I was you."

The moth had withdrawn! Over the next 5 years I put more than 50,000 miles on the bike; criss-crossing 2 continents, without a single repair except spark plugs. More than that, there seemed to be a blessing over it; I often found myself driving on wet roads but rarely in rain. At first I thought it was just my imagination – then perhaps coincidence – then... God? The very thing that I had coveted and then bought ended up being a symbol of grace in my life after repentance and forgiveness. It doesn't always work exactly that way; there are other times when we actually need to get rid of things and in all cases we certainly need to be willing to get rid of coveted things if we want to walk closely with God. Luke 14:33 reads: "*So therefore, whoever does not renounce all that he has cannot be my disciple.*"

Covetousness is idolatry – the coveted thing becomes an idol in our hearts; bringing impurity that prevents us seeing God; and double mindedness that prevents us from receiving from God. (Ref: James1:7-8) This motorcycle, though, after repentance and forgiveness became the opposite of that; an example of the Apostle's words that where sin abounds, grace abounds much more. (Ref: Romans 5:20) It was while riding this same CX500 Honda over the next few years that God began to teach me his supernatural ways; of seeing his hands at work and even ultimately his Face.

I should also mention here that when I gave notice to quit at the company I was working for at the time, in order to go to bible school in Germany I was immediately confronted with the temptation of a large salary with extra benefits and so on. This is a very common temptation to people who respond to God's call. If I hadn't just had the experience of overcoming covetousness on a small scale with the motorcycle I would have been much more likely to succumb to this new temptation – with the consequence of not entering my calling at all. This is often the way with

God and it is important therefore that we live each day to the full, not ignoring the little tests along the way, but remembering the words of Jesus when he said: "*He who is faithful in little is also faithful in much.*" (Luke 16:10)

Chapter 5

The Call of the Open Road

If a transtemporal, transfinite good is our real destiny, then any other good on which our desire fixes must be in some degree fallacious, must bear at best only a symbolical relation to what will truly satisfy.
(C. S. Lewis "The Weight of Glory")

Ordinary Miracles

My remaining few months in Canada, before heading to Bible School in Germany, were memorable for two long distance road trips by Motorcycle. The first was a trip to Vancouver, Victoria, Port Angeles and Seattle through the Rocky Mountains on the TransCanada Highway outbound and mainly on I-90 homebound; through Washington State and Idaho; before turning northward to the Crow's Nest Pass. Brenda and Clayton Mellsen were on their Yamaha 500 and I was on my redeemed CX 500 Honda. The accommodation was to be tents except in emergency – with friends and Clayton's sister in the Vancouver area providing a welcome relief from the discomforts of camping at that point in the trip.

The very first night was a test for me. Climbing higher and higher into the Canadian Rockies with their icy peaks was fine in the warm July sun but the temperature dropped as evening approached and by the time we decided to stop for the night the air had a wintery tinge. I thought it might be an emergency hotel situation but the Mellsen's thought it was ideal for camping. I moaned about the rocky ground as we pitched our tents and Brenda laughingly said she would pray for pillows under me. I can still see the Godly, playful joy on her face as she said it. Later, as I lay on the rocky floor of my tent twisting, turning and reflecting on the day's events I thought of her words again, and just as I was drifting off to sleep, I had a perception of many little feathery pillows coming under me – but I banished them as ridiculous and impossible in my grumpy unbelief. What followed was a long, cold, night with fitful sleep and little rest.

Unbelief is an amazing thing; even Jesus marvelled at it according to Mark 6:6 and was not able to do many mighty works because of it. Human beings seem to be disposed against the supernatural intervention of God even though they often lament its lack and talk as if they want it. Many times I had said and heard others say something like: "I wish I lived in bible times when miracles happened." But at a moment when God was trying to do something I found myself resisting and rejecting it. I can identify pride as one piece of the puzzle; I didn't want to be childlike, vulnerable and open hearted. Who wants to be foolish and gullible enough to actually think that little feathery pillows could miraculously materialize out of thin air? (Check your own heart right now!) Control is another resistant strand; I was a little miffed at having to camp in the first place and wasn't about to cooperate with the idea of camping being a good – even a God thing. Self-pity was a player; I was 'enjoying' my suffering too much to let go of it! Fear was there too; "I don't want to let my heart run down this path in case it doesn't actually happen." Perhaps the trickiest of all is the religious attitude that says: "I don't deserve this so I won't receive it." This has the appearance of honour and decency but is really just pride as well. I had been unkind to my friends by being moody and grumpy and so I felt distant from God. The honourable thing seemed to be to 'take my medicine' because I deserved it.

Another thing about the supernatural is that when it is happening or about to happen the atmosphere can seem very normal and ordinary – even boring sometimes. The Israelites became very bored of the manna that miraculously fell from heaven every day and the Pharisees asked for signs even after healing miracles had taken place right in front of their faces. The heart attitude is always the key issue; the impure heart; the unsound eye will simply never see God; for in truth, unbelief in God is actually a belief in something else.

In the story, told by Jesus, of the rich man and Lazarus, (Ref: Luke 16:19-31) the rich man, in his torment, begged Abraham to send Lazarus back to warn his brothers but Abraham said: "*If they do not hear Moses and the prophets, neither will they be convinced if someone should rise from the dead.*" (Luke 16: 31) As if to prove the exact truth of this story Jesus actually did raise a man called Lazarus from the dead with the result, as far as the Pharisees were concerned, that they plotted to kill him again. (Ref: John12:10) That is the nature of unbelief: it actively resists the activity of God; it is an aspect of the dynamic darkness of the serpent's kingdom. This was my first test with this kind of miracle and the result was a cold uncomfortable night. I had failed miserably but I was beginning to realise that God was using this motorcycle as a tool in my life to open my heart to the supernatural and to teach me the attitudes of heart and mind that are needed to live in that realm.

The whole journey was idyllic with the mountains, the coastal cities, the sea, the islands and the great wooded regions of western Washington and British Columbia but I was not yet awake enough in my spirit to drink more than a few tiny sips (like old pastor Knutson) of the beauty and glory liberally scattered there by the creator. I know that some people draw a distinct line between what they call aesthetic pleasures and spiritual pleasures but I am on a mission to blur the line. Even, 'immortal longings', found in the title of a previous chapter, is a term taken from a margin note in C.S. Lewis's *Problem of Pain* where he casts doubt on the connection of these longings with the Holy Spirit. I on the other hand am recklessly – but in company with Isaiah's seraphim – declaring that the whole earth is full of his glory. All that is beautiful, all that is true, and all that is good could not possibly have any other source than the brooding Holy Spirit of the opening verses of Genesis.

Lewis was necessarily wielding a hard, sharp sword against a literary corpus that had high-jacked and corrupted such pleasures into other systems of thought and so I'm not suggesting he was wrong – but not right for all times and places either. Certainly the list of sublime and subtle pleasures that the margin note about 'immortal longings' refers to would be called 'desires of the Spirit' in my way of thinking. (And I think Lewis would have agreed by the time he wrote, *The Chronicles of Narnia.*)

TRUE LOVE

The second trip was with my brother Duane. Another brother, Arlin, had found true love at last while on a Missionary training course in Los Angeles. The marriage was announced for August 8, 1981. I wasn't leaving for Germany until September so Duane and I decided to ride our motorcycles down to LA for the wedding; a round trip of almost four thousand miles.

We pitched tents in idyllic conditions somewhere in the hill country of Montana on the first night, got an early start and arrived next midday in Yellowstone Park. We had lunch near 'Old Faithful' and then pressed on to Salt Lake City where we stayed in a hotel that night because of the heat. I got a new rear tire first thing in the morning, from a man who assured us he wasn't a Mormon, and we made a late start southward on I-15 in oppressive heat. By mid afternoon the temperature reached 116 F and we put our leather jackets back on to shield us from the scorching air off the desert sand and the fierce heat of the sun. Only once – in the deserts of central Asia several years later – have I ever been hotter. We passed through the bright lights of Las Vegas around midnight but it was still so hot that we decided to keep on driving through the night rather than face the sun another day. I had been told and always believed that deserts got cold at night – and I was longing to experience this – but in fact this desert stayed hot. We arrived in LA mid morning where the discomfort and tiredness of our day and night drive were soon soothed away by the gracious hospitality of my brother's soon-to-be in-laws; the most hospitable people I've ever met to this day!

I was best man at the wedding with a bright red, sunburnt, peeling face... but of course it doesn't matter what the best man looks like. The last item of the ceremony was a love song sung by the bride to the groom. As the song progressed the tangible intensity of the love increased so greatly that I had to try and hide behind my brother because I... well only old ladies are supposed to cry at weddings... right? Surely not weather beaten bikers! I could see that the love was true love, and so it has proven to be – not just for each other but overflowing in blessing to many as true love always does.

Arlin and Ruth have spent their lives serving God in very difficult and dangerous third world and conflict situations, raising 5 beautiful children at the same time. On one of their trips back from Canada they had a stopover with us in London for a few days. We had just bought a house which the Lord had provided the money for. After they left, the Lord said to my wife: "If I had given you this house just to accommodate them these few days it would have been reason enough for me." Such is the honour and regard that the Lord holds for his servants and the relative low regard that he holds for houses and money and such things. When he calls us into loving and serving him rather than money we can be assured that he follows the same principle in his love for us; he loves each one of us more than his own life, as he proved on the cross. Certainly, he wouldn't trade the most sinful of us for all the money in the world.

Our failure to recognize or remember this fact is one of the reasons we find the Lord's action with money so confusing – such as when he got Judas to look after the money box. The truth is; Jesus loved Judas: not money and he loves us: not money – whatever lies may have been sown into our hearts by heavy teachings on tithing and stewardship and by the behaviour of some religious people and organizations. Let me be a little provocative just to drive that point home. It is more important to Jesus that people – including the poor and even dishonest and irresponsible people – are loved, than that the church finances are 'perfectly' managed.

To be a window to the light of our heavenly father's face, the church too must adopt his values; it must clearly show and demonstrate that it loves the lost men and women of this world far more than money or any other worldly power. People instinctively know, the moment they step into a church, whether they are loved and valued or whether they are way down the list below the building, the program, the vision, the giving of tithes and offerings, the leadership structures, the general pecking order of whose who, the doctrines and distinctive traditions, and all the other trappings of religion. When the church loves God and people more than all that: how great is the light; when it doesn't: how great is the darkness!

At a macro historic level, when the early church fell to the love of money and the other worldly powers, an age of darkness ensued that endured 10 centuries; the dark ages. How great is the darkness! After the Reformation many parts of the church (both Protestant and Catholic) once again became the 'salty salt', the 'light to the world' and 'the city set on a hill' that we are called to be; pioneering in education, medicine and social justice, for example, as well as preaching the gospel of salvation.

Arlin and Ruth and also my old friends Clayton and Brenda Mellsen are part of that genuine church. They have all lived for the best part of the last thirty years serving some of the neediest people in the darkest and most dangerous places on earth; their very lives – though often at risk – a light and a window to heaven.

Home by a Different Road

After the wedding Duane and I spent a day in Disneyland with family from both sides. We noted the sign at the entrance as we passed under it: "Here you leave today and enter the world of yesterday, tomorrow and fantasy." I like Disneyland, but there is nothing – other than the true spiritual world and the heart of God – that could actually fulfil the longing evoked by that boast. The slogan draws our hearts because we were made for the true spiritual eternal world and will never be at peace until we find it. I was on quest for it but I didn't yet know exactly what 'it' was.

For the return journey we decided to take the cool and scenic route rather than I-15 through the desert. The first leg was northward on route 1; a road etched into the cliff faces of the pacific coast from Los Angeles to San Francisco. The name Monterey (a town and district along that road) had always held a certain mystique for me because of an oil painting, by my mother's sister, of an ancient gnarled tree on the rocky coastline there. The painting held pride of place in the family living room and as a boy I had often starred at that wild, windswept scene and longed to be there – or at least desired to access it in some way. This day I was driving miles and miles in the very atmosphere that the picture had evoked; actually there at a place that had previously seemed like a fairy tale place and, because my spirit was beginning to wake up, the feeling was just beginning to rise above the disappointment that travelling to longed-for places sometimes can be.

We crossed the Golden Gate Bridge (in fog at night) and then pressed northward until the chilly coastal climate made us turn inland. The giant Red Woods of Oregon and Mt St Helens, which had had a volcanic eruption the previous year, were additional points of interest on the trip but we were hurrying to get home and, for the most part, only touched the surface.

MATTER BREACH?

One little event I will mention because it put something in my mind that had relevance a few years later. We had driven past Mt St Helens and toward Yakima in pretty desolate country and I was very low on fuel – easily done on a CX500 Custom which only holds a bit over 2 gallons. I was already on reserve which is probably only good for 20 miles at most. (The older bikes didn't have a gauge; they relied on a secondary supply or 'reserve' that you could switch too when you ran out on the main tank.) Mile after mile, another curve, another hill and not a service station in sight... praying for the fuel to hold out... it did. I can't claim it as a miracle because I had never actually run the bike totally dry but it was about 20 miles farther than I estimated it should have been able to go by the time we did find a service station.

Several years later I made a deliberate effort to push

into this kind of supernatural supply – literally driving without fuel – with the same CX500. I was in London then and very short of money when I felt an inclination – maybe even a challenge from the Holy Spirit – to try to break into this. As the miles beyond the safe limit passed the pressure on my spirit – an intense barrage against my heart and mind – increased until each time I would eventually give in and buy some fuel. At the worst: fierce reptilian beasts – visible to the eyes of the spirit – were attacking my soul and the pain in my heart would increase until I could no longer tell whether it was physical or spiritual. After two or three attempts I dropped the idea... or sort of put it off... just too painful. My experiment ended in apparent failure.

However, when reflecting on it some years later, a clear thought dropped into my mind: *It wasn't God who resisted; but the serpent.* This world and its usurping ruler (the serpent) resist this kind of thing just as surely as you might expect an oil company or oil exporting nation to. God, on the other hand, is quite open to it. It is the open road. Narrow and steep perhaps; but open all the same!

About Miracles

In conclusion: three comments about the miraculous. The first is that seeing God and seeing the activity of God are inseparable. The Pharisees were 'blind' as evidenced by their refusal to acknowledge that the miraculous works of Jesus were the activity of God; they could not 'see' God in them. This was because, rather than choosing God and his kingdom, they had chosen the powers of money, religion, politics and military that comprise the serpent's kingdom; 'the kingdom of this world'. And they had got into this state not by big decisions but by little day to day choices to indulge selfish appetites and desires; little abuses of power – perhaps some angry psychological warfare (like me against Mr Yamaha); a bowl of soup here (like Esau); a coveted motorcycle there (like me); or a posture of superiority here and there (as we've all done).

The second comment is that the miraculous is needed to get us out of the serpent's control. The serpent is 'the ruler of this world' and he's a bit like the mafia in Italy in that where ever you turn his power hems you in. He controls the banks, big business, the government, the police and the courts, so to speak, so where ever you turn you're trapped.

The temple council of Pharisees and Sadducees at the time of Jesus' first century ministry also controlled every aspect of worldly power and the only thing that rocked them was Jesus miraculous power; anything less and they would have ignored him. Similarly, Israel in Moses' day needed the supernatural power of God to get free from Pharaoh and Egypt; the supernatural plagues delivered them from the religious and political power of Egypt; the supernatural crossing of the red sea delivered them from the military; the supernatural supply of food and water were needed to keep them from going back to Egypt's financial power 'cap in hand'.

In a similar way, for us, there is no doorway out of the kingdom of darkness and into the kingdom of God without the miraculous help of God himself. This begins with our great 'Red Sea crossing': the death and resurrection of Jesus; and continues with a life in his miraculous grace; the continuous filling of his Holy Spirit. Anything less and you will find yourself in a religious cul-de-sac; a road to precisely nowhere; like the proverbial bear you will see only the other side of the same mountain; just the other side of the same old slimy, grinning serpent face.

The third comment: we need to broaden our understanding of the miraculous. One of my ongoing and recurring themes is that of discovering – of 'seeing' – God and his kingdom and his glory throughout the whole earth in everything from nature to history; from culture to astronomy; from architecture to mathematics; and so on. The man or woman with an impure heart will not see God in the obviously miraculous; neither will they see God in any of this vast array. I don't say this to condemn but to urge you to accept the miraculous grace of God, repent of the lukewarm dithering attitude and take up the healing eye ointment that Jesus offers. (Ref: Rev3:18) Get your treasure and your heart into the same heavenly place and the whole universe will open up its beauty, truth and goodness to your eyes.

Chapter 6

Of Castles and Cafés

It is not the physical objects that I am speaking of, but that indescribable something of which they become for a moment the messengers... . For they are not the thing itself; they are only the scent of a flower we have not found, the echo of a tune we have not heard, news from a country we have never yet visited.

(C. S. Lewis "The Weight of Glory")

When Heart and Treasure Meet

I parked my Ford F150 pickup at the side of a road that comprised the 4th boundary of an abandoned farmyard. On the other three sides were trees that had been planted in some distant springtime of hope as a buffer against the winds that swept relentlessly across the surrounding prairie. The family that had thus hoped were long gone as was their house and barn, though the trees were now of a size and density to provide useful shelter. It was a scene common in the countryside where I grew up though in some places a few derelict buildings remained.

More than once I had heard my grandfather echo

the lament of the old prophet: "*Woe to you... who add field to field until there is no more room and you are made to dwell alone in the midst of the land,*" (Isaiah 5:8) as he felt the loss of community that accompanied ever enlarging and increasingly mechanized farms. Today, however, the time line seemed to be in full reverse and the farmyard was a hive of activity. In front of rustic looking tents were rustic looking people cooking over open fires. I walked over to have a closer look.

A man in medieval attire and a sort of medieval voice shouted at me: "Yon carriage seemeth offensive to mine eye!"

"...ah... you don't like my truck?"

"Thou thyself art welcome but thy carriage offends for it hails from a different age". I moved it along the road out of view behind the trees and walked back.

Battle lines were forming by men dressed in rough leather and iron armour bearing spears, swords and maces. Battle was joined; bloodless, thanks to padding at all the sharp edges, but intense all the same. After the battle I mingled as best I could. These were members of a historical re-enactment society that had divided North America into chapters or 'kingdoms'. They regularly met for battles and general medieval 'hanging out'. The making – by ancient methods and crafts – of armour, weapons and the various other accoutrements of medieval life was all part of the experience.

It seemed like an interesting hobby but I already knew what they were after in their hearts and I already knew that it couldn't be attained – at least not at the level I wanted –

by re-enactment. I could identify with the longing evoked by all things medieval and there was a joy and a sorrow in dwelling on it – by re-enactment for example – but I already knew that those kinds of longings disappeared the moment you 'captured' them. For a plainer example: I had already noticed that homesickness faded away from my heart as I approached home so that the moment of actually arriving was never the pleasure that I had imagined it would be a few hours earlier; desire and fulfilment never quite connected.

We touched on this in a previous chapter in reference to C.S. Lewis's . Lewis describes the longing as joy that could equally be called a kind of sorrow. I used the term Joy-Sorrow to describe the feeling I had already been experiencing for some time. It was a desire – unfulfilled – but which the very desiring was joy and yet a longing for fulfilment at the same time. To be sure, there was sorrow with the joy – but it was a clean and cleansing sorrow with nothing at all of despair or hopelessness in it. Longfellow's lines from, *When Day is Done,* skirt the edge of it: "I see the lights of the village gleam through the rain and the mist and a feeling of sadness comes o'er me that my soul cannot resist. A feeling of sadness and longing that's not akin to pain, and resembles sorrow only as the mist resembles the rain." The sheer joy element is weak in these lines but C.S. Lewis with his Cair Paravel Castle in Narnia, beginning in *The Magicians Nephew* and climaxing in the rediscovered ruin scene in *Prince Caspian,* blends the joy-sorrow elements with perfection, like a master brewer of the richest coffees or a vintner of the finest wines.

But even before reading any of these, or tasting these coffees or wines, I was on the quest for the elusive longing and the presumably even more elusive thing that the longing ultimately pertained to. I have already, in an earlier chapter, given some definition to this as far as it relates to the desire and heart responses of God to his sons and daughters – how he loves them and longs that they should be wise and deep enough to be his true companions and friends; "*...deep calls to deep...*" (Psalm 42:7) There is no doubt that the Father's heart is the deep well from which the deepest sorrows and the deepest joys, travail and mingle to form the longings and desires that capture our hearts and that these are searched out and brought to us by the Holy Spirit; the 'desires of the Spirit' as Paul calls them. (Ref: Gal 5:17) But I did not make those theological connections until much later; at the time I was working with fleeting images and feelings coupled with an elusive sense of destiny.

I don't remember whether it was originally a dream or a day dream but I had an image in my mind of myself standing in morning sunshine. Behind was a tangled wood. My face was tear-stained and dirty and weary; but joyful with disbelief at the sight before me. Initially I was seeing the lower corner of an ancient stone wall just where it met the grass and ground. As the view widened I could see that I was a child leading a small group of other children. We were all weary pilgrims with the troubles of the woods and the night still clinging – though beginning to fall away; if only we dared let them go. As we looked up, there before us was a castle: strong and old, beautiful and new; its banners waving in the cool morning air; its battlements and turrets glistening in the fresh, living light of the morning sun. Here was joy beyond joy and hope beyond hope. "*Weeping may endure for the night but joy comes with the morning.*" (Psalm 30:5). Whether this image was the origin of my desire for castles – 'the joy of castle' I called it – or just a strong incidence of it, I can't say. Perhaps every boy who grows up in western Canada, where no building is likely to be more than a hundred years old, longs for something ancient, strong and mysterious.

Very quickly in my quest to fulfil this desire – this 'joy of castle' – I learned an important lesson. It's a lesson that applies to many, most or even all spiritual quests. The first way I approached this was to think of ways that I could own my own castle – perhaps build one or buy one. I saw a documentary about a man in Pennsylvania (or was it California?) who had spent 30 years building his own castle; a replica of an actual medieval castle. He had come a long way on his project but to me it seemed like a long and dreary route. I could see the weariness in his face and hear the pain of despair in his words at the cutting and carrying of stones as he laid a couple for the cameras. The kings of old had had thousands of peasants and craftsmen working on their castles; so what hope was there for him... or me? And anyway, as discussed above, I knew in my heart that this wasn't really what I was after.

Rich men occasionally buy castles and some build houses in a "castleley' style – these are called 'Follies' in

England. The 18th century architect John Vanbrugh built one that stands near Greenwich Park in London and I used to walk by it almost every day when I was living near it a few years later. The plaque on its wall designating it: 'England's first Folly' was a poignant reminder of this lesson which is so easy to forget. (Incidentally I went there a couple of days ago and noticed that they've changed the big sign that used to be there for a little round blue plaque that has no reference to 'follies'.) Rich men and rich churches are always tempted in this way and it is one of the aspects of Jesus' warning about how hard it is for the rich to enter the kingdom of heaven; surely wise poor men and poor churches are more likely to succeed in this quest.

But I didn't make those connections until some years later. At this point in my story, in the summer of 1981, I coined the phrase: 'the apprehensible reality of castle' – and pressed on. The idea was to find the thing that actually met the desire – that actually connected to the heart; where desire and fulfilment actually met. The list of 'non apprehensible realities of castle' was growing for me but I was undeterred. The lessons from coveting motorcycles which I discussed earlier were also happening at this time and it was by the grace of God that I was on my way to Germany that autumn and some wonderful castles were waiting. To that point I had never been near an actual castle at all. You might ask why, if I was so wise (or cynical a romantic might say), was I still interested? All I can say is that, even though I knew the quest wasn't about actual stone castles, they still were some sort of doorway – at least to the feeling – and I wasn't always confident that I was right anyway.

After the lessons and trips discussed in the last chapter, my new motorcycle was put into storage on dad's farm, as the old one had the year before, and I boarded a British Airways flight to London. From there I travelled by train to southern Germany. Jet lagged and burdened with luggage for a six month stay; the trip seemed more an endurance course than a romantic quest but some gray and dreary mornings give way to sunny afternoons. Besides I hadn't really collated all the strands I was seeking. At one level I was pursuing God and training for foreign mission work; at another I was a young man seeking adventure in the wide world; and at yet another, an inveterate romantic fighting against the dull practical world that seemed, eventually, to conquer every dreamer.

Bodenseehof Bible School was set in surroundings every bit as lovely as the brochure had claimed; the Bodensee (Lake Constance) was only a stone's throw away and the mountains of Austria and Switzerland were intermittently in view across its ancient misty waters. The countryside around was idyllic with its gentle hills; its woodland and meadow; and its quaint villages connected with curvy little roads.

The Gasthof Traube

I loved southern Germany. Almost everything I touched seemed to have a depth and genuineness to it that

I had longed for in Canada but never found. The coffee was strong and delicious – the very essence and reality of coffee which I had before only sampled a shadow of. The cakes were freshly made with real fruit and real whipped cream and served by waiters and waitresses dressed in traditional uniforms of real cotton and linen. My favourite restaurant, the *Gasthof Traube*, was only a short walk from the school. Its interior was framed with great oaken beams; not plastic; not fake plaster board; not veneered wood; but real solid thick oak beams that actually held the building up. The table cloths were several layers of heavy embroidered linen; the flowers on every table: fresh daily; the candles: traditional wax in a heavy metal lamp with a glass chimney. My soul craved such things and here was a case of finding something that could actually be appreciated, where desire and fulfilment met.

Though I didn't understand it at the time I can now see that I was beginning to enter into the blessing of having chosen heavenly treasure above treasures on earth; "*For where your treasure is, there will your heart be also.*" (Matt6:21) And I didn't need to own things, in fact it was better not to. If I had owned a restaurant, castle or whatever I would have to pay for it, clean it, guard it and pay taxes on it. And what pleasure is there in any of that? No it is much better to be on the road with God; to those persons: the whole earth is full of his glory and in his right hand are pleasures for evermore. Since God actually and ultimately owns everything only those to whom he gives it can ever really own it in their hearts where it is actually pleasurable. John Wycliffe, sometimes called the Morningstar of the Reformation, wrote a treatise on this idea: *On Dominions* which didn't gain him any friends within the Catholic Church. The Church at the time had huge wealth and they didn't take kindly to the idea that God hadn't really given it to them.

Like rich churches, rich men are liable to try to own the things they desire but are equally liable to end up with J D Rockefeller's lament: "The only pleasure left to me is to watch the dividend checks roll in." And even that, for him, as far as I can see, must have been about as much fun as watching paint dry – as they say. In the words of Jesus: "*What does it profit a man if he gains the whole world but loses his own soul?*" (Luke 9:25) James makes a similar point when he writes: "*...So will the rich man fade away in the midst of his pursuits.*" (James 1:11) Rockefeller's soul, by his own confession had faded away to the point that he had no pleasure except watching numbers on paper; numbers that actually had no effect on his life at all – heart and treasure in completely separate places.

This is the perennial problem with covetousness; the ungodly desire for things or money always ends with both the heart and the desire fading away. This is why covetousness, ever promising Eden, has in truth, always been the road out. The serpent who tempts us to covet is a liar of course but even if he tried, he is not able to fulfil what he promises. As for me, I am fairly certain that the Gasthof Traube gave me more pleasure than it gave the man who held its title deed; but I'm very thankful he managed it so well on my behalf!

When Jesus promised that our hearts and treasure would be in the same place if we laid up treasure in heaven he was not talking about 'heaven' after we die; he had already taught that the Kingdom of heaven was 'at hand' and he later taught it was 'in our midst' and moreover he taught us to pray for its coming on earth in the Lord's Prayer. When we experience the true heavenly pleasure of anything in God's creation – including the whole human cultural dimension – it is an answer to the Lord's Prayer, a fulfilment of his promises and also in keeping with the purpose for which our heavenly Father made those things in the first place. All the trees in Eden were good to eat from, except the one. The one bad tree was the serpent's kingdom; all the others were part of – at least potentially – the Kingdom of Heaven. Proverbs tells us that: "*...a desire fulfilled is a tree of life.*" (Proverbs 13:12) The true fulfilment of God intended desires, which are desires of the Spirit and the experience of having heart and treasure in the same place, are a tree of life and a facet of the Kingdom of Heaven right here on earth.

And these 'trees of Eden' are not just those things normally associated with Christianity or the Church, but every good and beautiful and tasty and fun thing on earth. Every set of meaningful and cohesive ideas and feelings that have life in them are trees in Eden. They grow from God given seeds but require watering from heaven and tilling by man to grow. (Ref: Gen2:5) The great nations and cultures are the big trees; the institutions and businesses – yes even restaurants – are the shrubs and bushes and so on. But of course there are many complicated caveats because many – maybe all – of the trees are under the shadow of the one bad tree since the fall and their true natures are obscured to some extent. They all wait for the day "*...they are freed from their bondage...*" (Rom 8:21); when the clarion call goes out: "*The kingdom of the world has become the kingdom of our Lord and of his Christ...*" (Rev 11:15)

However you can get a little foretaste of the coming kingdom at the Gasthof Traube – in English literally: '*Guesthouse Grape*'... nice vine that one! For a shrub that grows the finest cakes and chocolates try *Cafe Hoepker* just up the road a ways – it comes highly recommended by the staff and students of *Bodenseehof*. Then again, if you prefer a fast growing conifer with many branches, there's always *McDonalds*. And here's the verse: "*And the Lord God commanded the man, saying, "You may freely eat of every tree of the garden...*" (Gen 2:16) Of course I'm being slightly tongue-in-cheek because literal eating and drinking is only one small aspect of 'eating' and 'drinking' in the full spiritual and universal expansion of those words as metaphors – but it's not disconnected as far as the Kingdom of Heaven is concerned and I hope that is becoming obvious by now.

The city of Saskatoon, the city nearest my father's farm, was and is a city of many restaurants – many of them decorated in some nostalgic or other place kind of theme. You could eat in a cave or a crate, in an old railway carriage or a cow town saloon, in a Mediterranean garden or a metropolitan penthouse – and of course there was every nationality represented several times over – more than six hundred restaurants in a city of only a quarter of a million people. To a restaurant buff... well it's the Garden of Eden... so to speak.

Human beings have a great wish in their hearts to somehow connect to other places and – especially, I think – to other times. The restaurants of Saskatoon were really just a little appetizer and even if they didn't, in every case, hit their target place or period very well they never-the-less testify to the existence of such desires in the human heart. Let's move on with my story to some castles in Germany and, among other things, we'll try to understand this almost universal craving for other times and places and what that means in terms of the kingdom of heaven.

THE FORTRESS AT MEERSBURG

A few miles along the lakefront from Bodenseehof, on a steep, high slope was the village of Meersburg and at the heart of that village stood Europe's most ancient fortress. It was a castle to be sure but an older castle than I normally thought of. It dated from the 6th or 7th century with many later additions and adaptations; the most recent being late 19th century. A famous poet of that era, a woman I had never heard of, had lived in part of it and her apartments were still intact. These parts had that semi-insane, dark, romantic, sickly, deathly atmosphere common to the period, but which, I was not seeking. You can find some of it in Dickens, H G Wells, Conan Doyle and many others, with some redemption of it in George McDonald, if you wish to, but I side-stepped it on our first outing from the Bible School – nor did I ever look up the woman's poems. The more ancient parts of the castle provided a pretty good access (aesthetically – spiritually?) to the early medieval period with even a memory of the last strands of Rome. There was an armoury full of swords, spears, maces and other cruel looking iron forgings along with full suits of armour. There was a dungeon, a moat and drawbridge, a portcullis, a stone ringed well and even a secret tunnel that once led down to the lake front. The stone work was crude and thick; complex of memory, and laden with the burden of time.

I suppose I must have been aware of period atmospheres before this time and, I had had conversations with my cousin Carolyn about rooms decorated and furnished in the style of particular historical eras but the only source I would have had would have been cheap romance novels of the Harlequin or Mills & Boone genre. (Yes, I confess, I read a few of those!) Films probably didn't contribute much because I can't think of a single movie from the '70s (my teen decade) that had a very convincing sense of period outside of its own. Whether it was war, sci-fi, adventure, comedy or romance; none seemed to escape that '70s feeling – and what a vacuous feeling that was! (And who can forget the unbearable architecture of that period – utility on self destructive steroids!)

Star Wars was perhaps the one exception in that it

reached 'escape velocity' in part at least; not only in earthly historic periods but some sort of 'true' other world as well. I don't deny that Luke Skywalker and Han Solo had that goofy '70s hair but wasn't Princess Leah's hairdo truly transtemporal and intergalactic? And didn't old Obe Wan take us to a medieval hermit's grotto? And wasn't Governor Tarkan a perfect embodiment of that amorphous evil that seems to stalk the nether regions of Eastern Europe – occasionally even penetrating Germany? And wasn't that the perfect foil to Darth Veda's truly transcendent evil? And didn't all those partiers, drop-outs, hippies and flower kids from the sixties – fuzzy headed, hung-over and weary of irresolute wars – rejoice to see some clear genuine goodness kicking some genuine evil into touch? In San Francisco the number of *Star Wars* tickets sold in the first few months surpassed the population of the city. We shouldn't wonder that it was so popular – people thirsting to death in a desert might well be expected to go crazy over a few drops of water! Something of the same ilk was here at the fortress in Meersburg; something real; real access; a glimpse of Eden and I liked it.

Schloss Heiligenberg

On a high hill about an hour's drive from Bodenseehof stood the grand castle Heiligenberg. This was one of the early tours organized by the staff to introduce new students to the area. It was more a mansion or stately home by some measures but it had been built on a castle site and some medieval parts remained.

The chapel was part of the old section and beneath its altar, in a stone casket, one of the castle's treasures: the legs of a medieval pope. A dispute had arisen over who should have the dead Pope's body and in the end it was chopped up and spread around in the belief that any part of it would bring blessing wherever it might be placed. A number of other relics were housed in the chapel and this was my first encounter, apart from school history lessons, with that trade in relics that was 'important' in the medieval church scene. There were splinters of wood supposedly from the cross, vials – some claiming to contain Christ's blood and others the breast milk of the Virgin Mary – and various items of 'Apostolic' origin. The sheer folly of trying to access spiritual realities by such means was strongly reinforced for me in this castle but ironically there was also a sense of access through it to that strange era of church history. And is there some sort of poetic justice in a Pope being cut up and added to the collections?

The other aspect of this castle was the connection it had to the aristocracy of Germany. At the time I was there it was still owned and lived in by descendants of an old family and historically the castle had been a meeting place of princes and kings including the Russian Tsars. At the time I didn't recognize that this had any connection with the Kingdom of Heaven or 'Eden', but I loved the feeling.

HEIDELBERG

The first break in the school year came in early November and all students were required to leave for about 4 days as the dormitories were being used for a house party. I rented a car and drove up to Heidelberg in search of castles along the Neckar River. When I got to Heidelberg I turned in at the first hotel I liked the look of and found myself in an ancient inn dating back to the fourteenth century. The atmosphere was richer than any hotel I'd ever been in and the ambiance of the evening dining room with its off season cliental of German business men was itself an access point to something in the heart of the nation. It's a hard thing to define and I doubt it can be grasped at all apart from travelling away from your home land because the atmosphere at the heart of a nation is taken for granted if you grow up in it and always live in it. People who have never lived in another country usually assume that many of their nation's unique aspects are the norm for all people everywhere – and sometimes they take a shallow negative view of the many beautiful 'other' things they touch as tourists, cinema goers or readers.

Next day it was straight down to Heidelberg Castle. Vast and sprawling, and ruin though it is, it gave me a pleasant access to later medieval periods and also a touch of the Napoleonic age. A great round guard turret stands split in half, with the outer half leaning outward but still standing. Napoleon's troops had managed to throw one of those classic black round bombs with a fuse into an archer's slot with the resultant graphic testimony to the end of the age for castles as a serious military defence. For a fleeting moment in time, I was standing among those troops on that fateful day, watching the turret fail.

Neuschwanstein Castle

On a gray January morning in 1982 the students and staff of Bodenseehof boarded a charter bus for a day trip to the most frequented tourist site in Europe; Neuschwanstein Castle in Bavaria. It's the famous castle you often see on posters and vaguely copied on a much smaller scale in Disney Land. Strictly speaking, Neuschwanstein is a 'folly' built, not as a real castle, but by a king in the mid 1800's obsessed with the idea of castles. Our first stop was the castle Hohenschwangau just across the valley from Neuschwanstein.

Hohenschwangau was the boyhood home of Ludwig II, (sometimes called mad king Ludwig) of Bavaria, who built Neuschwanstein as well as other grand houses or castles. I remember three things about Hohenschwangau. First, it was full of secret passageways that had been designed to allow the servants to tend the fires in the ceramic stoves without a hint of disturbance, spark, dust or ash to trouble the royal family.

Second, in one of the display cases was a sealed bell jar containing bread and salt; the ceremonial gift of a Russian Tsar on a visit some hundred years earlier. The sight of a loaf of bread older than communism made a strange impression on my soul which I can't fully account for to this day. Perhaps at one level it cut communism – that exalted menace of my grandfather's after church conversation – down to the little historic blip that it really is; just another passing tyrant; another 'Ozymandias' in the sands of time. (From the poem, *Ozymandias* by Shelly)

And third, on the ceiling in little Prince Ludwig's room, a mural of a starry night sky had been painted. A small hole had been drilled in each star and in the chamber above a long suffering servant had tended a candle at each hole to provide living light for the little prince's pleasure. Here then was a prince groomed to romanticism on a scale that qualified him, or at perhaps drove him, to build the castle of all castles.

The climax of my castle quest, as far as literal castles are concerned, came when we crossed over to Neuschwanstein itself. I jokingly linked arms with another student – a pretty Bavarian girl who had grown up not far from there as it happened – and we began pretending to be King and Queen. Arnold, a Dutch Canadian took on the job as a very gracious courtier and a few other students joined the game as we walked through the great halls and turrets of what must be the very spirit of 'castleness' itself. The architecture had been inspired by the grail legends and was modelled on *The Wartburg* at Eisenach (A real medieval castle where Martin Luther had been kept for his own protection and where he had famously 'thrown an inkwell at the devil') and *Pierrefonds* (a French Château of some glory).

The walls were adorned with murals from Wagner's *Twilight of the Gods, Tristan and Isolde* and others. The lavish interiors and furnishings and even a fully integrated romantic grotto provided the perfect backdrop to the whole fantasy. If I had somehow married the girl, been transported back in time and been the actual king of this castle it would have come no closer than what I got that day to the thing I was after.

In fact I suspect such an actualization would have cumbered and fettered the thing I longed for with all sorts of dreary administration I was definitely not seeking. This theory was borne out by a booklet I picked at the Castle gift shop that day. Ludwig II had bankrupted himself and borrowed heavily to build this and several other castles or grand houses. He too was searching – perhaps for the 'joy of castle' or the 'apprehensible reality of castle.' The booklet's writer suggested that it was 'the validation of his kingship' that he was seeking. Once he even took bread and wine ceremonially (a private Eucharist) seeking to access that grail blessing so earnestly sought by King Arthur's Knights of the Round Table. Perhaps I could say that he was seeking the 'apprehensible reality of kingship' to put it in my own terms. And so it is obviously possible to be an actual king and own a castle or two or even four and still not really possess it in your heart – and it may drive you crazy or maybe you'll just 'fade away' if you try to get it into your heart.

Now of course it was a sobering thought to me that some people considered Ludwig II to be quite mad and my fellow students teased a bit in the way students do but I was undeterred. My quest was for reality – the very spirit of reality – and I was beginning to feel that it was quite near. Mad or not, Ludwig II was searching for a kingdom that was absolutely real – more real than what he already had – and in my heart of hearts, so of course was I. The kingdom of heaven is more real than any earthly kingdom; more real than physical stone castles; more real than actual political and financial power over people. The desire we have in our hearts for a kingdom will only be met by the Kingdom of Heaven because the Holy Spirit brings the reality of that Kingdom right into our hearts so that the desire and the fulfilment actually connect; a tree of life. (Ref: Proverbs 13:12) The apostle Paul tells us in Romans 14:17 that the Kingdom of God is "*...righteousness, joy and peace in the Holy Spirit.*" But that doesn't mean it is purely mystical or just a feeling; quite the reverse: it's more real than the physical; indeed the whole physical universe is just one facet of the kingdom of heaven. The Holy Spirit takes the natural world, that became inaccessible (meaningless and subject to decay – Rom 8:20) after the fall, and restores our proper connection to it so that we can not only experience the joy of it, but so that we can have headship over it as was intended in the beginning – in Eden.

As far as the access to other times are concerned this is one of the most pleasant aspects of knowledge – I'm not saying we can actually travel in time physically as in Sci-fi but we can access the spirit, atmosphere and significance of different eras and through prayer even effect changes to the consequences of those eras. This great and joyful work of intercession involves healing of individuals, families, institutions and nations by bringing resolution and forgiveness and by casting out darkness with the redeeming light of Calvary. This work goes beyond living memory into the roots and foundations of human life and society; to the souls of people and peoples; to the removing, from the little trees, the shadow of the one evil tree. (This theme will be taken up in more detail in the next book in this series: *The Gates of Eden.*)

CHAPTER 7

CITIES AND CIGARETTES

By ceasing for a moment to consider my own wants I have begun to learn better what I really wanted.
(C. S. Lewis "The Weight of Glory")

THE QUEST FOR A CITY WITH FOUNDATIONS

Abraham, we are told in Hebrews, sought a city with foundations whose builder and maker was God. He had already – or rather his father had – rejected Ur of the Chaldeans and gone on a quest for a different kind of city. The writer of Hebrews suggests that if it was a normal city that he was after, he would have been able to go back to the one he had left behind. (Ref: Hebrews 11:8-16) It was a better – a heavenly city – that Abraham was seeking. This quest is the same as the quest we have been grappling with from a number of angles for several chapters now. At this point we are bringing in 'city' as a term to define at least some aspect of 'Eden' or the 'tree of life' or the 'matrix of heaven-earth integration'. 'City' on the positive side is a metaphor for the society, nation or church that God is seeking to build and it is also used throughout scripture as the opposite; the serpent's society or kingdom. The serpent builds 'cities' with his loveless powers and false loves whereas God builds his 'city' with love and service and humility.

Against the general eastward flow of the world: Abraham journeyed west. He was a man submitted to God; a servant characterised by sacrifice of worldly power and security, in pursuit of heavenly power and life. This is the life of faith – faith not in the world but in God. The life of sacrifice was symbolized and integrated in reality in Abraham's life by four alters. Sacrifice is that act by which we take faith or confidence out of one thing and place it in another. When Abraham burned an animal on his alter, it was an act of sacrifice that demonstrated and declared: "my faith is not in my wealth and ability to prosper on this earth but in God." It was a rejection of the serpent's system of ever increasing getting and an acceptance of God's way of giving. Abraham took his faith out of Ur of the Chaldeans (Babylon) and placed it in a promise of God for a different kind of land and city; Jerusalem in the Promised Land – though Abraham didn't get this clear in his own lifetime.

Abraham's fourth alter at Mt Moriah, where he was tested to the point of being willing to offer his son Isaac, is particularly instructive in understanding the foundation of God's city. (Ref: Gen22: 1ff) Hebrews 11:19 informs us that in this experience Abraham reached the point in his faith that he was able to believe that God would raise Isaac from the dead. In this spiritual faith sense Abraham did reach the city he was seeking – he was there shoulder to shoulder with God in actually laying the foundation. Jesus said of this moment that Abraham saw all the way into the future to Jesus himself (Ref: John 8:56). This 'resurrection believing faith'; this 'love stronger than death' is the absolute bedrock and foundation of the city God is building. We are called to be sons of Abraham in being those who have faith in the God whose love even raises the dead. Therefore we should be willing to sacrifice worldly – that is serpentine – power, wealth and position... and beyond that even too hold with pure hearts the blessings that God himself has given us.

To a degree I was on that 'Abrahamic pathway' having left behind a lucrative job in Canada but I wasn't coping overly well and I had a lot to learn about submitting to God as I will illustrate with a little event that happened in Lucerne. As with the second motorcycle and the job back in Canada, a small choice preceded a larger more consequential one, and once again I was a slow learner – but God is very kind, patient and gracious.

A Lesson in Lucerne

One of the day trips out from the school in the first term was into Switzerland. We went by charter bus to Lake Lucerne, took a lake boat the length of the lake to the city of Lucerne itself where we had a few hours to shop before rejoining our bus at the central train station for the journey home. The beauty of the mountains and the lake were overwhelming but for some reason I was in a depressed state of mind that day and by the time we reached the city I was positively self piteous for no concrete reason. Most of the students scattered in little groups searching for gifts and souvenirs but I walked off into the city by myself. I decided to buy some cigarettes – my normal self pity-come-rebellion outlet at that time. (I had quit smoking several years earlier). When I had made this decision, the Lord said to me: "If you buy cigarettes you'll get lost."

So I said: "No I won't". I walked up to a street Kiosk and asked for a pack of cigarettes and some matches.

"What kind?" said the man with a heavy Middle Eastern accent?

I didn't recognize any of the brands so I said: "anything, I don't care". He gave me a pack branded: 'LORD' and I thought: "This isn't a good sign... a coincidence? ...Hmmm... probably not!" I was grumpy enough to light one up anyway but determined to be very careful not to get lost. I looked back at the station and made a mental note of some of the streets and buildings. Then I walked farther along the street the kiosk was on until I came to a T-junction. I turned left. At the next street I turned left again. I'll play it real safe I thought – just go around the block and back to that main street leading down the hill to the station. Then I'll wait at the station.

I'm sure it is possible for God to actually intervene in the layout of streets and change them to keep his promises but I suppose he doesn't need to do anything so drastic to confuse a Canadian prairie boy in a strange European city

– even the idea that there was such a thing as a 'block' to go around was probably enough. In any case I was soon completely lost. Not even a sign pointing to the station where I had to rejoin the group on the bus. I tried to back track, then I tried to back track back to the first back track point and try an alternate. I never seemed to ever find the same spot twice. This was serious. I chucked the cigarettes and muttered a repentance – that felt totally hollow and fake.

I had no alternative but to start asking directions (and we all know how much men hate that!) with my one line of German: "Wo ist der Bahnhof?" (Where is the train station?) The only problem was I couldn't understand the answer and could only go by whatever direction they happened to point. For non English speakers: don't ever try this in England because usually the English point totally randomly and purely to illustrate the idea of movement when giving directions. With a farmer in western Canada on the other hand you could safely set your compass by the exact angle he casually points his finger. I assume the Swiss are somewhere between these extremes because I had to ask about ten different people before I finally rushed breathlessly up to the bus, minutes before it departed. I bore the lesson in mind.

A Vision in Vienna

The idea of not going home for Christmas was one that had never crossed my mind. Mum and Dad had, whatever the financial or other pressures they were under, always created a beautiful heavenly atmosphere in our home at Christmas time – especially Christmas Eve. It was an event never to be missed and I hadn't in the previous 22 years. However, I had a strong feeling in my heart that God had something different in mind for this Christmas of 1981. Here was a case of being asked to give up something good that God had given in the pursuit of something further. I realize this was a small thing compared to what Abraham was asked to give up, but small isn't necessarily proportionally easier.

I wrestled around the decision for a while but could not escape the sense that Jesus, on that first Christmas, was spending his first time away from home and that somehow this truth was at the heart of what Christmas really is. I tried to pretend this wasn't perhaps true, but of course with certain truths, no amount of deliberate fuzzy and loose thinking can get you out of it once you've seen it. My choice was stark: make the sacrifice or disobey. And besides the Lord had clearly said to me that if I went home I would not return to Europe – and he said this in a way that implied a great loss. Because of the lesson in Lucerne I knew that when God speaks this way he really means it, for sure.

Christmas then, was to be spent away from home this year, in Europe, but where? As the time to decide came near the only opportunity that interested me, was a winter outreach team in Vienna with OM (Operation Mobilization – a youth mission organization). It was for German speaking people but perhaps some immersion would aid my language learning I thought. The event didn't begin 'till after Christmas but I had to be out of the school so decided to spend the Christmas week in Vienna on my own. An Austrian lecturer had stirred my interest in the city with tales of Turkish siege and how coffee had fallen for the first time into western hands. The name 'Vienna' always stirred something in me as well – I ignored the German name for it because that wouldn't have carried any mystique at all. 'Wien' and 'Wiener' wouldn't have and still don't do what 'Vienna' or 'Viennese' did for my sense of romance.

I arrived on a very snowy December night; so snowy in fact that the last few miles were by bus because the rail lines were blocked. The bus shuttled us into West Bahnhof and I wandered out into the cold and snow, found myself on Maria-Hilfer Strausse (literally: 'Mary-[our?]-Helper Street' – clearly a very Catholic place this!) Then like all cold, lonely and insecure North Americans, went into McDonalds for a hamburger and to collect my thoughts. I had no hotel booked but after continuing down the street for a mile or two I found one that, though full, had a very small and rarely used single room that they offered to me for half price – still well off the scale for my budget. I took it with a view to finding something cheaper the following day for the remainder of the week. Next morning was cold and clear and I continued down Maria-Hilfer Strausse towards the town centre. I turned left at a rough looking area and found a poor very 'untouristy' hotel that was as cheap (and nasty) as I suspected it might be. It was the worst hotel I've ever found in Europe but the bed seemed clean enough and I booked and paid for three days. Breakfast was included; dry, stale bread and terrible coffee – actually turned a dull greyish colour when the milk went in. It was the worst coffee I've ever had apart from the time several years earlier when we were snowed in for a week during a blizzard and my uncle Lorry had reheated a thermos of frozen, mouldy coffee that had been left for several months in a tractor cab; my absolute rock bottom zero on the coffee scale.

After my first night and this shocking breakfast, as it was Sunday, I decided to attend a church nearby; an ancient stone Catholic Church just around the corner from the hotel. The interior walls and ceiling were richly adorned with stained glass, stone sculptures and wood carvings – all the work of skilled artisans and craftsmen. By contrast the people were given rough and plain wooden benches to sit on. The oaken pews of Skudesness Lutheran, back home, were luxurious by comparison. What a strange contrast I thought – as if the people were an irritating afterthought. I had plenty of time to reflect on this because the combination of Latin and German that the service was conducted in were so far beyond me that I could barley identify those languages much less understand them. And then, though I wasn't a 'Charismatic' at the time, I had a little vision or 'picture' of a crowd of people trapped at the mouth of a cave by gates made of iron bars. Behind them the cave extended back into darkness – the past I thought. This image settled into my imagination and remained heavily upon my soul for some time.

After church I found a little café and, still somewhat traumatized by the morning's coffee, decided to try a cup of tea. It came on a nice little china platter with milk, a lemon wedge and squeezer, sugar and even a nice little lacy doily. Ah... I've never tried tea with lemon... so I squeezed in a generous amount... very bitter indeed... perhaps milk would smooth it out... no just turned it into a curdled mess. My tea was ruined. I quickly paid and exited leaving a perplexed waitress standing and staring in bewilderment. Clearly my romantic quest was in trouble and I was on the brink of settling into grim endurance mode; wishing I was at home.

Somehow I was spurred on by remembering how Jesus on that first Christmas had come into such an unwelcoming and uncomfortable situation. In the end I settled into a sober joy. I was beginning to learn, what I now consider a basic truth of all quests: that the early days often feel like the exact opposite of what you are seeking. Soon my eyes and heart began to open to the beauties of the city. I had previously known nothing at all of the Austro-Hungarian Empire or the Habsburg dynasty that had ruled for more than five centuries prior to World War I. The palaces and houses and even the apartment blocks were grand, ornate and beautiful beyond anything I had ever seen. It was the first time I had ever encountered what could perhaps be called 'majesty' or 'glory'. The atmosphere was full of it. It was as if a bell had been struck centuries earlier and the faint resonance continued – not loud or clear enough to get into focus and understand but still present enough to find some depth to awaken and call forth in my soul. I had the little line: "...as though of ancient glory still ringing faintly there..." but I've never been able to write the rest of the poem. In one sense – a particular type of knowledge – this book is the filling out of that ancient resonance as I've come to understand it. Truly the whole earth is full of His glory.

Christmas was still a few days away and I busied myself touring the city; streets, palaces and museums. One museum – the museum of modern art – stood out because its atmosphere was so utterly in contrast to the prevailing beauty of the city; it served only in making the other seem even more to be desired. The highlight of the few days I spent at the grim hotel was one evening that I attended a performance by the Prague Symphony of Handel's Messiah in the 'Konzert Haus'. The beauty and glory of the city seemed to swirl into the beauty of Isaiah's prophecy and the majesty and romance of the Christmas story. I cannot adequately describe the complex panoply of life that alternately buffeted and comforted my soul during that concert. Perhaps some of it was no more than culture shock. There I was among Vienna's elite; they: clad in evening attire; I: dressed in scruffy jeans and a puffy Canadian goose down – bright red – parka. I don't suppose many of them had any idea of what was happening in my soul as they glanced my way disdainfully.

The hotel seemed especially lonely and dreary after

that, so next morning I went to the OM office to check in for the after Christmas team though it was still two days before Christmas and four days before the event was due to start. They informed me that my event was not based there but in an outlying district. However, they welcomed me to join them for lunch and said that the team leader of my team would be in the following day. The OM office and base in Vienna at that time was in a very large apartment of the type common in the old city. A grand spiral staircase swept gracefully upwards in a central atrium; marble pillars, potted plants and ornate banister railings all bathed in natural light from the glass dome above. The apartment itself was huge with 16 ft high ceilings and windows to match. OM had roughed in some mezzanine floors in the bedroom areas to accommodate more people but the main rooms were original. In the drawing room stood the largest Grand Piano I had ever seen. It had come with the apartment as it was too big to move – apparently it had been brought in through one of the windows by crane some hundred years earlier. My imagination could easily project back to that era; see the room as it once was, among well dressed gentlemen and ladies in beautiful gowns on their way to a concert or dance in that most musical of capitals – perhaps a classic concerto of Mozart or Beethoven or a new waltz by Strauss. Glory and beauty were everywhere. Even the scruffy clutter of all the OM presence and their shabby building work could hardly detract from it.

A complicated Afrikaans girl who was working with OM invited me to join her that afternoon for a trip to the Christmas market. In spite of failing all her 'husband criteria' – that took less than half an hour – I enjoyed the market. Everything seemed so much more real and genuine than what I was used to; so much less plastic and so many more things made of real wood and glass and metal. Traditional craftsmanship had somehow managed to live on here in a way that it hadn't back home; it was like stepping back a few years in the direction of the days my grandfather yearned for.

I stayed that night with the OM team in central Vienna and met my own team leader, Mark Parsons, the next day. Mark was a true Christian leader; humble, hospitable and kind but still decisive, diligent and focused – a balance much harder to attain and rarer to find than one could wish. He insisted against my mild protests that I come out to the team base and spend Christmas with him and several people from the year team who were staying on to help with my short term team after Christmas. Their base was an old house in the village of Kritzendorf on the outskirts of Vienna. This part of the village was on an island on the Danube River and only accessible by a small stone bridge. It was a rustic little place rather than grand, but not the western rustic of wagon wheels, horseshoes and rough pine rails that that word conjures up in the North American mind. Rather it was a Germanic form of rustic, from which the poorer side of the old European empires could be accessed; the plain wood and stone work of humble honest craftsmen. Externally the whole scene could have been on a Christmas card with the river, the little bridge and the surrounding woods all covered in a heavy blanket of snow. The house itself was very primitive with no hot water, no shower and no bath and only a small wood stove for heat. There was one cold water tap in the kitchen from which water for washing could be taken and heated over the wood stove. The 'bathroom' was just a small closet with a toilet in it. Only the sofa was available for me to sleep on; but the welcome was so warm that none of the practicalities mattered.

I soon settled into the team and had a great time doing outreach and evangelism in the 'can do' OM style. We had open international evenings at the house, a huge bonfire with youth and did door to door carolling – always assuring our prospective host: "Wir sind nicht der Heiligen drei Konnige", in translation: "We are not the three Holy Kings". (Apparently these 3 Wise men, as we call them in English, frequent the Germanic regions of Europe in the Christmas season, collecting gold rather than giving it!).

I returned to Bodenseehof in mid January, my mind swimming with new ideas and feelings. For one thing, it was the beginning of understanding the longing and desire for cities, which, most famously, Hebrews ascribes to Abraham as we discussed above, but which is also a perennial theme in the scriptures generally. This theme has a positive side with the focus on Jerusalem; a beautiful and gracious mother; and a negative side in the dominating promiscuous Babylon. The literal cities of the world are a complex matrix of these two metaphysical realities:

locked in constant battle. A person on the road with God will occasionally get an over-arching glimpse of the big picture but most days we must simply join the battle at which ever detail we have been assigned.

Conflict across the Generations

In a way of speaking it is ironic that Vienna was such a blessing to me on my journey and it wasn't until many years later that I began to get a vague mental grasp of the extent of that irony and of the majestic wisdom of God. By majestic wisdom I mean something like words, meaning and feeling all interacting at multiple complex levels with actual events, places and knowledge over time – past, present and future... for some purpose and destiny. Since an earlier chapter we have been exploring that which a philosopher (of some ancient school) might call 'the unified field of reality' – an integrated matrix of beauty, truth and goodness to which I have sought to also add the dimensions of space and time; that is: history and territory – but I don't pretend for a moment that all that I write of it, is more than the briefest, vaguest shadow of that 'thing' which from another angle is called the Kingdom of God.

Vienna of old, an imperial capital, was the very seat of power that militated against what I understand now to be my spiritual forefathers. These were the pre-reformation Christians in Bohemia who were forerunners to the protestant reformation and also to the pietistic renewal within Lutheranism in later times. The genuine spirituality of these humble pilgrims brought them into conflict with the apostate but politically, financially, militarily and religiously powerful Catholic Church. Jan Hus from Prague, who along with John Wycliffe from England is considered a forerunner to the reformation, was burned at the stake in Konstanz and famously proclaimed from his flaming pyre: "Today you burn a 'goose' (meaning of 'Hus') but in a hundred years time a swan will arise whose song you will not be able to silence." This prophecy and a dream from another quarter about a feather knocking the hat off the pope in Rome came true when the feather pen of Martin Luther wrote words that shook the Catholic hegemony to its core. (The 95 thesis were posted by Luther 102 years after Hus was martyred.)

It was from this Bohemian stock too that a small band, fleeing Habsburg Catholic persecution, came to live on the estate of Count Zinzendorf in Saxony and became known as the Moravians. The fiery missionary zeal of these Moravians propelled them to the ends of the earth and, closer to home; they were instrumental in John Wesley's conversion and release into fruitful ministry in England. The Anabaptists and Mennonites are also from this same counter establishment stock; all true hearted pilgrims who, like Abraham, dwell in tents and seek for a city with foundations whose builder and maker is God.

The church is called to be that city but sometimes falls to being a Babylonian unity of rebellion against God rather than his dwelling place on earth. Jesus warned against this spiritual dynamic in comparison with an unclean spirit having been cast out of an individual returning to its old home. "So shall it be with this generation." He said. Unless a nation (Israel in that generation) or a Church gives place to Jesus and keeps the Holy Spirit dwelling in it, it will have a religious, political, military and financial spirit – the serpent – instead. Again we could apply the warning of Jesus that we discussed earlier: "If the light in you is darkness, how great is the darkness!" A vacuum is not an option; there is no such thing as a neutral unity or solidarity – people scatter if there isn't something spiritual holding them together.

The Catholic Church, with its military, financial, political and religious power throughout the dark ages and into the reformation perfectly illustrates the tragic truth of Jesus' warning words. A historical biblical key to understanding the Babylonian unity is the tower of Babel where men decided to build 'a city' and 'a tower' with a view to having power – "its top in the heavens" – and identity together: "*make a name for ourselves*" and "*lest we be scattered*". (Gen 11:4) God intervened and confused the languages, resulting in many separate nations. He did this in order to prevent a rebellious solidarity from becoming too strong; "*...nothing they now propose to do will be impossible for them.*" (Gen11:6) Throughout world history this kind of rebellion has persisted as men seek to unite nations together in empires in their relentless pursuit of unity and power apart from God. As with the scattering at Babel, God continues to resist this serpentine plan.

Pentecost, where a new supernatural language was given as a sign of unity in the Holy Spirit, is the antithesis to Babel in God's long term plan for all the nations. In the final analysis there are only two kingdoms; one a loving unity of people and nations in the Holy Spirit – the New Jerusalem; and the other: a loveless conglomerate of people and nations under the crushing dominance of the serpent with his four loveless powers – same old Babylon.

During the long years of persecution many of the wandering pilgrims rightly viewed the Catholic Church, not as 'Jerusalem' but 'Babylon'; she had been captured by the enemy and was now a stronghold fighting for the wrong side. My heritage was to be at war with Babylon from generation to generation. When I stepped into that Catholic Church in Vienna, Christmas of 1981, I had unwittingly stepped into a date with destiny; the God of the ages was about to settle some old accounts and I was caught up in his coat tails. The little vision I had of the people trapped at the cave mouth was just the opening thought that led into a full scale encounter with religious strongholds. More on how all that unfolded in the next book in this series where I take up the spiritual warfare of the kingdom under the title: *The Gates of Eden.*

To avoid misunderstanding now though let me just say this: when God settles accounts or brings judgement it is always for the purpose of freeing people and nations and even churches from the serpent's kingdom. Neither the Catholic Church nor any other church – and many so called charismatic churches can and have fallen as well – is an enemy of God. He does, however, war against the serpent's loveless powers wherever they manage to set up shop.

I was once at a conference when Francis Frangipane was asked: "What is the biggest spirit attacking the church today?" His slightly tongue in cheek answer: "Yahweh is the biggest Spirit attacking the church today." Truly our God is at war with Amalek (a son of Abraham, Isaac and Jacob but through Esau: thus Amalek is understood as 'flesh' that gives place to the serpent) from generation to generation. (Ref: Exodus 17:15) Where churches have given place to the loveless powers and false loves of the serpent's kingdom they can be sure – and it's because of his love and grace – that God will be at war... they might even get lost for a while until they repent; like I did in Lucerne... and there is an epilogue to my Lucerne story which shows among other things that God has a great sense of humour and is having fun as he weaves his beauty, truth and goodness into the fabric of this world across the generations.

Swiss Epilogue

About twenty years after my experience of getting lost in Lucerne I was travelling through Switzerland by car with my wife Sarah and the four sons that we had at the time; the youngest being less than a year old and the eldest about eleven. I had studied the map carefully and besides I was no novice now to the confusing structures of European cities. This time maybe God did move the streets around... well... probably not... maybe the boys distracted me...? Anyway we got seriously lost right in the middle of Zurich even though I had been trying to simply skirt the edge of it with the view to crossing into France before nightfall. Not only were we lost in the middle of the city but somehow we managed to get on the wrong side of a barrier and became an unwilling vehicular participant in some kind of anarchist or 'sexual liberation' march. People were steaming past and all around us dressed most inappropriately – some in nothing more than body paint – for the young eyes of our four sons. Sarah prayed hard – and she may have rebuked me strongly for getting us lost as well, but I only vaguely remember that bit. We were surprised that the boys didn't seem to be able to see what was all around the car but only commented, as we finally found our way out, that the streets of that city weren't very clean and that there was a lot of rubbish left lying around.

When we got out of the city and had driven some miles toward the French border we came to a restaurant on a hilltop that had views across the countryside for miles in every direction. It was called the *Fier Linden* (The Four Oaks) and we thought it looked perfect for our evening meal. It seemed quite crowded but we drove into it anyway and as we parked we noticed that a wedding party were there. In a treed garden, just in front of where we parked, the Bride in a beautiful wedding dress, was playing some sort of 'hide 'n seek' game with the groom – apparently for the photographs. It all felt so beautifully glorious and enchanted in a Holy Spirit sort of way that we just sat and watched for a while. It seemed the perfect antidote to what we had just come through in Zurich but beyond that the symbolism woven into reality and the joy in it all... and I knew God was teasing me a little all the way back to Lucerne twenty years earlier. But the serious side is that he is always working to free his bride from the rebellion that the lost cities of the world have walked in since the days of Cain. And on a twilight evening not far in the future, around the four trees of life: Jesus will play hide and seek with his bride just for the fun of it. And at another level he already has been for some time – ever since he came looking for Adam and Eve, hiding amongst the trees in Eden.

Chapter 8

Of Ships and the Sea

We are told to deny ourselves and to take up our crosses in order that we may follow Christ; and nearly every description of what we shall ultimately find if we do so contains an appeal to desire. If there lurks in most modern minds the notion that to desire our own good and earnestly to hope for the enjoyment of it is a bad thing, I submit that this notion has crept in from Kant and the Stoics and is no part of the Christian faith. Indeed, if we consider the unblushing promises of reward and the staggering nature of the rewards promised in the Gospels, it would seem that Our Lord finds our desires, not too strong, but too weak.
(C. S. Lewis "The Weight of Glory")

Desire, Ships and Leadership

I had always had a fascination with boats but growing up in land locked Saskatchewan had provided little opportunity beyond rafting on ponds and motorboats on lakes. Such trifles to me as a boy only increased the desire for what I imagined to be the real thing behind the desire. Like other desires of the Spirit, sown so liberally throughout life and the universe, this desire associated with ships, the sea, sailors, the navies of old and even pirates is rarely understood as something spiritual. (Nor thankfully: religious) Sometimes it's just frittered away in theme parks or dismissed until weariness of life and old age take their toll and one resigns one's self to the dreary view that such longings are only youthful whims. If you dive in and take the adventures life affords, however, you will find that every longing leads to something meaningful, significant, instructive, joyful, and – like everything the Holy Spirit gets involved with – the kingdom of heaven.

In the summer of 1982 I was back in Canada at Tero Welding and Machine where I worked all hours to repay my debt from the previous year in Europe and make enough to get me back again. This time I was taking my motorcycle and I had booked passage from Montreal to Rotterdam on the Stephan Batory – last of the scheduled transatlantic steam ships.

Ships are often understood, prophetically, to be symbolic of leadership – and vehicles generally are taken as pertaining to a person's ministry. So for example if someone came with a dream in which they were riding a motorcycle, car or truck a foundational idea of interpretation would start with seeing this vehicle as symbolic of the person's ministry or calling. Likewise with a ship the idea is of leadership – and obviously there is an overlap because a ship is a large vehicle and its captain and crew would be leaders perhaps over a church or ministry organization. Now of course all these are subjective ideas and should not to be taken as biblical doctrine – more of a rule of thumb that the Holy Spirit uses as a kind short hand. Each of us may develop our own unique set of symbols with the Holy Spirit; but some, like vehicles and ships, seem to have come into common use by many different people. In this chapter I will continue to merge the various themes I'm weaving together with a particular focus on boats or ships and of course my motorcycle was now going with me to Europe as well.

I see this period as training for leadership and ministry that God was preparing me for. Of course this wasn't a dream or vision but real life ships, trips and an actual motorcycle. This is the nub of Eden and the road to it – that God integrates meaning and reality in a way that no human mind could possibly contrive. This integrated life where meaning and symbol and reality are woven into one is life in the Holy Spirit, the Road to Eden, and surely what we are all after in our heart of hearts.

The Mackenzie River

Before my return to Europe with my motorcycle on the Stephan Batory, I arranged to take a trip into the Canadian North West on a much smaller boat. My uncle Lorry had built a 36 foot steel hulled river boat and fitted it with a V8 truck engine. I had helped him with some of the drive line shafts and he invited me to accompany him on a trip up the Mackenzie River to Norman Wells; a trip of some 300 miles by road and then another 800 by boat. I would fly back from Norman Wells to Edmonton and then travel by motorcycle to Montreal and catch the Stephan Batory in mid August; a trip of some 3000 miles by road and then another 3000 by boat.

The Mackenzie River trip began on the south shore of Great Slave Lake. We would first slip across the lower corner of the lake for about 50 miles to the mouth of the river.(The lake is not called 'great' for nothing; it's about 300 miles east to west and 200 miles north to south and its choppy waves are famously deadly for shipping.) Being fearful landsmen we tried to stay close to the shore – an almost fatal mistake. We hit shallows about half way. I would have previously considered great depths to be more frightening than shallows but the sound of rocks on a steel hull when the shore is several miles away is a more terrifying sound than the imagined monsters and darkness of the deep. The prop was damaged and jammed by drift wood. I dropped the anchor and watched in horror as it hit bottom and flopped over only half submerged. The lake bottom had risen up to attack us! With every wave we were crashing hard against solid rock. My uncle Lorry got out of the boat and standing on the rocks managed to get the log out of the prop. We restarted the engine and headed for deep water, frequently crashing against rocks and with a one blade propeller vibrating heavily. Thankfully the welded hull held.

The lesson is this: the deep is safer than the shallow. God is deep, reality is deep, and the answers to all the important questions of life are deep. The human desire for simple, one dimensional answers to deep multi-dimensional questions is based in fear. This fear is a suspicion of things that aren't immediately obvious, of things that require diligent searching and of things that require serious thinking. This fear clings to prejudices and tired old patterns of thought like the landsman clings to the shore and it prevents many people from ever finding truth.

Do you want truth? Then launch out into the deep with God and abandon the fearful mindset of the landsman because ultimately truth is found within the dynamic relationships of God: Father, Son and Holy Spirit. And all of that living truth flows from the great unfathomed depths of the Father's heart – even the Holy Spirit remains forever a searcher! (Ref: I Corinthians 2:10) The Holy Spirit will not multiply laws and rules – or even over many 'facts' – but rather draw you into his own adventure and search.

The bible too, is no science, math or history text, but a chapter in the great book of life that God is writing across eternity. To the fool it's all fiddle-faddle of course, and to the honest landsman it will contain more questions than answers; but to sailors it will chart the seven seas... so to speak!

After reaching the Mackenzie River itself we had to travel another 50 miles before we reached the first village where we could make the needed repairs to the boat. The area is a vast scrub forested wilderness; empty but for the hoards of insects. There were flies by day and mosquitoes by night in such numbers that local workmen wore beekeepers nets and even these had to be sprayed with repellents to keep the visors clear. The beauty of the wide river with its wooded islands and meandering valley through rolling hills eventually giving way to mountains was awesome in the old sense of the word but the struggle against bugs, mechanical difficulties, bad weather and food shortage left us little time for reflection.

One evening, however, just before sunset we rounded a bend in the river and the beautiful panorama of the Franklin Mountains spread out before us. The vision of those gentle purple-green peaks against the last light of the sun in the western sky was painted forever on my soul. Here was untamed, untouched nature full of longing potential; the whole matrix of joy-sorrow in undiluted form. It was not a heavenward vision though; rather a vision into an ancient past; those mountains still waiting for the man from Eden to arrive and 'name' them. These kinds of things in nature speak symbolically of the past and the contingencies still open there. If you want mountains that take you closer to heaven (mountains already 'named'); go rather to the Alps where humble cowmen have yodelled and lived the forging cry of the Spirit over long centuries.

As we travelled northward we edged ever nearer the land of the midnight sun and I remember finishing C.S. Lewis's *Miracles* in natural light at 3 AM on our last night on the river. We weren't far enough into the north for actual sunshine at that time of night but rather a lingering twilight as the sun dipped just below the horizon. As I looked at the shadowy cliffs on the starboard side I noticed small fires and embers casting an eerie glow up the cliff face. I discussed it with uncle Lorry and we decided we had better report it to the police or fire service when we arrived in Norman Wells next day because of the risk of forest fires. We duly fulfilled what we considered our civic duty and were met with laughter. Apparently these petroleum fuelled fires had been noted by Sir Alexander Mackenzie when he first navigated the river in 1789. It still seems strange to me to think of a fire burning for hundreds of years. Is it another example of the ancient and fresh? The fire is old but each tongue of flame is new second by second.

The batteries in my cassette player seemed to share some property with those fires. I had brought with me about 40 hours worth of lecture tapes by Roger Forster who was a London church leader and one of my Bible school lecturers. We listened to them all; an in-depth study on Romans and a series on apologetics, with only 2 sets of AA batteries. I'm reasonably sure that each set of those batteries would normally only last for 2 or 3 hours at most – certainly never 20 hours. But some miracles seem very ordinary and unremarkable when they are actually happening. Looking back now I can see that this miracle was a sign of that heavenly fire that has continued to burn within the true church – those tent dwelling pilgrims like Abraham – and that Roger, whose work in London I joined a year later, was one of God's carriers of that fire.

Norman Wells to Montreal

The oil wells, after which the town of Norman Wells was named, bore crude so light that it could be burned as gasoline without refining but the town itself was a grim work camp needing refinement and in keeping with arctic contradictions it was Canada's hot spot that late July day with temperatures reaching 33 degrees Celsius. Some of the men my uncle knew had gone ice fishing, wearing no shirts, on Great Bear Lake and caught a bad sun burn as well as some fish. (The ice on Great Bear Lake breaks up in late August and starts freezing again in October). Because the northward leg of my journey had taken longer than planned I had to catch a plane back to Edmonton the day after we arrived in Norman Wells but I had no wish to stay longer in any case.

Watching the river and surrounding lands from the plane window restored to me that 'modern man sense of power' after feeling so much at the mercy of the elements for the previous two weeks. Cathartic though that feeling may seem at times, my quest for truth has always prevented me from seriously indulging it. Surely man – modern or not – could truthfully afford himself generous portions of humility at every turn, in the light of the natural universe alone. (Whatever reality those words: 'the natural universe alone' could possibly correspond to!??)

I picked up my redeemed and seasoned CX 500 Honda in Camrose and headed eastward, spending a night at the family farm in Saskatchewan, and the next couple in

Winnipeg catching up with old friends from Bodensee-hof. I rode hard for the next two days and covered the 1400 miles to Toronto where I spent another day with friends before completing the road journey the following day at the docks in Montreal.

The Stephan Batory

To see my motorcycle beside the massive hulk of the Stephan Batory only amplified that sense of amazement that such a small machine could propel me so far in such a short period of time. I think it's one of the many little feelings bikers have that compensate for the less desirable elements of motorcycling. I left the bike with the loading crew to prepare and secure into the cargo hold while I went for a wander around the old city of Montreal where I purchased from a second hand book store the only memento I still have from that trip: a cloth bound copy of the complete works of Shakespeare. I know that this great volume is hardly the handbook of the motorcyclist but I fancied myself a bit of a renaissance man at the time and would never eschew anything simply because it didn't seem to immediately suit a particular scene. I have often found that the synthesis of apparent contradictions is the most fascinating doorway into truth (and the whole summer so far seemed to be shouting this at me) though to be honest I don't make any claim of insight from this particular connection. But I still like Shakespeare and motorcycles and in that Renaissance-come-Victorian romantic 'synthesis' one could say there must be a connection somewhere. (Well that's the kind of thing I might have said at the time anyway!)

I returned to the ship, found my cabin, and met my cabin mate for the journey; a Dutch American bound for Holland who was frightened enough of flying to spend most of his holiday on the journey. He already had claimed the bottom bunk which suited me fine though the roof was too close to the top bunk for my liking. The cabin was tiny with just enough room to stand beside the narrow built in bunk. I sorted my stuff as best I could, exchanged a few pleasantries and then returned to the deck to watch the send off.

I found myself standing next to a dignified looking man in his mid sixties, and being in the habit since Bible School of being generally friendly with a view to sharing my faith when appropriate, I struck up a conversation. He was a retired civil engineer from Spokane, Washington and though well spoken and polite he stiffened slightly on hearing my occupation. Besides that he seemed vaguely familiar and reminded me of someone but I couldn't immediately place the connection. The conversation moved to discussing our cabins and I mentioned how small mine was and he agreed that such a small cabin so far below deck and with no window was indeed unpleasant to say the least. He on the other hand had a large deck cabin but of course had paid many times what I had paid for the space, windows, comfort and extra service. He even had

2 separate beds instead of a bunk and had paid extra so as not to have to share. He suggested that if I found my place unbearable I could perhaps use the extra... and his voice trailed off. I was tempted but something in my heart urged me to stick with my own cabin and overcome that cramped, claustrophobic feeling.

That battle was harder than I thought it was going to be. For one thing it was completely dark at night and I was being assailed by fears I had never before encountered – which I now realize may have in part originated with my fearful Dutch cabin mate. The first night I gave up after a short while, got dressed, went back up on deck, paced back and forth, prayed and even tried to sleep on a chair for a while but it was cold and damp. One thing did occur to me as I was thinking of the elderly gentleman I had met earlier and his 'sort of' offer of a more comfortable cabin. I realized it was Somerset Maugham the novelist and short story writer that he reminded me of. Maugham of course was dead and I had never met him but through his writing I had an image of him that was strangely proximate to my new acquaintance. I also knew that Maugham was a homosexual and I was thankful to God that he had steered me clear by the gentle warning to my heart. The discernment proved accurate in conversation the following day. He also turned out to be a misogynist to an extreme that I have never encountered since. Having begun with a mild and casual attempt to explain his switch to homosexuality after 30 years of traditional marriage he ended up uncontrollably raging against womankind, salivating heavily, until he was suddenly embarrassed and made a hasty exit. He avoided me for the rest of the trip. For my part I had simply tried a bible school student's version of the gospel; but he had already heard and dismissed much more eloquent presentations.

After a few miserable nights – the first being spent about half time on the deck – I managed by prayer, mental exertion and persistence to overcome the claustrophobia and the fear of fire and violent crime enough to endure my cabin and sleep reasonably well for the remainder of the nine day crossing. I could now appreciate the pleasant luxury that transatlantic travel once was. The Stephan Batory was a liner operated by the Polish Ocean Lines when Poland was still deep in iron curtain days and it may have been an attempt by communists to impress westerners as well as a generator of much needed hard currency. Apart from the tiny cabin, which would have been typical of any of the passenger lines, even in the halcyon days of the great liners like the Queen Mary, the Stephan Batory was very luxuriously appointed. The upper decks offered bars and lounges with live entertainers, a library, cinema, swimming pool and sport areas. The dining room was excellent and we were waited on by one waiter to two tables of four people each. Gourmet multi course meals were served three times a day. There was a set seating plan for the whole journey and I found myself at a table with a young honeymooning German couple and another German woman about 5 years older than me. I don't remember the conversation flowing easily but everyone was on their best behaviour and very polite.

Between meals I met other people in the lounges or on deck. There was a retired literature professor who had several times crossed the Atlantic on the Queen Mary and had once been caught in a storm so severe that he likened the great ship to a little cork bobbing around in the waves. I had been on the Queen Mary, now berthed as a hotel in Long Beach California, and had read about the storm where waves were crashing over the bridge. The bridge of that ship, to a prairie boy, looked about the height of a grain elevator (80 Ft). Such waves sounded terrifying indeed and several passengers shifted uncomfortably in their seats at the thought of it. We never had any bad weather on our trip though and only on one day were the waves severe enough to warrant raising the table edges to keep the tableware from sliding off.

One other man on board made an impression on me. He was a Methodist itinerant preacher returning to England from America. He was every bit as dejected and discouraged as John Wesley, the founder of the Methodist church had been on a similar journey home from America two centuries earlier. He grumbled about the Christians in America suffering from 'tapeitis' – apparently a disease that caused people to want to listen to tapes of other preachers rather than come to his meetings. I helped him put on a Sunday service in the cinema and, though I could sympathize with his plight, I could understand why he was not in demand as a preacher. Sadly, whatever gifting he may have had was negated by bitterness. The parallel of John Wesley and Peter Bohler's encounter two centuries earlier, with this Methodist, and me – being in the spiritual heritage of, if not exactly the Moravians, certainly the Hussite root of reformation and the Pietistic renewal streams – went over my head until I began to write this account of it all. Unfortunately, though this Methodist preacher may have been as dejected as Wesley, I apparently wasn't as blessed as Peter Bohler! The parallel, however, is still an example of how God operates in time and across the generations... and besides that, the outcome of parallel things in different times is not necessarily the same. One lesson to take away, by even a very plain interpretation of this event, is a warning about the destructive nature of bitterness.

By the 7th day I was seriously bored and longed for the trip to be over. To make matters worse we were ahead of schedule, because of a constant tail wind with its corresponding forward thrusting swell, but could not dock any earlier. And so the great turbines were throttled back to about ten percent and we crawled painfully around the south coast of England and up the channel toward Rotterdam. I spent most of my time on deck starring toward the land and wishing for a way to get there. One day I managed to slip past a sleepy guard and make my way down to the lower cargo deck to checkout my motorcycle. It was parked and tied right up in the front point of the ship way below the waterline. I could hear the swish of the heavy steel hull ploughing through the Atlantic waters and on the inside the surface was cold and covered in condensation.

We docked in the early morning and I got off as quick as I could which wasn't very quick because the crew had to check each person off and this process didn't begin till after breakfast. I was standing on the deck as my motorcycle was lifted out and lowered by a tall quayside derrick. Very small it looked swinging on the end of that cable high above the dock. Eventually I had possession of it again and had to push it to get it started because the battery had lost its charge in that cold damp cargo hold. After that I tried to buy European insurance for my motorcycle from an office in the dock area but the price was way beyond the money I had and extortionate by any standard. I felt God say that he would look after the insurance so I headed for the open road. I know some people think that God would never lead a person to break the slightest law of any kind... except maybe in communist countries or Nazi Germany or Arab countries but my experience would suggest that he isn't overly concerned about every little traffic and insurance law even in the 'righteous west' though I wouldn't presume to build a great teaching around it. But is there after all a better underwriter than God? Maybe it's his little response to insurers calling all the things they don't like: 'acts of God.' However you wish to judge me on this point, the test came when I arrived at the German border. The bike had an Alberta licence plate and I had my Alberta driving licence but of course no insurance. The border guards scrutinized my documents and searched all my cases and then just when I had repacked did it all over again and when I had repacked again they stopped me just as I was leaving and asked for my licence and passport again. They finally let me go. Later in the year, when I had insurance, the only document any German border guard ever asked to see on several occasions was an insurance card.

Bodenseehof Staff

My second year at Bodenseehof was for the most part as painful as the previous one was pleasant. During my year as a student I had idealised the staff unity and I suppose that a young man generally is not aware of the subtle discords that occur between people around him unless it affects him directly. Such youthful self centeredness is probably no more sinful than a baby crying for milk and may even be a protection for young open hearts in the gentle wisdom of God's providence. It seems that second year was a time for growing up and waking up to the realities of what I still consider the darker side of leadership; when domination and control creep in and love grows cold. I don't remember many details and it would be pointless to go into them if I did, but by the time I packed my meagre possessions onto the carrier of my CX 500 in late March of 1983 my heart was pained and weary.

The weather was as my heart: gray and dreary and cold. I was heading for London for an interview with the hope of getting onto Ichthus Fellowship's Network Training Course (Inner city evangelism and church planting). On this occasion I decided to take a route through Freiburg and then travel mainly through France rather than the Belgian route that I later came to prefer. I mentioned earlier that I hardly ever travelled in rain on this motorcycle which seemed to carry that blessing but this dreary day threatened to be one of the exceptions. I was also caught off guard by the tolls on French Auto routes, and because money was scarce, I had to take secondary roads. I remember a seeming unending series of scruffy gray villages and torturous twisty roads that might have been nice on a sunny day but on this day I had to hurry on in the drizzle and fog (but very little actual rain) to catch the last ferry from Calais.

On a motorcycle, as opposed to a car, the rider is much closer to the actual machinery of the engine and gear trains and my CX500 seemed so diligent and constant and untiringly loyal as we trudged together through that long, long day. When I'm weary of people I'm tempted to this kind of feeling but I don't indulge it much because several times I've had a distinct suspicion that God has thrown a 'spanner into the works' to warn me against what could potentially become a form of idolatry.

The ferry crossing provided a chance to warm up and dry off and catch an hour or so of sleep on a seat before completing my journey to London, early the following morning. After my interview, which didn't seem very encouraging but got me in the door, I stored my motorcycle and took the train to Heathrow for the flight back to Canada.

I had bought a return ticket from London with a visa credit card at the travel agent in Friedrichshafen before leaving Germany. I had originally wanted to stay in London but one of the leaders of the course had told me that he thought I should go back to Canada for the few months until July when the course would start. Against my own desire and thoughts, I took this as wisdom from God that I should submit to – a relatively new idea to my independent Western Canadian soul at this point. Anyone can submit to what they want to do or have thought of themselves but the real test of submission is when you don't like what you are being told. On this occasion God honoured my decision in an unusual way. I never got billed for the airline ticket which was about $700 at the time. Even when I wrote to Visa, I only received a reply pointing out how rarely banks ever make mistakes and that I should recheck my own records. How God did that I don't know to this day, but I wish it would happen more often!

Normandy Epilogue

I'll now skip ahead a year to stay with the ship and leadership theme to make a final point. In the summer of 1984 I went on holiday with friends to Normandy France. One day my friends went off for the day touring the WWII sights and I walked down to the beach alone. My mind was swimming with the complex experiences of the previous year in London and my heart was still smarting from the reluctant, hesitant, acceptance I had finally received onto the Ichthus full-time team of workers for the next year. Of course being accepted or not into the scene that a person does ministry training is not always a measure of success or failure, but it is hard for students to believe this. I probably knew then but can see clearer now that it was only my own pride that was wounded by being at the sharp end of such deliberations.

In the midst of these thoughts and feelings the Spirit showed me something from his beautiful heart. I was standing on the beach looking out over the water when the Father said: "What should we do with the sea?"

Before I had time to think, a cry rose from my heart with a deep, longing sigh: "Oh, let's just have sailing ships." I felt the depth and beauty of that desire and agreed with it but I didn't really understand it. One thing that was nice was the feeling of being at home with Father God and being his friend to the extent that he would ask me the question in the first place. When you know that place it really doesn't matter much whether the various inner circles in churches or on earth generally, include or exclude you.

As far as the ships themselves I thought how good it would be to have sailing ships only and I remembered how the gritty diesel smoke swirling down from the billowing smokestacks of the Stephan Batory used to irritate my eyes as often as not when I was on deck for fresh air. I even fancied for a moment that the *Cutty Sark*, the famous tea clipper dry docked in Greenwich, might sail the high seas again and I asked God if that could be arranged... if it wasn't too much trouble. I suppose that could be equivalent to Peter offering to build 3 booths on the Mt of Transfiguration. It occurs to me now that God might find these kinds of responses quite cute just as I might find my 3 year old daughter's answers cute if I were to ask her a question about Nuclear Physics.

I may not be much further now but let me offer something from my current thinking. Picture a huge diesel powered steel ship with its great churning propellers and belching smoke stacks alongside a beautiful sailing ship with its crafted wood – rope and canvas taut before a freshening wind. And instead of sickening, gritty diesel smoke; sheer, clean and bracing salt spray. If we take ships as spiritually symbolic of leadership or ministries we can begin to understand the Spirit's desire. How weary the Spirit is of those leaders and ministries driven by the oily, smoky engines of men; how delighted he is with humble sailors. What are those oily engines? They are pride, self will and self importance along with their greasy friend: money. These are the energies of religious, political, military and financial power by which the serpent has built his kingdom.

And the freshening wind? It's the Spirit of God who blows where he wills; drawing, as he does, from the great depths of the Father's heart. To be born of the Spirit and to be blown only by the heavenly winds of the Father's great heart is the very best and the only absolutely essential qualification for true Christian leadership.

When I first began to see this, I thought it applied to the people who weren't treating me as I wished to be treated, but I can now see that in that moment on the beach in Normandy, God, in his very loving gentle way, was addressing my pride and my desire to be honoured by men (pharisaic religion). However long it may take us to learn it, this lesson is essential because in the end there will only be sailing ships; all the rest are doomed to join the Titanic – that symbol of human wealth, power and pride – in its watery grave. In that day the desires of the Spirit will finally be fulfilled on earth as they already are in heaven. And here's a verse to go with it: "*...Not by might nor by power but by my Spirit says the Lord of hosts.*" (Zech 4:6)

CHAPTER 9

CHRISTIANS AND COMMUNISTS

Almost our whole education has been directed to silencing this shy, persistent, inner voice; almost all our modem philosophies have been devised to convince us that the good of man is to be found on this earth. And yet it is a remarkable thing that such philosophies of Progress or Creative Evolution themselves bear reluctant witness to the truth that our real goal is elsewhere.
(C. S. Lewis "The Weight of Glory")

Just after the Atlantic crossing on the Stephan Batory, and a few days after arriving back at Bodenseehof by motorcycle from Rotterdam – before the students arrived and the school year started fully – I had the chance to accompany a fellow Canadian on a trip into East Germany; still behind the 'Iron Curtain' at that time. I had met Martin when I was a student the previous year because he had come in one day to acquaint the students with his work behind the Iron Curtain. His mission supported the underground church in whatever way it could; primarily with Bibles, books and teaching materials, which of course had to be smuggled in. Our cover for the trip was that we were Canadians touring the Martin Luther sites. It was the

500th anniversary of the posting of the 95 theses and the East German communist government – with ironic hypocrisy – were using the occasion to attract western tourists with hard currency, to aid their crumbling economy.

Of course the romance of smuggling, spying and generally sneaking around behind the iron curtain greatly appealed to me. This was the stuff of legend in my grandfather's Sunday after church talk. I had read: *God's Smuggler* by Brother Andrew but also the more sobering story of Romanian pastor Richard Wurmbrand's long and torturous incarceration. The general wisdom for us was that a worst case scenario of getting caught would result in confiscation of our vehicle and a short internment of perhaps a few days with some uncomfortable interrogation, but not outright torture. Several people known to Martin had been caught, stripped naked, hosed down with cold water, and then faced into the glare of bright lights and interrogated aggressively for 24 hours by a rotation of men they couldn't see. Any westerners caught so far, had, after this treatment, been given their clothes back and dropped at the nearest border crossing with no vehicle or valuables. The purpose of the interrogation would always be to uncover contacts and collaborators in East Germany, who would of course, face serious consequences: long term imprisonment with torture.

I agreed to go but decided not to tell mum and dad until after I got back – because I thought they would worry too much. Late the next afternoon I packed my CX500 and set out for the mission base in Stuttgart under grey and threatening skies. I had hardly ever been rained on while riding that motorcycle but this was to be the greatest exception ever – before or since. It rained heavily every inch of the 100 mile way and with the spray off the busy autobahn I was totally soaked and freezing cold by the time I arrived. My travel companion, Martin, a Canadian Mennonite, lived with his German wife in a very spacious old apartment near the mission base. They were the perfect hosts for a cold and wet biker. By the time I set my bags in the entry hall, the largest cast iron claw foot bath I had ever seen was already filled with almost scalding water – the very thing I needed most. After that: 'Gluh Wein' (hot spiced wine) as only the Germans can make it – something I now always desire on a certain type of cold day. I was fully restored but hoped my wetting wasn't foreshadowing any future unpleasant wettings behind the iron curtain.

We got up early and Martin showed me around the mission workshop while last minute preparations were made to our vehicle; a classic VW mini bus. In the workshop various VW vans were being modified for smuggling – secret compartments, false floors and even an extra but concealed sliding door on the opposite side from normal on one of them. I couldn't see much point in that but Martin assured me that it could be very useful indeed.

"Could you hide stuff in the tires?" I asked.

"Not much and they'll cut them open if they suspect anything"

"Where is the stuff hidden in our van?"

"There isn't anything in our van."

"Oh"

"Yes, we will search over it again right now to make sure there is absolutely nothing – and check all your pockets – a little book of matches with a bible verse on it once betrayed a team."

Martin was a stocky little man in his mid thirties; moustached and goateed; round wire rimmed spectacles and a demeanour most serious and measured. He spoke no needless words and even those he did speak had obviously been carefully rehearsed internally with full analysis before finding tongue or breath. His eyes searched my face as he spoke and, I imagine, found manifold follies and frivolities there which he factored into his complex considerations. Such were the skills of God's smugglers in that legendary era. We took our vehicle and began the journey toward the East German border. The several hours we had driving in the west were spent briefing me on the things we couldn't talk about once we had crossed the border because after the border we had to assume our van would be bugged. I would never hear any of the actual names of the people we would meet and even Martin himself knew few of them and he had disciplined himself to forget them as soon as possible. Also it was better if I did not take note of any specific locations such as streets or villages – Martin remembered locations but only minimally used street names or numbers; "they can never get from you what you don't know."

"So we are picking up the stuff on the way before we get to the border" I suggested trying to sound a bit "with it".

"No."

"Are we meeting people then?"

"Yes, we are meeting people."

"Oh, I thought we were bringing them some bibles and stuff."

"Yes we are, but we won't get it for a couple of days"

"I thought we were going in today."

"Yes we are."

"Where's the stuff then?"

"It's still at the mission base in Stuttgart."

"Will we take a trip back for it then?"

"No."

At this point Martin must have read a look of: "Maybe these guys are actually crazy," on my face because he finally offered a bit of information. "We'll pick it up from another team on the 'Transit Route', but for the first few days we will do exactly what we tell the border guards – visit Luther sites and museums – we're likely to be watched for the first day at least. On the third night we will drive half way across the country to the meeting place with the other team. They'll be 'driving transit' to West Berlin."

"What's that?"

"The transit route is the autobahn that runs from the border to West Berlin, which, though part of West Germany, is an 'island' within East Germany. People who only want to go to West Berlin don't need a full visa for East Germany but can get a transit permit where they are given a set period of time to drive straight to their destination. They aren't allowed to leave the autobahn – there's police at every exit – and they are clocked at the beginning and end. It's called 'driving transit'."

"But we are going into East Germany farther south and with full entry visas – right?" I clarified.

"Yes" said Martin "But we will meet the other team – driving transit – at a rest stop on the transit route."

"Oh, so we can go on the transit route too?"

"Yes the East Germans also used the same autobahn and we can go on it. The critical point is when we leave it again after we pick up the stuff – being a western vehicle we are likely to be checked when we leave the transit route."

"But we'll have the visas that allow us to do it, right?"

"Yes but pray we don't get stopped, all the same, because we'll have the van full of stuff then."

"OK"

"Of course the rest stops have surveillance cameras on them as well – the East Germans are the most advanced of all the Iron Curtain countries. That's why we have the concealed door on the side of the other vehicle – so we can transfer the stuff across even with a camera on us – we park the vehicles side by side and quickly hand everything across."

"Sounds kind of risky" I said. "And with the vehicle bugged as well..."

"Yes – we do everything we can to prepare and then rely on God to help us" said Martin with one of his rare little smiles. "And of course we don't talk about anything we are doing – just keep talking about Luther and Canada and our parents and friends and stuff like that – and pray under your breath."

"OK"

We crossed the border near Eisenach and the Wartburg Castle, one of the inspirations for Neuschwanstein in Bavaria, and where Luther had been protected centuries earlier. We were thoroughly searched at the border. Every inch of the van was carefully analyzed for contraband; the tires, the wheel wells, the thickness of the floor and every corner inside and out underneath and on top. Meanwhile, we stood idly by, looking as sweet and innocent as the proverbial lamb. Of course there was nothing to be found.

As it was already late afternoon we went straight to our hotel which had been recently built for tourists and wasn't affordable to local people even if they had been allowed to enter. We had to assume it was bugged and so kept our conversation as bland and banal as the hotel itself. In all the annals of human history I doubt if anyone has been able to create anything so bereft of life, atmosphere, culture, goodness or beauty – all the things my heart sought – as the communists. In their effort to remove God from their lands they had instinctively tried to tear the very souls out of man, woman and child as well. Most truly did the Psalmist say: "*The fool says in his heart: 'There is no God*". (Psalm 53:1) This was my first encounter with communism in its actual outworking and I guess you get the idea I wasn't very impressed. Here truly a beast had arisen: '*without royal majesty*.' (Oblique reference: Dan 11:21)

The next couple of days we toured the Luther sites including, the Wartburg castle, the door upon which the 95 theses were posted, and the grave of Luther. Nothing was in very good repair, not even the secular museum in the area. Many of the buildings still had war damage from WWII and with the prevailing smell of raw sewage and the general rustic atmosphere I felt as if it could be 1946. This slight feeling of time warp had a certain kind of spiritual pleasure to me which I did not at that time realize was the beginning of the Spirit's leading into prayer and intercession for issues arising from that wartime era a year later in London.

The Wartburg Castle, where Luther had been accommodated by the Duke of Saxony for his own protection, was of course of great interest to me. Here Martin Luther had done a lot of writing and had famously 'thrown an ink well at the devil'. Being Lutheran, this was a story I had heard since boyhood. I had seen a picture in a Sunday school book of a large ink spot on the wall with ink running down and as a boy I had imagined that Luther had seen an apparition of the devil and literally thrown his ink container. I could never get a straight answer from adults when I questioned them about this incident.

"Was the devil actually standing there?" I had asked mum. And she only mumbled a response in a back and forth sort of way.

Thinking of it now, it seems more likely that when Luther said this he was referring to the battle to express his knowledge of the truth clearly. 'Throwing an inkwell at the devil' could be a figure of speech meaning that he used a whole inkwell full of ink in writing out cutting edge truth against the fallacies attributable to the devil.

But he may also have had a consciousness of the direct personal resistance to writing this truth. Perhaps, like the apostle Paul in Romans, he found himself caught up in an argument with a personification of error – 'the straw man' as it is sometimes called. Paul talks in another place about the 'doctrines of demons.' Presumably, the writing out of true doctrine would encounter resistance from the demons behind the false doctrines. If Luther was actually 'seeing' these demons or the devil in his spirit as he wrote – and it's hard to judge this sometimes – then the analysis does not so much centre around the literal vs. symbolic; but the spiritual which is literal but not necessarily physical. The beauty of God's ways is that simultaneously many things are: literally physical (Jesus physically bled and died at a point in history): symbolic (his death had great symbolic meaning): and spiritual (the physical death and its symbolic meaning are actually made real in our lives by the Holy Spirit).

If Luther then was functioning on all cylinders we could say he was; using ink to write truth against demonic error; that he perceived the presence of the devil's resistance – maybe even saw a vision of the devil – and physically picked up his inkwell and threw it as a symbolic act of defiance (the ink hit the wall and ran down it leaving a stain like the Sunday school book showed); and that he refilled his inkwell and continued his argument against error in the anointing of the Holy Spirit – enhanced by his 'prophetic action' or 'acted out prayer.' To search out, understand and live this integrated body, soul and spirit way in the Holy Spirit is, in my view, a great and noble quest; it is the quest for the kingdom of God, and of course, the 'road to Eden.'

After a couple of days visiting the Luther sites and eating for pennies (by western currency) in the nicest restaurants in the country we made our rendezvous on schedule and loaded up the Christian literature and Bibles (our inkwells so to speak) as planned. As we were leaving the transit route – the most vulnerable moment in our plan – the dreaded thing happened; the flashing blue lights of the East German police appeared immediately behind us. Martin, even then, said nothing about prayer but gave me a meaningful look as he calmly said: "Now we will show the police our visas because they will be checking that we have the right papers."

We stopped and handed our passports to the officer. He studied them carefully with his flashlight in the semi darkness and then sighed with evident disappointment. No doubt he was hoping to at least fine us for leaving the transit route without authorization – and of course to collect the immediate cash payment that was required for western vehicles at that time. Then he began to shine his light into the rear windows of the van and look inside. In the back, under a loose blanket, was a large stack of boxes full of books and magazines. They were pretty obvious unless God hid them from his eyes – or 'blinded their eyes' as Brother Andrew used to pray on such occasions.

We prayed. What must have only been a few moments, seemed to be a suspension of time itself as he pried with his light around the side and rear of the vehicle. We stayed calm on the surface but were desperately praying on the inside. At last he seemed satisfied, handed us our passports and waved us on. Now I was in the spirit of it and Martin's warning look as he saw that I was about to speak was unnecessary. "It's a good thing those visas were all in order or we would have had to pay a fine," I said, just as blandly as any communist could.

We drove another half hour or so to a village whose name I never saw or heard and then stopped on a street outside a two story house. We watched and waited; Martin was looking up at an upstairs window. A tiny flash of green light was our signal that all was well – the date and time had been arranged months earlier on a previous trip. We drove around to the back of the house and into an open shed door that was quickly closed behind us. A tall, gaunt and slightly stooped man greeted us warmly but quietly in German and escorted us into the house – his name was never used in our presence. We stepped into a bright spacious kitchen – all blinds drawn – with a big round table in the middle of it. Five or six adults and at least as many children, were gathered around with eager and excited faces.

The table was covered with food – not great food to our eyes; a small amount of tinned meat and a lot of bread and pickle type things. Martin had warned me most emphatically that I should graciously and thankfully receive what I was given and that any reluctance would deeply offend our hosts – this food was the very best they had and they wanted to share it with us to show their appreciation for the risks and effort we were taking. Any reluctant superiority on our part would be to humiliate them. Martin dived in and ate with more gusto than I had seen him do anything to that point and I did my best to follow suit.

The meat, such as it was, was pushed over to us while the children looked hungrily at it – and us while we ate it – with big, beautiful eyes. Even some of the adults glanced furtively... to ensure that we were enjoying it. Martin said later that the amount of meat they gave us would have been about a month's worth of their normal supply. That meal stands forever as a monument in my mind and a memorial in my heart that it is often the poor of this earth who are most generous and hospitable. I've met the rich who will reluctantly spare a scrap or two out of their excess but I have never met one who would give anywhere close to this – nor have I ever given to that level myself.

Besides the hardship and poverty these Christians suffered direct persecution by the secret police and the communist authorities. The tall man who first met us had been imprisoned several times and still bore the scars of beatings and torture. He didn't want to talk about it though – just waved it off as if it was no more than a little discomfort.

I was out of my depth for sure and perhaps it was as well that, with my limited German, I only understood a fraction of all that was said on the several days that we spent driving through the country to various underground churches. Martin sat up front – chatting freely at last – while I crunched up in the back of the tall man's smoky and noisy old Trabant. (An iconic East German car with 'communist finesse')

The advantage of this car was that we could move around the country as natives without a western vehicle to attract unwanted attention. Everywhere we met poverty, hardship and suffering; everywhere we met generosity, hospitality and love. Such were the islands of the Kingdom of Heaven in a vast sea of communist darkness and atheist foolishness.

When we had finished our travels and said goodbye to our hosts we took our own vehicle and headed for 'Check Point Charlie'; the notorious crossing through the Berlin Wall into West Berlin. We had to zigzag slowly through huge concrete blocks that had been placed to prevent a ramming escapc by bus or tank. Our vehicle was searched again and the underside studied with mirrors – on wheels with a handle – in case any East German was trying to escape by hanging onto the under carriage.

The final leg of our journey, now that our visas were spent, was to drive transit from West Berlin back into 'mainland' West Germany. Once again – but as if for comic relief this time – we were stopped by the police; apparently we had failed to signal a lane change. A very podgy policeman came puffing importantly up to our window and wagged his finger as he emphasized the gravity of our failure: "Heir muss man immer blinken!!" Then his greedy eye gleamed as he demanded the ten Deutschmark fine. Martin, barely able to conceal his delight, drew from his wallet an East German ten mark note and handed it over with a smile. The policeman's cheery red face turned immediately sour. His appetite was for hard currency and what it could buy on the black market, but, to adhere to the official line that the eastern mark was worth the same as the western mark, he had no choice but to accept Martin's payment. He remonstrated briefly to no avail and then mumbled and grumbled to himself as he walked back to his car.

The trip back to Stuttgart consisted of a few final exhortations from Martin to keep quiet about everything we had seen and done – even in the West... because of spies? Well, yes... but there was a spiritual dimension to secrecy as well... We left it at that. I picked up my CX 500 at the mission base and rode back to Bodenseehof in late afternoon sunshine.

CHAPTER 10

WORD AND WILDERNESS

We are to shine as the sun, we are to be given the Morning Star. I think I begin to see what it means. In one way, of course, God has given us the Morning Star already: you can go and enjoy the gift on many fine mornings if you get up early enough. What more, you may ask, do we want? Ah, but we want so much more – something the books on aesthetics take little notice of. But the poets and the mythologies know all about it. We do not want merely to see beauty... we want something else which can hardly be put into words – to be united with the beauty we see, to pass into it, to receive it into ourselves, to bathe in it, to become part of it.

(C. S. Lewis "The weight of Glory")

THE DEEP END

In late June, 1984 I returned to London to begin my course in inner city evangelism and church planting. To join Ichthus Fellowship in its halcyon days was to be pushed straight into the deep end of the word of God in action. The word of life preached and lived is the great purifying and cleansing detergent from heaven. Every meeting and teaching was full of the Lord's power and presence or so it seemed to me who had been searching since childhood for that thing which I perceived my forefathers had experienced. My course began with the 'July Project', a one month 'shock treatment' in theology and evangelism. It consisted of about 5 hours of lectures per day and 'n' number of hours of evangelism within a team assigned to open up a new area for church planting. In my case I was sent out to Peckham where a small group was already replanting an old Baptist congregation. My team leader, Paul, had just completed the course I was beginning and was, to my eye, embarrassingly bold – maybe even brash and crazy. But it was all very contagious and my old comfort zones were crumbling to the left and the right in both the lecture hall and on the streets and doorsteps.

I became conscious of hearing from God at a much clearer level than before especially in 'pictures' or visions and sometimes dreams or just knowing through a thought from an unexpected angle. "My sheep know my voice..." and "the sons of God are lead by the Spirit of God..." were forged into my spirit in the lectures and lived in reality the same day. Faith explodes in such a context. It was during this time that I had the encounter (mentioned in an earlier chapter) with the young man who was part of an occult coven in Greenwich, when I experienced being one with the Father and Spirit in loving Jesus; and when the beauty of the stars and the heavens swept through my heart for a fleeting moment.

One thing I resisted was the teaching on adult baptism but the historical testimony of the Anabaptists overpowered my objections in the end and I felt the Lord was leading me to get baptised again in water but also in the Holy Spirit. The Lord showed me, in a dream, the difference between the two pools in John's gospel; Bethesda was stagnant but Siloam had a fresh stream running into it. Jesus healed a man and took him away from Bethesda but sent a man to Siloam for his healing. Bethesda corresponded to the old covenant, the ministry of angels and the baptism of John (right and good in their time) whereas Siloam was the new covenant, the ministry of the Son *sent* (Siloam) from heaven and the baptism of the Holy Spirit. The miracle of Siloam was the opening of eyes born blind and I asked the Lord for the gift of knowledge during the baptism service and he said: "I will most certainly give you that."

I had, since before I left Canada the first time, been seeking knowledge because the Lord had spoken to me from Proverbs 2 where the promise is given: "...and knowledge will be pleasant to your soul." Even that promise had been the culmination of a quest that began in school when I had first studied the renaissance and I had longed for a comprehensive and unified knowledge as I imagined Renaissance men pursued. This has been called 'the unified field of knowledge' though I can't remember where I first heard that term. Einstein pursued this in a scientific way as regards time, space and gravity resulting in his theories of relativity and also spent his last days seeking a unifying theory of gravitation and magnetism. All these considerations were open questions I held in my heart at the time. The night of my baptism I had a vivid dream in which I was in a room with Leonardo Da Vinci, Michelangelo and Albert Einstein. Einstein was pulling out white paper from a large roll attached to the wall against which a wooden table stood and laying it out on the table; ready for me to write or draw on it. The paper was an intensely pure white and he gave me a look that seemed to say: "... and mind that you keep it absolutely clean."

"*...as Christ loved the church and gave himself up for her, that he might sanctify her, having cleansed her by the washing of water with the word.*" (Eph 5:25b-26)

A Tent in the Wilderness

The July Project finished with a Love Feast which was a semiformal and beautifully presented banquet with all the students, lecturers and staff in attendance. The atmosphere was rich in love, goodwill and promise for the future – and to my eye and heart at least – free from that dreadful pretence and religion that so easily creeps into events like that. I then had the month of August off before beginning the main body of my course in early September.

I decided to take a motorcycle and tenting trip into the countryside for the first week, then come back as a volunteer to help with building work at Ichthus House for a couple of weeks and to spend the final week on a motorcycle trip to Germany to visit old friends at Bodenseehof. The idea behind the tenting trip was derived from the Old Testament 'Feast of Booths' or 'Tabernacles' (literally: Tents). I understood that the essence of this feast was to leave your normal secure comfortable dwelling and live in a tent or leafy booth – an altogether insecure dwelling. This is an exercise of putting faith in God rather than in earthly securities and is, besides being an official feast from the time of Moses that continued to celebrate the wilderness journey, a facet of Abraham's life that the writer of Hebrews notes in saying: "... *living in tents...seeking a city with foundations...*" (Hebrews 11:9-10) Historically 'wilderness' has also been one of the classic ways that men and woman of God – mystics, prophets and intercessors – have sought a deeper spiritual life with him.

Such were my thoughts as I packed my tent and a few clothes on my CX500 and headed out into the English country side in the direction of Cambridge. Because in normal times I would have driven all the way to Scotland the first night, I decided to set a limit of about 100 miles per day to curb that busy nomadic tendency. I arrived in Cambridge, resisted the temptation to take in all 3 episodes of Star Wars that were playing in the cinema there that day, discovered that English mustard is very different from American mustard and actually destroys hamburgers, and then drove out to Trumpington Camp ground to pitch my tent. The sky threatened rain by evening time and a friendly Dutch neighbour offered me a fly sheet as my tent was a very basic single sheet type.

"That's OK" I said, "I'll just pray."

Not a drop of rain fell that night and my neighbour was eyeing me very studiously the next morning as I packed and loaded the bike.

"See – prayer is better than a fly sheet," I said.

"Yes, maybe... but I think you have ein relationship with God" he said, in a heavy Dutch accent – starring hard, as if to see beyond my scruffy jeans and black leather jacket. I got onto my motorcycle and was wondering if I should say anything else to this man when a woman appeared to me in an open vision. She was so beautiful and majestic – neither young nor old – walking straight towards me with so much freedom and joy in her face that I lost track of my surroundings for a moment.

"Come and Play" she said in a way that flooded my heart with the same freedom and joy that was in her face. Then she disappeared. I said good bye to my Dutch neighbour – now very perplexed indeed – and drove away; resuming a north-bound route. My mind was occupied with the vision I had just seen. The woman resembled 'Liberty' (and the statue of that name) as far as dress was concerned – though the crown was lighter and more gracious and the robe richer and more elegant. Her face was full of life, freedom and joy and her movement was light, easy and delightful – so playful – and her words: "Come and play" were perfectly in keeping with her whole being. I thought it was a vision of the Holy Spirit and have never had any reason to change that thinking – though in the back of my mind I could, and still can, hear all kinds of accusations and objections.

20 Angels

My hundred mile quota brought me into Nottinghamshire and I could not resist pulling in at the Sherwood Forest visitor centre. I've always liked Robin Hood and I was pleased to walk through the ancient woodlands to 'Major Oak' and take in an open air theatre with full costume presentations of some of the classic tales that have somehow captured the hearts and imaginations of many generations. The ongoing fascination with Robin Hood arises, I believe, from the deep longing we have in our hearts for justice and righteousness – oppressors overcome and the oppressed freed. The 'robbing of the rich to feed the poor' strikes a chord with the prophetic mandate to pull down mountains and fill up the valleys to make a level and righteous ground. This is the work of destroying the serpent's kingdom and bringing in the kingdom of heaven but unless it is kept within the guidance of the Holy Spirit the 'Robin Hoods' are likely to end up as oppressors themselves. The kingdom of God we are told is: "*...righteousness, joy and peace in the Holy Spirit.*" (Romans 14:17)

I continued northward late that afternoon but stopped after about half an hour in a little woodland clearing just off the road. It wasn't an official campsite but there were no signs prohibiting camping so I pitched my tent and then sat down on my motorcycle with my bible open across the fuel tank. A playful little breeze began to flip the pages and I decided to let it continue until it stopped on its own. It stopped at Luke 4 and on reading the account of Jesus' wilderness temptation I was struck with how the devil had quoted scripture in saying: "...he will give his angels charge over you lest you dash your foot against a stone..." I looked up the cross reference to Psalm 91 and was surprised to see that the verse continued: "...the young lion and the serpent you will trample underfoot..." – the devil sure didn't quote that bit! I read all of Psalm 91 and was very comforted by the words: "...because you have made the Lord you dwelling, the most high your habitation... no evil shall befall you, no scourge come near your tent..." This was a wilderness Psalm for the feast of booths. It was probably what Jesus was meditating on while he was in the wilderness which fits in with how the devil was trying to twist it in the temptation. The condition for all the promises is making God your dwelling place – your security, trust and confidence must be in him.

I looked around at my lonely surroundings and began to feel a bit nervous as the evening grew darker. Here I was, a stranger in a strange land, beside an ancient wood, alone, with only a tent. I thought of criminals, wild animals, witches and just the plain unknown – could there be crazy insane people living in these woods? Did evil covens have midnight rituals here? As these thoughts began to press in on me I heard God say: "Why don't you ask me for some angels?" I briefly wondered if this really could be God but then it occurred to me that, even though the devil had quoted the promise of angels to Jesus in an attempt to trick him, the fact remains that the promise is there in the scripture.

"Ok" I answered "I'll have some".

"How many?" he said.

I had no idea how to answer this so I said: "ah... ok... um... twenty?"

I knew the answer was immediately: "yes" and I sensed, but did not see physically, a group of twenty angels arrive and form a ring around my tent. They were laughing and joking in surprise at being sent in such a number for this assignment – more than enough for a whole city seemed to be the general sense but they were always pleased to be surprised by the Father's extravagance. I am always surprised when I reflect on these kinds of supernatural experiences later because of my own state of mind during them. I was very calm and logical – something like: "oh that's good". And then: "I think I'll move my motorcycle inside the circle of angels". I backed the bike right in beside the tent but part of the wind screen and front wheel were protruding just beyond the circle. I couldn't see any angels but I could "see' exactly where they were. I then crawled into the tent and had a good sleep. When I woke in the morning the first thing I noticed was that the tent was absolutely dry – no dew or condensation at all – a thing that never happened before or after this one time. (Usually the tent was dripping wet with dew in the early morning – unless you were in a very dry climate). When I poked my head out of the tent flap I noticed that the angels were gone but the little bit of the bike that was outside the circle where they had been was wet with dew. There was literally a diagonal line across the windscreen –

one side dry the other wet, exactly where I had perceived the outer edge of the angelic circle to be. "...no scourge will come near your tent..."

Wilderness Grace

I have heard people say since that immediately after supernatural experiences is a time when we are most vulnerable to temptations and weakness and I can testify to that. It's not that I fell to some obvious sin of the flesh – just went against the leading of the Spirit and chose to deviate from my path and drop in on an old friend I had met the previous year in Germany. I had a pleasant and comfortable time but knew in my heart that I had missed something God had in mind. I drove away the next afternoon under gray skies and with a heavy heart. The sky never cleared and there was even an occasional spit of rain as I made my way northward on the M6 toward the Scottish border. I could get no clarity at all about where to stop and had passed several campsites by the time I neared the border.

It was getting late now so I turned off the motorway at the next campsite sign and followed the narrow road to a campsite near the village of Kirkpatrick-Fleming. The office was closed and I drove a wide circle around a central grassy area that was surrounded by Caravans and tents. It was dark and raining and not at all pleasant for setting up a tent much less sleeping in one. A woman, perhaps in her mid twenties stepped out of one of the caravans and waved me over. "We're bikers, and any biker is welcome here" she said waving me towards the caravan. "Come on in for a coffee." I hesitated. "My boyfriend will be back in a minute", she said, as if to clarify the whole situation. The Lord seemed to give the nod and I was glad he was still speaking to me and also to be getting in out of the cold wet night.

The woman made some coffee and we began exchanging stories. I of course had grown up on a farm in Canada and gone to bible school and was now on a special church training course. She had grown up in a poor area of Cleveland in the north east of England and was now riding with her Hell's Angels boyfriend and living semi permanently on this campsite. This same boyfriend was still nowhere to be seen and she breezily said "He just went out for tobacco but he must have stopped in at one of the other caravans – there are a few other bikers living here." Then perhaps noting a little concern on my face she added "He won't mind you being here – bikers are always welcome" I wasn't quite sure if he would consider me a biker and I remembered the time a group of serious bikers had called me and my brother: "Boys on the scooters." In the context of our stories I had managed to share a basic gospel message and she seemed very interested and almost tearfully said she would like to go home and see her parents and family again.

At this tender moment her boyfriend stepped in looking like a scrawny crazed Viking; his long blond hair the victim of years on the open road. He was at least ten years older than her (and me) and stood for an awkward moment assessing the situation with wild bloodshot eyes. I tried to look like a nice guy. Whether it was the element of surprise at having a Canadian appear in his home at that late hour or some direct intervention by the Holy Spirit I can't say but he relaxed and listened to his girlfriend explain my arrival. She assured him he would be most interested in what I had to say – apparently he had spiritual interests of his own. It turned out he considered himself to be a druid and was interested in ancient occult and pagan spirituality. He claimed to have been instantly supernaturally enabled to read ancient Norse occult books in their original languages without ever studying. We talked until 4 AM quite amicably and I shared the whole gospel to the best of my ability. I gave them my address in London before we finished and invited them to stop in if they were ever down that way.

They insisted I put my camping mat on their floor rather than pitching my tent in the now heavy rain. The Lord was there with me and so I agreed. The young woman already had her night dress on under her robe and slipped into their double bed which pulled down from the wall. I camped a few feet away on the floor of the small one room caravan – fully dressed in my sleeping bag. The man stripped stark naked and got into bed without a stitch on... oh dear what would mum and dad think of this situation I thought? We all apparently slept well and awoke just before 9 AM. The woman asked if I wanted a coffee as she got up and the man sat on the side of the bed looking quite weird and grumpy. The Holy Spirit then spoke with absolute clarity and force: "*Get out of this place right now.*" I looked at my watch. It was exactly 9 o'clock. I said thanks and no thanks and left immediately.

The young woman stood on the doorstep and watched me load up and drive away. She sent a letter a few weeks later and told me that she had left her boyfriend the following week and returned home to Cleveland. I sent her some books and tapes and she asked Jesus into her life and began her own journey. I never saw her in person again but wrote a few times and helped her connect with a local church in Cleveland. Of course rescuing distressed damsels from the clutches of fiery dragons has always been the work of our great fisher king, Jesus – and a most noble occupation of his knights. (I was in the reality of this but didn't connect it all intellectually until writing the story years later.)

My wilderness tenting trip concluded with a ride to Edinburgh that morning. On a high hill at the city's edge, overlooking the sea, I spent the afternoon reading, praying and reflecting on the events of the last month and especially the last few days. Next day I returned to London.

WILDERNESS: PSALM 91 AND THE FEAST OF BOOTHS

When Jesus was in the wilderness, as we noted earlier, he was evidently meditating on Psalm 91; one of the

Psalms associated with the ascent to Jerusalem for the feasts – particularly the feast of booths. This Psalm is a treasure trove of some of the most frequently quoted and desired promises of scripture; protection from sickness, accident, deception and danger, from both men and demons. And to top it off: Honour, love and friendship with God. But there are some adventures that are prerequisite to receiving the full benefit of these promises: "*He who dwells in the shelter of the Most High...*" (v1) and: "*Because you have made the Lord your refuge...*" (v9). The adventure is in making God our dwelling, our security and the stronghold in which we trust.

If our main confidence and trust are in worldly, human or natural things then we are likely to receive only the boring fruit of those pursuits rather than the promises given in Psalm 91. While it is true that God in his kindness and mercy may even yet bless and protect people, it is clear from experience and observation that he doesn't always do so. Consider verse 10: "*...no evil shall befall you, no scourge come near your tent.*" The promises are to tent dwellers! Those whose hope is in God, who dwell in him and who, on this earth, are part of that great company dwelling in tents and seeking a city with foundations. (Ref: Heb11:9-10)

Of the three great Old Testament Feasts, which were first established in the wilderness journey in the days of Moses, only one remains unfulfilled. The Feast of the Passover, initiated on the night of Egypt's final judgement with the death of the first born, was fulfilled when Jesus became the ultimate 'first born' and 'paschal lamb', and died on the cross. The Feast of the first fruits was fulfilled fifty days later with the coming of the Holy Spirit on Pentecost. This feast was the time when the early first fruits of the harvest were celebrated and given thanks for in themselves, but, beyond that, as tokens of more to come in the final harvest. The Apostle Paul applies this idea to the Holy Spirit in the church age as the guarantee or 'down payment' of our full inheritance still to come. This, full inheritance still to come, corresponds to another feast which was the feast of final ingathering or harvest. It was also called the feast of' booths' or 'tabernacles' – in English: 'Tents'.

At the feast of the first fruits or Pentecost 3000 people from many nations put their trust in Jesus and found salvation in response to Peter's preaching and the anointing, power and presence of the Holy Spirit. The feast of final ingathering or booths will be an exponentially larger harvest of people from every nation responding to an exponentially larger outpouring of the Holy Spirit and a radically loving and supernaturally living Church. We will look at the Old Testament roots of this feast as well as some of Jesus' teaching and action in connection with it, in order to help prepare us all to be participants in the great harvest at the close of this age.

I believe we stand at the very brink of the great final harvest of humanity, the fulfilment of this last great feast and the fullness of the Spirit corporately and individually; our longed for and promised inheritance. But I am not seeking to labour any hype about timing and I categorically state one certain truth from the outset: even if the great harvest does not begin for another century or two; the principles and choices that I will be advocating are the right way for us to live, at any and all times.

This feast of booths was given in Moses day and was meant to be celebrated yearly; though in fact it seldom was – to Israel's great loss. The underlying principles of this feast are the way we should live day by day and even moment by moment – whatever the prophetic weather of our times. This of course is true of the other feasts as well. We need to live daily in the reality of the cross; Passover. We need daily to be filled with the Holy Spirit; Pentecost.

What then is the unique underlying heart of the feast of booths? In Lev 23:42-43 we read: "*You shall dwell in booths* ["tents" made from palm and other leafy branches – v40] *for seven days; all that are native in Israel shall dwell in booths, that your generations may know that I made Israel dwell in booths when I brought them out of the land of Egypt: I am the Lord your God.*"

At its plainest basic level, the feast of booths was a memorial feast commemorating the wilderness journey of Israel. God wanted every generation to remember that event, but not only to remember it, to actually experience something of it. Even before the days of Moses, God had called his people out of a heathen land and into the Promised Land through a process of wandering and living in tents. Abraham, Isaac and Jacob lived mainly in tents their whole lives. Of Abraham the writer of Hebrews declares that he was searching for a city with foundations whose builder and maker was God. If it had been an earthly city he was searching for he would have had opportunity to return to the one he had come from: Ur of the Chaldeans. The call to God's people in every generation has been to come out of our earthly 'dwellings' and seek heavenly dwelling places. On this earth we are an exile people with no enduring security – "*...no lasting city.*" (Heb 13:14). Never-the-less we are constantly surrounded by people seeking to build as much security as possible on this earth and we ourselves are constantly tempted to seek earthly security as well. It's uncomfortable and inconvenient – the best adventures always are – to live as aliens in a hostile world and therefore very tempting to try to 'blend in' and 'go with the flow'. Against this pressure God gave the feast of booths; an annual literal reminder to Israel of its roots in the wilderness and its destiny for something heavenly – The Promised Land.

When Israel crossed over the Red Sea and Pharaoh's army was swept away they were free from the power of the Egyptian kingdom. For us as Christians this corresponds to salvation through baptism into the death and resurrection of Jesus which we accept by faith as a free gift, and through which, we are delivered from the kingdom of darkness. The wilderness journey, however, corresponds to our sanctification or purification and, though this may take longer, the length of it is partly in our hands. If Israel had cooperated fully with God their journey would have been about 2 years or maybe even less before they entered the Promised Land. Others have succinctly put it this way: "Crossing the Red Sea got them out of Egypt, but the wilderness journey was about getting Egypt out of them."

Like Israel of old we today are constantly tempted in our hearts to gravitate back toward Egypt; the world with all its comforts and securities. Sometimes our hearts prefer slavery and bondage because we are like prisoners who have become institutionalized and want to get back in prison to the concrete; the familiar; the structured; and the controlled. The scary big world of the wandering pioneer and explorer seems wonderful to talk about by the comfort of the evening fireside, but in the cold chill of morning light we may choose rather to pull the blanket over our heads and forget about it.

The true Christian life is nothing if not a major stretch to all our natural inclinations and yet deep within our hearts there remains those ancient desires, born of the Spirit of God, that will never be satisfied with a bit of half hearted religion on the side. Like Tolkien's Bilbo, even the cosiest home lover amongst us has a dragon slayer in his heart of hearts; and like Bilbo he will have to leave his cosy home in order to release that dragon slayer. This is the inner essence of the feast of booths; leave the 'earthly' behind and step out into the vast and lovely – and scary – world of the heavens. Reach for the stars. Live the dream; the real dream.

Abraham's father had got his family half way out of Babylon, to Haran. Abraham went into the Promised Land living in tents but even beyond that – which had became an earthly comfort of sorts – God called him out of his own tent of anxiety and worry, into the great tent of the heavens. "*Count the stars, if you can,*" he said, "*for, as many as they are, so shall your descendants be.*" (Genesis 15:5) Such is the life of faith. Even what we pioneered in the past can become an earthly rather than a heavenly security and so to truly celebrate this feast we must remain in the place of listening and obeying. Moses fell to this snare when he struck the rock the second time as he had done the first time rather than obeying what God had actually said; "Speak to the rock."

The end time's connection of the feast of booths with the coming of Moses and Elijah is foreshadowed at the transfiguration of Jesus as recorded in Matt 17:1-13. Peter, who Luke tells us in his account (Lk9:33): "*...didn't know what he was saying*", offered to make three booths. Usually this is interpreted negatively of Peter, but let me suggest rather that Peter was making an appropriate connection, but didn't have full understanding of it. God interrupted from heaven saying: "*This is my beloved son, listen to him.*" (Luke 9:35) Not that Peter's interpretation was wrong; only his application and timing. The lesson is this: though there is value in literally keeping the feast by going out into wildernesses in tents or booths, the ultimate reality and fulfilment is a life lived listening day by day and moment by moment to the beloved son, Jesus – which of course means obeying. This is how we make God our dwelling place.

And God may from time to time test us to the limit in order to reveal to us (not to him – he already knows) whether the deceitfulness of worldly riches and all the houses and other securities it can buy are stealing our hearts from him. "*Beware*" said Jesus "*of the leaven of the Pharisees, which is hypocrisy.*" (Luke 12:1) They loved to be seen to be very godly and spiritual but in truth they were lovers of money with confidence in the physical temple and the structures of human power. Their dwelling place was not God but in truth the devil – for in his lies they had placed their confidence. In truth they did not dwell in Jerusalem but in Babylon – Ur of the Chaldeans that Abraham had abandoned. And their tents... well they were actually whitewashed tombs. Again and again in Israel's history and in church history, those who began well – trusting in God – when they were successful began to trust more in the structures, the institutions, the systems, the 'houses' they had built. Even churches, movements and ministries themselves can, in this sense, become an idol in which we find our security, over against God himself. Abraham was tested to the point of giving back to God what he had received by faith; Isaac the son of promise. Are we willing to hand our ministries and positions within the church back to God when he asks?

We are so prone to gravitate toward the earthly whereas God desires, and will have in the end, those who truly make him their dwelling place. In every generation he has had some of these, but according to Zechariah all the nations will, in the last days, come to Jerusalem and celebrate the Feast of Booths. This is the day of the great harvest of the earth; the coming of the Kingdom of God in fullness;

that Kingdom which we are called to seek first before all else. (Ref: Matt 6:33) If our hearts are right in this we can absolutely depend on every promise of Psalm 91 and all of Jesus' other promises of provision and care as well.

I am not putting this forward as some new religious activity that will finally get everything to 'work' for us – the exact opposite in fact. The appeal here is to faith, to trusting God, to accepting his saving gospel in the death and resurrection of Jesus and to receiving the gift of his Holy Spirit. The sacrifice required of us is not one of religious activity or discipline but of taking our confidence out of the serpent and his kingdom and placing it in God and his kingdom. It is to reverse the choice made by Adam and Eve; it is to reject the serpent's power and plan and to receive the power of God to be transferred into and then to continue living in the kingdom of his son Jesus. If camping in some literal wilderness in a literal tent helps you get your head and heart around that... well then I highly recommend it. And of course it was God's idea in the first place.

CHAPTER 11

FATHER'S FACE

To please God... to be a real ingredient in the divine happiness... to be loved by God, not merely pitied, but delighted in as an artist delights in his work or a father in a son – it seems impossible, a weight or burden of glory which our thoughts can hardly sustain. But so it is.
(C.S. Lewis "The Weight of Glory)

A TWILIGHT ENCOUNTER

I pressed my knees as best I could up against my CX500's V-twin engine for whatever scraps of warmth I could draw from it as it propelled me through the twilight toward the deepening gloom of night. The cheerful little engine seemed as oblivious to the darkness and the chill blowing in off the North Sea as to the last remnants of my slowly resolving frustration. Nor did the bleak, muddy flats of western Belgium, which were disappearing into the darkness, dampen its spirited diligence as we pushed mile after mile, first northward from Dunkirk, then westward toward Brussels.

I had always loved twilight and my mind drifted back to

an evening years earlier; a soft, warm evening when I was a child playing in the farm yard back home. Through an open window, I could hear my mum playing her piano and singing. The beauty of her worship flowing through the evening air caressed both soul and body, and merging with the interminable distances of the twilight drew me into one deep undefined longing; joy-sorrow, as unexpected as ever.

A sudden extra chill in the air jolted me out of the remembrance, and I tried to tighten the bottom of my jacket around my waist. There was no question of stopping for the night – I only had enough money for fuel and food. I looked again across the shadowy landscape in the last light of that September evening and wondered at the barren, cold, dampness of it. This was a different twilight and I thought of the battles that had been fought here in two world wars.

The grisly image of a young German soldier trying to gather up his intestines and put them back in his abdomen flashed across my mind. It was from stories told by my father's old friend, Bill Perrot, who had fought in the First World War when ammunition was in short supply and the fighting frequently reverted to knives and bayonets. Most of the soldiers never talked about what they had seen; but Bill was different (for reasons unknown) and this particular story had traversed a generation and was fixed in my imagination. I could even see the ghastly terror on the young German's face as the final blow was struck; ostensibly in kindness at that point – as the story went. My father hated war.

My thoughts drifted back to the present day and my current situation. I would have been well into the warmer regions of central Germany by now if not for an annoying delay in getting out of London. Someone had asked me to wait outside a meeting he was in – then forgetting me – exited the meeting by a different door, leaving me to wait 'patiently' for several hours. The plan for a leisurely late morning departure had become an irritated rush mid afternoon. And because of that, I had left my gloves on the hall shelf – a fact which was now painfully presenting itself to me. How pleasant would those generous motorcycle gauntlets have been, I thought, as the cold, damp air found its way up my sleeves. Still it was my own fault for being so grumpy when I left.

Then, my mind began to wander down a path that it had never previously travelled. My hands were really very cold by now and I began to wonder if a person could ask God for something when it was his own fault that he didn't have it already? And, when he could see no possible way for God to answer? After all, there didn't seem much chance of bumping into a generous Christian with an extra pair of gloves on this forsaken road – and I had no family on this continent, and not a penny to spare for extras. As I was reflecting on this theological point about prayer, but before I had a chance to pray – and I hadn't even decided if it would be appropriate to pray – two ancient looking figures in black hooded capes, resembling the grim reaper, appeared about a meter in front of my hands each holding a large block of ice. They hovered above the road and moved backwards, facing me, at the same speed as I was travelling. There would have been times that such apparitions would have terrified me, but on this occasion I simply thought: "Humph, they've been there awhile and I just didn't see them."

Then a memory began flooding across my mind of a day when I was about three or four years old and riding with my father on the tractor; a McCormick WD9 – with no enclosure for the driver or his tag-alongs. It was twilight, the temperature was dropping rapidly and we had probably another hour to go to finish the field. I remembered my boyish thoughts exactly: "If I tell dad I'm cold he'll say: 'I better take you into the house to mum while I finish.'" (The field was right next to the farmyard). "But... I want to stay with him... but, my hands are freezing."

I looked up into my dad's face, sort of trying to pretend I wasn't cold but, hoping for some help as well. I saw in his face the perfect understanding of my dilemma – he knew I didn't want to say I was cold and he knew why. And besides that, I could see that he was pleased that I wanted to be with him and he was pleased to have me with him. This whole complex thought was held in one glimpse of his face in an instant.

Then the memory merged with my present situation – as I saw that look on my father's face in my memory I was seeing the same look on my heavenly Father's face – now – in Belgium. Instantly my two hooded tormenters dropped to the road and disappeared beneath my feet behind me.

Back on the farm my father had said: "Put your hands on these hydraulic pipes." As he said this he gestured toward the hydraulic control valve that was bolted on to the fender of the tractor near where I was standing. Delicious heat from the high pressure oil lines flowed into my cold hands as I did this, and our faces met again with smiles.

My heavenly Father, on the road in Belgium, then said: "Put your hands on my hydraulic system – where my Spirit is moving under pressure."

I understood this as the call to intercession and the whole incident, though it took only a moment, revealed the heart of all kingdom work. But Jesus answered them, "*My Father is working still, and I am working.*" (John 5:17) He is out in the cold working and he makes it possible for us to join him and be with him.

This is the first, the deepest and the greatest desire of the Holy Spirit. The Holy Spirit brings to our tangible experience, from the depths of the Father's heart, the desire to be with us. This desire to be with us goes right back to before the foundation of the world – in fact it is the reason for the foundation of the world. The Father imagined us, just as a man or woman might imagine having a son or daughter; and having seen us in his heart's eye: he cannot rest until he has us tangibly with him.

As far as the work is concerned: one place you will always find God is with the cold and hungry and needy and with orphans and widows and lost sheep. "Why are you looking for me here?" he once said to me when I had gone away for a weekend to a nice hotel to seek him. I'm still often tempted to expect to find him at a conference or in the inner circles of the religious elite, but in my better moments, I know where he really is.

THE HYDRAULICS OF AN OPEN HEAVEN

I was only a little warmer after that experience but I had a different heart about it and I spent the rest of the night praying for people that were cold. I won't say it was fun or easy but it was joyful work, in good company and much came of it in future days and years.

Mid morning the next day I arrived at Bodenseehof, spent a good week with friends and had plenty of time to rest. If you have never been to that part of southern Germany, near Meersburg and Friedrichshafen, it would be hard for me to describe the idyllic beauty and surreal pleasantness of the place. In all my travels I have found no place on earth more pleasant than the shores of the Bodensee (Lake Constance); with views to Switzerland and Austria; with restaurants that have tables right up to waters edged by beautiful stone work; and with swans in those waters that glide happily past as a person samples delicious cakes, ice creams and coffees... But I had better stop, lest I tempt you beyond your strength! Well... actually, I think that most of the things I just mentioned are not wrong desires to resist at all but desires or 'temptations' of the Spirit to be given in to. Beauty in creation and even culture – 'the glory of the nations' in its myriad forms –

are overtures from God designed to draw us to him and open our eyes and hearts to his goodness and love.

And in general: I'm no advocate of asceticism. It is often a religious spirit that bids people to take up deliberate pains in intercession. Be assured: whatever pains you need, will come. Use the pains you already have rather than seeking new ones. I would hate to see a bunch of zealous motorcyclists freezing themselves trying to find God – and blaming me! Let the Spirit lead you by his beautiful desires. I had come on this trip because of all the above mention delights and incidentally 'Eden' means: 'delight'. If you would take the road to Eden or be an intercessor: never seek suffering or open any door to trouble or sickness, but rather: always seek what is good and beautiful and true.

The week passed quickly and I packed up for the return journey. I was taking a slightly different route than the one I had come on because I wanted to pass through Freiburg to see an old friend. The trip to Freiburg was pleasant, taking me through the Black Forest, an area famous for its Cuckoo Clocks, among other things. Ever since I was a little child I had longed for that wooded Germanic atmosphere that my grandfather's cuckoo clock and certain fairy tales evoked. And here it was. The road wound its way through rocky wooded valleys with quaint stone villages and little workshops selling cuckoo clocks and other crafted things.

I had driven this road before but my eyes had only been half open. People with eyes half open can access atmospheres in the way that films might access the old fairy tales, stories, places and time periods but this is considerably less than what the Holy Spirit can reveal. (I still enjoy films that do it well though and I find modern films do it considerably better for the most part than films from my youth and childhood era of the '60s and early '70s.)

Since this transitional summer of 1983 I felt that I could access the atmosphere of different time periods and situations in a deeper way; even the distant and ancient – I don't mean time travel in the scientific sense but that I could feel, experience and 'taste' the actual atmosphere, and not just an imagination or memory of it. As far as the longing for such things being satisfied this is better than going there in the scientific way (space ship or time machine) and finding that the atmosphere you crave is not actually there.

All of the treasures of wisdom and knowledge are hidden within Jesus as the apostle tells us (ref: Colossians 2:3). The treasures of wisdom and knowledge are not just information about times and places but the realities that knowledge corresponds to; it is the actual times and places. These are brought into our experience by the Holy Spirit who is the Spirit of reality or truth; in accordance with Jesus' statement that: "*He will take what is mine and declare it to you...*" (John 16:14). Such is the fulfilment in Christ of the pleasantness of knowledge coming into your soul – promised in Proverbs 2 – and which, in various ways, I had been seeking all my life. The reality behind the desire for other worlds, places, time periods and places at other times – in fact: every true, good and beautiful desire – is found fully in all possible depth in Christ.

This is a far cry from the bit of religion that people normally imagine and associate with Jesus and the church. Relative to what should have been – to actual truth – nothing and no one in all of human history has been as under-valued as Jesus. In him are hidden the glories and beauties of every time, place, nation, culture or art that ever has existed or ever will exist. He is: the hope of earth, the joy of heaven, the pearl of great price, the lily of the valley, the bright morning star, the desire of the nations and the treasure in the field; in all ways the true consummation of every human longing. I had often heard words like these in church, spoken or in song, but never accessed their real meaning and value – even now I have barely scratched the surface.

"*The kingdoms of the world and the glory of them, in a moment of time...*" (Matthew 4:8 & Luke 4:5) which the serpent showed Jesus (and Adam and Eve) was deceptively portrayed to mirror this actual thing. That hollow and empty promise has had people scurrying around after money, sex and the other loveless powers all their lives. The genuine thing that will actually satisfy is the kingdom of God. And that kingdom is only available in Jesus and the depths of the Father's heart in the Holy Spirit. In the heart of God, all times, all places and all places at all times can be accessed... but this cannot be reduced to plain fact or logic – rather tenderness of heart and love – and so the statement pertains to something that remains a mystery.

Meanwhile, back at a village near Freiburg, I saw my friend and was back on the road sometime after lunch. Rain was forecast and the sky was already almost fully overcast. Soon I was driving on a wet road and expecting to drive into the rain at any moment. But, the rain was staying just ahead... and I began to think it was a bit odd... but then I had rarely ever been in the rain on that motorcycle... it seemed to have a blessing over it. After about an hour there was rain behind as well, then rain to one side, and then to the other side; but still none on me. This continued throughout the afternoon as I headed northward up the Rhine Valley to Karlsruhe, then Mannheim. At one point it was raining right up to the sides of the road on both sides and almost as close front and back; at other times the area was larger. There was always a little patch of open sky and sometimes even sunshine just over me and I began to know it was in answer to my prayers for all cold people a week before on the outbound journey. At Mannheim the route was to take a sharp left toward Kaiserslautern and Luxembourg and I began to wonder if the little bit of open sky would turn as well. It did. It continued until nightfall when I decided to get a few hours sleep in Luxembourg airport – the one and only time I used an airport as a free hotel. I drifted off to sleep, lying across some chairs and watching the heavy rain through the windows; pleased to be dry.

The next day dawned bright and clear and my return to Dunkirk, the ferry crossing to Ramsgate and the short drive to London were spent pondering the events of the previous week. The 'Network' course in church planting and inner city evangelism with Ichthus Christian Fellowship began the following day. (Incidentally, the motorcycle gloves were still there on the hall shelf when I came in, looking quite as ordinary as any gloves do.) The experience touched on here, of walking in the revelation of the Father, seeing his face and placing hands where the Holy Spirit is moving under pressure became the doorway into prayer, intercession and warfare in London over the next couple of years. Most of it was over my head and it is only now, through chronicling in detail, that I have begun to see more than the slightest inkling of what was actually happening. The story of that period will continue in the next book in this series: The Gates of Eden.

For glory meant good report with God, acceptance by God, response, acknowledgment, and welcome into the heart of things. The door on which we have been knocking all our lives will open at last. (C. S. Lewis "The Weight of Glory")

EPILOGUE: KRISTOFFER'S STORY

[**Note**: What follows is fiction, which, though intended to convey and illustrate truth, is not to be taken as technically precise theology. My intention rather is that a theologian should regard it as a botanist might view Monet's garden through his paintings of it – perhaps we could call it 'Impressionist Theology'. Kristoffer Larsen is my grandfather's great grandfather but the only details that are historically correct are the dates of his birth and death, his wife's name, the number of children and the geographic location. It is pure speculation that this was the father that I encountered prophetically for the brief moment mentioned in chapter two but that is the notion that lies behind the germ of the story.]

Kristoffer Larsen set down his cup on the pine table and gazed out the kitchen window. His eye followed the coarse grassy slope and then the rocks beyond to where the restless sea, just out of sight, could be heard worrying the shoreline with such constancy that its sound had become, to Kristoffer, what silence was to men who lived inland. Farther out he could see the blue waters of the bay, two small bays just beyond, several distant headlands and at last the open sea as far as the horizon, which today was a blend of whites and blues.

"Do you want some berries and cream for dessert?"

"I didn't know we had any berries."

"I picked some wild blue berries with Johanna Aadland this morning" answered Kristoffer's wife Elen.

"Well mama, it would be a shame not to eat them I suppose" said Kristoffer with a little teasing twinkle in his eyes.

"I could put them in a pie for Sunday."

"I wouldn't want you to go to all that work – better eat them now."

"Humph"

"You have to admit I've saved you a lot of work over the years since you married me" said Kristoffer with mock seriousness. Elen feigned a tired smile as she set a bowl of berries and cream in front of her husband – she had heard all these little jokes a hundred times before, but they were the light banter of tender hearts, where each played the expected part and any deviation would have been discordant.

"Nils and Johanna are having their fortieth next month" said Elen presently.

"Long time with the same woman" answered Kristoffer; eyes relishing the sharp response he expected just as a man might like a piece of strong cheese after dessert.

"What kind of silly talk is that for a Christian" scolded Elen "I can't think what's got into you today?" She finished her own bowl, got up from the table and started clearing away the dishes. Kristoffer sat quietly looking out toward the sea lost in his own thoughts. Presently he looked back to his bowl of berries and cream. The particular swirl between the blue of the berries and the white of the cream seemed to form a shape like the sea, sky and headlands he had been gazing out at. A little nudge with his spoon perfected the picture and he fell to musing about the connections between such parallels. After a few minutes he spoke.

"Actually I had a very strange experience today – just after praying this morning – we need to pray for the children". He was referring to their three sons and two daughters who were all married, and to the twelve grandchildren who had so far come into the family.

"Is there something wrong?"

"No – something very right I think – at least if we pray. You know I've been troubled since the special meetings last spring?"

"When Jakobsen was preaching? I thought you liked those meetings – many of the young people were converted – and our Ola went forward too."

"Yah, yah they were good meetings. But Jakobsen warned us about the new German theology – terrible – twisting the whole Bible so there's no miracles left in it. A lot of pastors seem to be taking it pretty seriously – can't think what's getting into them."

"Well I don't think we should worry about pastors like that – the Lord will take care of them" said Elen, as ever steering her husband away from the worries she sensed in his face.

"Yes he will, mama, but some things he does through prayer. I saw terrible troubles coming because of that teaching – all of Prussia and Austria seemed inflamed with hatred and greed – I've never heard of anything like it."

"What? Visions?"

"Sort of pictures and thoughts from the Lord, I guess. And just after the meetings I had a dream one night where I was right in the middle of it and people were burning bibles and other books in huge fires while soldiers with cold hateful faces stood listening to preaching of some kind – angry and hateful – I woke up terrified and haven't been able to shake the feeling till today."

"What happened today then?"

"Well you know how I've had a real burden to pray into all this – today I felt the Lord had really heard my prayers – or that the burden was complete or something and just as I was leaving the wood shed I saw many years into the future to a city somewhere that had stood against the worst of the evils that are coming – and suffered much loss because of it. One of our sons – many generations forward was there with praying people who were clearing away... I don't know what... the scars maybe."

"What did he look like?"

"Reminded me of Ola... a bit taller with funny hair and strange clothes – the really strange thing was though that he could see me as well and our eyes met for a moment – and the Lord was there – so much joy I didn't know if I would cry or laugh."

"Well I guess the Lord can do what he wants" said Elen, not being able to think of anything else to say but feeling a need to make a positive response.

"The Lord likes to hear our prayers – though sometimes I can't think why," said Kristoffer "I have a great wish to pray for the children – forward through the generations – and I must tell the boys to keep praying too, even after I'm gone – it's a long time in the future – years and years – more than a hundred I think."

"A hundred and twenty" said Elen with a little laugh.

"Really?"

"I don't know – it's just the number I thought of – I was reading about Moses being 120 years old this morning and it just came back to me as you said 'more than a hundred'... I'm not saying it's the Lord."

"Could be – that would be 1958 – has a future feel to it, that's for sure... and I was born in 1758 – but I don't suppose that has anything to do with it. It's in the Lord's hands but we will pray too. Strange about Moses because yesterday I was reading about his encounter with the Lord – how God said:... visiting the iniquities of fathers on the children to the third and fourth generation... doesn't seem very fair, does it?"

"But that was old covenant – I remember Hans Nilsen Hauge teaching about that bible passage, years ago," offered Elen, "He said that in the New Testament times since Jesus came, blessings go down the future generations instead of curses."

"Then we take those words from Moses and pray them over the children – and tell them to keep on until... well more than a hundred and twenty years anyway," said Kristoffer with a little twinkle in his eyes. "Right to the end actually."

Then the two of them bowed their heads and Kristoffer prayed: "And Father we do pray for the children and the children's children even unto the third and fourth generation. That you would bless them Father and keep them in purity of heart and faith until your return at the close of the age."

Kristoffer followed through his intent over the next few years, frequently charging and reminding his children of the sacred burden he felt over the coming generations. Early in the spring of 1842, Kristoffer and Elen had come in late after an evening meeting in church. The fire had gone out so Kristoffer picked up an old newspaper he had kept for lighting the fire. He got the fire going and then sat down with the left over paper in his hand and drifted into a light doze. He dreamed, with a sense of great foreboding that he was reading an article in a newspaper about the Prussian victory at Koniggratz:

"*I cannot recover from the impressions of this hour. I am in truth no devotee of Mars; the goddess of beauty and the mother of the graces appeal more to my understanding than the powerful god of war, but the trophies of war exercise a magic spell even upon the child of peace. Involuntarily the look is fettered, and the spirit dwells with those countless masses who are hailing the god of the moment, success.*" (Gustav Mevissen, 1866)

When he was finished his eye scanned upward on the page to the date where he was startled to see the year 1866. The shock of seeing a paper from 24 years into the future jarred him from his sleep with such a jump and shout that Elen rushed in from the kitchen.

"Oh – just you – falling asleep in your chair again!"

"I had the strangest dream about reading a paper from the future – terrible things coming to Prussia..."

That night he dreamed of great troubles coming upon the earth. In the dream he was a warrior like his Viking ancestors before him, fighting not men but horrible beasts and creatures he had never seen before – trolls maybe, he thought. As the battle drew to victorious completion he found himself standing next to a great warrior, and looking up into his face saw those eyes he knew so well – "Jesus, it's you" he cried.

"Who else?" chuckled the warrior.

"I know this is a dream... I suppose I'll wake up soon," thought Kristoffer.

"No Kristoffer, this time you're staying here with me."

Kristoffer turned to look around and saw in the distance his little cottage, his room and his bed upon which his body was laying. "What about Elen," he asked?

"She will remain there and my Spirit will comfort her. She has a journey of her own to finish... I promised something to her when she was a child. I'll take you to our Father now – He's expecting you."

"Does He have time? And you too...? I thought angels looked after new arrivals and visitors."

"Ha! Ha! Ha! Our Father's house is more hospitable than that... and we don't have the time issues here that you have on earth."

"Oh, I never knew."

"No of course not, let's go," said Jesus as he swung

around and began to walk with clear purpose across a wide plain; undulating with windblown grass. "We can talk on the way."

"What is this place?" asked Kristoffer after they had walked, what seemed to him, a few minutes – and he couldn't think of anything else to say.

"Well it could be called 'Hades' I suppose: 'the place of the dead' because – well – here you are!" chuckled Jesus. (Slightly amused by his own joke, Kristoffer thought.) "Sometimes I call it paradise."

"Am I dead then?"

"Of course – look at that." Kristoffer looked back to where his cottage was and saw in the distance a little white church, its graveyard, and a grave with the inscription:

Kristoffer Larsen Styre
1758 – 1842

"That was quick!"

"Two weeks by that time."

"Uh?"

"Of course it depends whose looking from where to put it into your terms."

"The time you mean."

"Yes, but we will have to talk of other things first or you will become more confused rather than less; this walk is, after all, for your benefit. You can ask whatever you want and I will always answer but the answers will be older or younger, larger and smaller than you expect. In my

Father's presence you will see things that will help you fit the answers together in a way more pleasing to you – but we are smart up here so it will stretch the grey cells all the same." This last comment Jesus made tapping his head and with such mirth in his eyes that Kristoffer wasn't quite sure what to make of it. For one thing, he had for years now, considered himself to have a clear definition of humility but he couldn't remember it any more – try, as he might, to recall it.

"Are you always so happy?" Kristoffer asked at last, beginning to feel he was getting into the flow a bit.

"Ha! Ha! Ha! You've got me there – after all that build up too – the answer to that is simply: 'Yes' whoever or wherever you are."

Kristoffer began to wonder if he was still, after all, in a dream – could Jesus really be this light hearted about death. Wasn't that a bit disrespectful to the grieving – however the time thing worked?

"I have completed my sorrows," Jesus said in answer to this thought before Kristoffer could frame it with words. "That grace is carried by my Spirit to the bereaved on earth but here perfect faith and knowledge leave no place for any sadness either small or great as you imagine it." Kristoffer felt deep kindness in this answer and the sorrows of his own heart began to surface with tears and yet before any tears could fall they were swept away and replaced with solid joy; making his heart feel full and clean. "Is that too quick for you?" asked Jesus after this process had recurred many times, "Some like to savor their tears and it is a choice I give to all who walk this way with me."

"Does that help?"

"Kindness and love are greater than answers – I cannot choose for you."

"I think I would let a few tears fall for Elen if that's ok." In a moment Kristoffer found himself standing by the fire in his old cottage where Elen sat knitting. She looked older and the lines on her face spoke of many tears but her demeanor was solid and she had a strength Kristoffer had never seen before. She felt his presence and looked straight over to the place he was standing but nothing reflected light back to her eyes. Then for a fleeting second their eyes met. She sighed deeply and tears flowed freely over her cheeks as Kristoffer drew near and let his own tears blend with hers. She drifted off to a deep and peaceful sleep as the moment passed and Kristoffer found himself on the grassy plain once again looking into the eyes of the Savior; so free and joyful that every worldly pain seemed like a vague and distant memory.

The two friends walked together in a rich quietness for what seemed to Kristoffer perhaps an hour. Nothing needed to be said just then and also there was a feeling that there would always be time to say whatever anyone desired to say.

Kristoffer broke the silence: "That rock there," pointing to a large bolder that was partly submerged in the earth and covered with lichens "we have walked past that same rock three times now. Are we going in circles?"

"From the point of view of that rock I suppose" answered Jesus "but there's no reason for us be troubled by that. The lichen growth on it isn't exactly the same each time we pass it though – it's on a different timeline from us – not in the same dimension as some people like to say."

"Are we going straight... sort of... then?"

"Us, here, now – yes: perfectly straight."

"Oh, I see."

"No you don't – you just thought I might get tired of trying to explain it" said Jesus without a hint of condemnation or superiority.

"Ok, sorry."

"Why don't you say what you are thinking now?"

"Well... it's just that I'm sure that rock is the same rock that I used to walk past every day on my way to work in the copper mine back home on Karmoy" said Kristoffer scratching his head.

"It is the same rock Kristoffer. It is on a time line that ran across that dimension in Karmoy and also this dimension where we are now."

"I don't see that at all."

"Ah, ha now you are going too far – you do see – you just can't get it all into one simple thought pattern in the way earthmen always want. Because you can't picture it in one frame you think you don't understand. Let your mind expand Kristoffer – the human mind was designed to think multi-dimensionally, but on earth, man has taken pains to suppress all but a few of the dimensions." When Jesus had said this Kristoffer felt a refreshing across his itchy head and lightness in his mind – no need here to control thoughts in the heavy grinding way he had learned at school. Then Jesus continued: "What did you think of when you used to walk by that rock on the way to the copper mine?"

"It always made me think about time – it looked so old – so ancient... and then I thought about: 'Our God the rock' like Moses talked about."

"Exactly – your thoughts ran along the dimension in which the rock exists so to speak" said Jesus with a look of satisfaction. "The dimension of the rock is no more mysterious than your thoughts – you walked along a straight path to the copper mine but your thoughts moved along a dimension that ran across that path."

"OK... but where is the rock actually – really – then? Is it up here? There it is again!" cried Kristoffer as he pointed to the same rock just ahead of them. "Or... is it still down by the copper mine?"

"It is where ever you see it."

"So if no one is looking at it... it's not there?"

"If no one can see it it's not there."

"So does it move then?" asked Kristoffer – his head beginning to feel itchy again.

"Oh no – it is always where it is and my father always sees it and I see it. It seems to be here and there to you because of the dimension in which you walk. From the rock's point of view you appear to be moving in a circle which advances upward at each turn like a spiral staircase with the rock on a line parallel to the centre pole. From the viewpoint of the dimension in which we are now walking – on a straight line across a flat plane – the rock moves like a wave which intersects every few miles. As regards our Father, the rock is a ray that beams out from his heart and shines that light which came to your thoughts when you saw the rock – it reveals the ancient and mysterious from the depths of his heart. It is a ray of his memory – 'time' you might say."

"If time shines out from him, what do the theologians mean by God being outside time?"

"Those words may not have any meaning" said Jesus with a distant look "unless my Father chooses to do something that fills them. Even here there are mysteries in our Father's heart which are so deep and beautiful that no eye can yet bear them – not even mine." At these words Kristoffer was so taken aback that he stopped walking and stared at Jesus – for a moment he felt he couldn't breathe and looked away – but then a lightness of spirit came over him and as he looked back at Jesus he saw a depth of humility which he had never imagined before this moment. "There will always be actual reasons to be humble" said Jesus in response to this thought "my Father is and always will be greater and he is the creator of all dimensions. And yet He walks humbly in them all so no one has ever had reason to be anything other than humble as well."

"But why would he make the words of theologians true?"

"Our Father hates to prove someone else wrong and only does so as a last resort. His humility and love are one and his greatest pleasure is when his creatures receive his love – he never cares to be right for its own sake. Many times I have seen him create entire dimensions just so someone else's words could be true within that dimension."

"You mean like when you said from the rock's point of view we are walking in circles kind of thing?"

"Yes that's it – and many of the surprises we experience here in heaven arise from the extravagance of Father's humility... I'm quite sure he is surprised himself sometimes – by the things the Spirit searches out and reveals from the depths of his heart. Yes, I'm sure I've seen surprise on his face at his own ideas and of course his children surprise him too."

"Really?" asked a very puzzled Kristoffer.

"Oh yes – love always has surprises."

"So you think he might create some dimension where He is outside of time?" said Kristoffer getting back to the original question "Just to be kind to a theologian?"

"No I can't see how He ever could and I would be very surprised if He did!" said Jesus mirthfully. And Kristoffer couldn't help but feel that he liked Jesus very much – what a great friend – so much room for everyone else around Him!

"When I was born onto the earth it was a great miracle from heaven's point of view and very surprising because it filled a human wish to be outside of God – to have a God that could be fit into human systems of knowledge and thought. Our Father, in his humility, was pleased to give man experience of that, but it was also inevitable that man would seek to kill me – human knowledge could not contain me."

"That seems strange to me" said Kristoffer "If He knew how it would go...?"

"Those who loved truth saw me for who I was and freely chose a place inside me – they rejected the outsider's view that human pride sought; thus vindicating our Father's extravagance. The proud were judged by one who is humble, not trumped by one prouder – this is always our Father's way; humility always triumphs over pride just as mercy always triumphs over judgment." At these words Kristoffer could not help feel how prideful his own thoughts were and was beginning even to drift toward self-pity about it. Jesus stopped walking and Kristoffer took a couple of extra steps before he realized. Then he also stopped, turning toward Jesus as he did with a questioning look. Jesus said nothing.

"Ok – how can a person ever be right about anything if truth is so... flexible?" asked Kristoffer moving now from self-pity to self justification.

"Here is a truth that never changes: our Father is gentle and lowly of heart, just as I am – if you take our yoke you will find it always light. The yoke of always being right, even theologically, is heavier than men can bear." Then Jesus paused and looked intently at Kristoffer. "Love and kindness are always greater than answers and explanations, just as the acts of love are greater than the words about them." These last words Jesus added with such kindness that they flowed over Kristoffer's soul and settled like deep peace upon him.

"I'll take that yoke of yours then" he said at last.

"Thought you might!" said Jesus as he strode forward

putting his arm around Kristoffer, sweeping him into his own stride.

"Sorry not to surprise you!" Kristoffer quipped.

"I don't want those kinds of surprises – you've given me enough of the good kind and I'm sure there'll be more."

"I can't think of any time I could have done that" said Kristoffer

"Well, when you saw 146 years into the future, you saw something that none of us knew anything about – that was a secret in my father's heart that you found together with the Holy Spirit as you were praying. It caused surprise and rejoicing among the angels because, though they could see the wars coming, there remained many unanswered questions. We weren't sure if the city you saw would hold faith through its darkest hours."

"I find that hard to understand."

"Our Father is greater than all others and it is only in His heart that the deep secrets of the future can be found. Many things are fixed by His authority and choice which is also shaped by the prayers of His children. Your prayers in the Holy Spirit and our Father's heart established another timeline through many difficult years."

"But... why me, other people had better prayers and greater ministries?"

"Each of His children touches areas of His heart no one else can – it's not about greatness – I have made access to the Father available to everyone."

"Was the descendant I saw, already there in His heart then?"

"Your descendant was there when you saw him and that is why he also saw you."

"I've heard it said: 'seeing is believing' but this 'seeing is being' stuff is pretty strange" said Kristoffer thinking back to their earlier conversation about the rock.

"Within our Father's heart this is part of the creative ruling dynamic which all His children can share" said Jesus and after pausing he added: "And it is truly a great mystery and full of surprises but under it all, His humility and gentleness are unchanging just as the love between our Father, Me and the Holy Spirit can never fail."

"So if I hadn't been praying that day...?"

"Things would have been different, but of course I cannot see them and they don't exist."

"So what is there in Father's heart to start with?" Kristoffer persisted.

"There are deep waters of love and desire – potential, by one way of thinking, which comes into existence when one of His children enters His heart and sees it – the seeing makes it come into being – such is the dynamic within our Father's heart. The whole universe came into being as the Spirit revealed the deep desires of our Father's heart to me in the beginning. As I saw those treasures and spoke of them, the heavens and the earth and all they contain came into being."

"So creation was like prayer – you touched something deep in the Father's heart and spoke it out... to Him?"

"Not 'like prayer' it was prayer. The Spirit searched the deep things of our Father and it was my joy as a Son to form those ideas into words. Then it is the Father's pleasure to fill the words with meanings in many dimensions – one of course being physical things; earth, stars and planets. This is the same work as artists, musicians and craftsmen; they start with an idea; then put it into words or sketches and finally make it; a physical painting, sculpture, sound or whatever."

"Does the original idea always and only come from the Father?"

"He is always the source, though He gives more than anyone can actually develop and often in such basic forms that human imagination is stretched rather than limited. It is not possible to be otherwise because nothing created will ever be greater than the creator – to wish it to be otherwise is a great evil; the inner heart of the serpent and the origin of the one forbidden tree. It is the desire for outer knowledge; above and beyond Him – but of course it is a great folly and deception – that is to say: there is no such knowledge, just as there is no rest for the wicked. The wicked ever gropes in darkness; rejecting the one who alone is light and the Father of lights."

"And yet as you said before – when you came as a man on earth you allowed men to taste this unreal outer position of knowledge."

"Yes, that was the choice in Eden coming to its full maturity" answered Jesus "but on the cross all that outer knowledge, so called, and its fruit, which is sin, was swallowed up into my heart and our Father's heart. The outside is judged; the old tree condemned; my cross has become the tree of life; and our Father welcomes every prodigal who wants to come home."

"I see that" said Kristoffer "the old tree was a door out but the new tree – your cross – is a door back in... only of course there was no 'out'... just a fake... a dark closet... no great free outdoors; but only the confines of a coffin."

"Preach it Kristoffer!" said Jesus joyfully and Kristoffer's face broadened into a wide smile.

At this point in the conversation the two walkers arrived at the crest of a ridge they had been climbing for some time and they paused to take in the view. Before them spread a wide gentle valley and beyond that wooded lands and at the horizon, which seemed very far away, a bluish haze with a jagged top which Kristoffer took to be mountains. They stood – still and silent – just looking.

"What is this place?" asked Kristoffer in a quiet, awed voice; not wishing to disturb the stillness.

"It is many things" answered Jesus "but in the way you are asking now it is a land on earth in which your descendants will walk for several generations. As we see it, however, Adam is still in Eden. The lands of the earth also exist on timelines that shine out from my Father's heart. We have walked this one and sown many seeds as we spoke."

"What... like prayer or something..." asked Kristoffer?

"Yes, actual prayer – you have been talking to me... right..." said Jesus with a playful smile.

"Hmmm"

"From here" Jesus continued "our Father's throne is there where you see mountains but we will walk another way and come to Him freer and sooner – from here on this path it is almost as far as on earth in Adam's day. If we turn left or right, up or down or any other direction from here we will walk another path in another dimension."

"Like the rock's dimension..." Kristoffer offered.

"Yes, but we will not walk that path either – other paths will be more interesting for you," said Jesus. "But first we'll rest a while."

As Jesus said these words Kristoffer noticed a house; palatial, ornate and beautiful and yet so perfectly in keeping with the surroundings that he wasn't sure if it had been there all along and he simply hadn't seen it – or whether it had recently appeared. "I saw it," said Jesus with a slight smile and sparkling eyes.

"Dinner's ready!" called a voice.

"And I heard that," said Kristoffer turning to Jesus. Then it occurred to him that he recognized the voice. As he looked back toward the house he saw a man and a woman standing on the doorstep. They were so beautiful and bright that at first he struggled to focus... and then... "Mum and Dad!"